The American Romanian Academy
of Arts and Sciences (ARA)

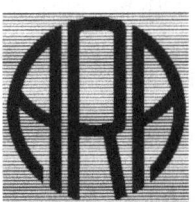

The 40th ARA Proceedings

July 28-31, 2016
Montreal, Canada

Proceedings of the 40th Annual Congress of the American Romanian Academy of Arts and Sciences, July 28-31, Montreal, Canada

Editors: Ruxandra Vidu and Ala Mindicanu

Front cover: "Tree of Life" by Carmen Doreal

General Chair: Dr. Ala Mindicanu, President of the Moldavian Community of Quebec, Vice President of the Romanian Cultural Association, Montreal, Canada

Advisory Committee: Ruxandra Vidu, Marius Enachescu, Catalina Curceanu, Isabelle Sabau

Board of Reviewers: Prof. Ruxandra Vidu, PhD, University of California Davis, USA; Prof. Marius Enachescu, PhD, University Politehnica Bucharest, Romania; Dr. Catalina Curceanu, PhD, National Institute of Nuclear Physics, Frascati, Roma, Italy; Prof. Isabelle Sabau, PhD, Northern Illinois University, IL, USA; Prof. Ileana Costea, PhD, California State University Northridge, CA, USA; Prof. Dumitru Todoroi, PhD, Academy of Economic Studies of Moldova, Republic of Moldova; Prof. Ioan Opris, PhD, University of Miami, FL, USA; Prof. Eduard Harbu, PhD, Chisinau, Republic of Moldova; Prof. Vasile Staicu, PhD, University of Aveiro, Portugal.

Organized By: the American Romanian Academy of Arts and Sciences in collaboration with Embassy of Romania to Canada, Embassy of Republic of Moldova to Canada, The Romanian Academy, Academy of Sciences of Moldova, Romanian Academy of Medical Sciences, Academy of Romanian Scientists, Moldavian Community of Quebec, Romanian Cultural Association, Montreal, Canada.

Local Organizing Committee: Ala Mindicanu, PhD, Journalist; Ruxandra Botez, Dr.Prof.Univ.; Jacques Bouchard, Dr.Prof.Univ.; Liliana Surugiu, Phd., Dan Fornade, PhD.; Simona Pogonat, Journalist, Marc Marinescu, Designer; Eva Halus, Painter

Student Track Chairs: Isabelle Sabau and Ioana Onica

Editorial services: Iulian Gherstoaga

ISBN: 978-1-935924-19-7

Published by ARA Publisher

ARA President's Wellcome Message

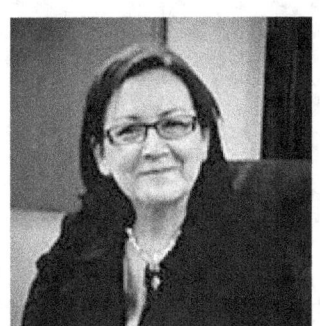

Professor Ruxandra Vidu, PhD
President
American Romanian Academy
of Arts and Sciences
University of California, Davis

With great pleasure I welcome you to the 40th ARA Congress in Montreal, Canada. We are excited to be in the city of Montreal once again, as the Congress has been previously held here in 1990 and 2001. This year's Congress promises to be a highly rewarding experience on an intellectual and social level. Speakers are expected to present revolutionary ideas in technology, medicine, education, arts/culture, and other fields that mark our lives and shape our future.

This Congress marks 40 years since the first ARA Congress was held in Berkeley, CA. The congress was originally designed to provide a forum where intellectuals of Romanian descent from all around the world and friends of Romania of all nationalities, meet to discuss recent contributions to art and science. Over the decades, the mission of the congress evolved into a forum where comprehensive overviews of innovative developments in major academic fields are presented.

Since our last Congress, the Academy has been growing and moving forward, gaining international recognition. New collaborations with Romanian academies are signed each year as we continue to gain new members. As our membership grows, our ties with the academic communities are strengthened. In addition, the ARA Journal that's been inactive for 12 years will be resumed this upcoming October.

I would like to express my gratitude to our dedicated and loyal members, and to those of you who attend the Congress each year. The success of the ARA Congress depends on our distinguished scholars and guests. Additionally, I would like to extend our sincere appreciations to our sponsors and local organizing committee, without whom our congress would not be possible.

I look forward to seeing you in Montreal for the 40th ARA Congress, undoubtedly a special backdrop to our anniversary event.

Ruxandra Vidu

ARA President

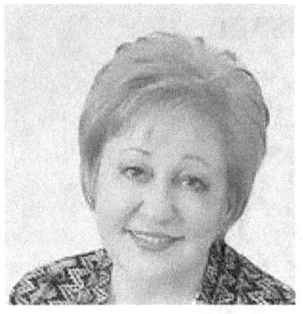

Congress Chair's Wellcome Message

Dr. Ala Mindicanu
40th ARA Congress Chairman
President of the Moldavian Community of Quebec
Vice President of the Romanian Cultural Association Montreal, Canada

Dear friends, the 40th Congress of the American Academy of Art and Roman Science is an enthusiastically awaited event in the world. The anniversary of four decades of ARA is a source of joy and pride for its members and for the entire scientific community of Romanian origin in the diaspora.

Community Resource Centre in Montreal will host the event with great hospitality and will offer spacious rooms with the entire technical assistance to the participants at the ARA Congress. A special session is dedicated to the history of ARA and its personalities, and a unique photographic exhibition will evoke the last four decades of activity.

Scientists, researchers and PhD students from the US, Canada, Romania, Moldova, etc. will participate in the most famous and appreciated forum diaspora. In Montreal, where the ARA Congress was held 15 years ago, come with a world of welcome Her Excellency Mrs. Maria Ligor, the Romanian Ambassador to Canada, Her Excellency, Mrs. Ala Beleavschi, Moldovan Ambassador to Canada, Mr. Victor Socaciu, Consul General of Romania in Montreal, representatives of universities in Montreal, leaders of associations of Romanian diaspora in Montreal, and other guests of honor.

ARA President, Prof. Ruxandra Vidu, and her team have an extensive experience organizing substantial events revolving around ARA intellectuals, scientists, and researchers of Romanian origin around the world. Their efforts are now focused on evaluating and digitizing the ARA archives, and publishing literature that recalls the Academy's past and present.

Romanian diaspora communities in Montreal and around the world eagerly await the meeting with the representatives of science, art and literature of Romanian origin who attend ARA Congress - in 2016. The local team lead by Prof. Ruxandra Botez, journalist Simona Pogonat, Marc Marinescu, designer and community leader, Dr. Dan Fornade, author of numerous works about the Romanian Diaspora in the West, Dr. Nicolae Margineanu, Master of Arts, Eva Halus, painter, numerous students and volunteers, welcome you to the 40th ARA Congress. Media representatives are accredited to Congress to inform ARA members and supporters around the world about the unfolding event.

On behalf of the local organizing committee are very proud to be hosting the 40th ARA Congress and look forward to welcoming you and wishing you success.

Local Organizing Committee:
Dr. Ala Mindicanu, Journalist
Prof. Ruxandra Botez, University of Montreal
Dr. Liliana Surugiu, University of Montreal
Dr. Dan Fornade
Mrs. Simona Pogonat, Journalist
Mr. Marc Marinescu, Designer
Mrs. Eva Halus, Painter

Table of Contens

CARMEN DOREAL

2016 Donation

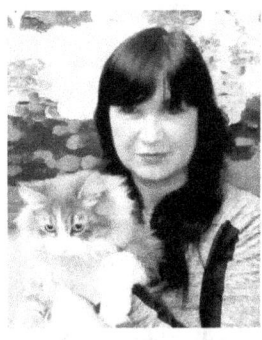

Carmen Doreal is the artistic name of the poetess and the artist painter, Carmen Tuculescu Poenaru, of Romanian origin, living in Montreal, Canada, since 2001. She is a member of The Association of Romanian Writers from Quebec, and The Association of Romanian Writers from Canada, of Romanian Cultural Foundation Constantin Brancusi, of International Academy of Arts from Montreal and The Circle of Plastic Artists and Sculptorsfrom from Quebec, Canada. She was published in literary magazines and anthologies in Romania and Canada: Luceafarul, Bucharest, Branches, Craiova, The Column, Targu-Jiu, The Tribune, Cluj, Astra, Brasov, The Canadian Romanian Toronto, The Capital, Ottawa, The Candle, Romanian Pages, Our Tribune Montreal, Agonia (Wordpress, Culture inside), The Online Sphere. She contributed to several TV and radio shows in Romania and Canada.

Carmen Doreal is the author of three volumes of poetry **"The love opening"**, 1999 Ed. The Danube Currier Press, Bucharest, **"Poems in colors"** 2010, Ed.Nemesis, Montreal, and **"Dating without arguments"** 2013, Ed. Fides, Iasi. All these three volumes represent the double artistic credo of the poetess-painter. She is also, the author of hundreds of paintings in oil and acrylic. Her works are exhibited in Europe and in North America in art galleries and particular collections.

At Montreal, Carmen Doreal have had three personal exhibitions, in January 2009, and June 2010 at Inter-Pallas Art Gallery, and in 2012 August- September, at the Municipal Library, Deux-Montagnes.

"Her creation manifests the music of poetry and the poetry of colors in an original fashion."

The writings and the paintings of Carmen Doreal fits smoothly into the creations of artist in accessing her emotions. Through her eyes the ordinary becomes extraordinary and each color becomes more vivid. She likes to feel the words and the musicality of the colors, she likes to feel the blues and the horses of our dreams.

"Watch what you listen and listen what you watch." Each piece tells a unique story and is a visual record of events both real and imagined ... Here is a series of materials depicting basic human instinct with harmony of the passion, balance of feelings and rhythm.

Art is not just art, but is a therapy. It's a therapy to lighten people's mind and make them see insight of beauty, the expression of life. Who will read the poems of Carmen Doreal, or will visit the artist's works gallery, will feel happier, because her "joy of life" is infectious and therefore, her art is unforgettable!"

Prof. Arnie Greenberg

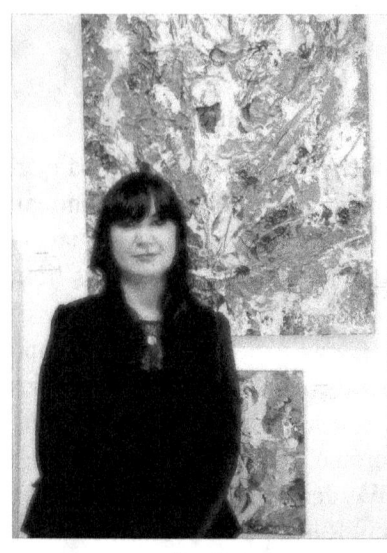

CARMEN DOREAL
(Presentation in Romanian Language)

Carmen Doreal este numele artistic al poetei şi pictoriţei de origine română Carmen Ţuculescu Poenaru, stabilită din 2001 în Montreal, Canada. Carmen Doreal este membră în Asociaţia Scriitorilor Români din Canada, Asociaţia Scriitorilor de Limbă Romană din Quebec, Fundaţia Culturală Constantin Brâncuşi, din România, Academia Internaţională de Arte, Quebec şi Cercul Artiştilor Plastici şi Sculptori din Quebec, Canada. Publică în diverse reviste literare din România şi Canada: Luceafărul (Bucureşti), Ramuri (Craiova), Columna (Târgu-Jiu), Tribuna,(Cluj), Astra (Braşov), Constelaţii Diamantine, Românul Canadian (Toronto), Capitala (Ottawa),. Destine Literare, Candela, Pagini româneşti, Tribuna noastră (Montreal), Agonia,Wordpress, Culture inside, Sfera online. Colaborează la diverse emisiuni de radio şi televiziune în România şi Canada.

Carmen Doreal este autoarea a trei volume de poezie Vernisajul iubirii, 1999, Ed. Curierul Dunării, Bucureşti Poeme in culori,2010, Ed. Nemesis, Montreal, si Întâlnire fără argumente, 2013, Ed.Fides, Iasi, toate aceste volume reprezentând dublul crez artistic al poetei-pictoriţe. Lucrările sale artistice sunt expuse în Europa şi în America de Nord în galerii de artă şi colecţii de particulare.

În Montreal, Carmen Doreal a avut trei expoziţii personale, în 2009 şi 2010 la Galeria de Artă Inter Pallas, Montreal şi în 2012 la Biblioteca Municipală din Deux-Montagnes. În creaţiile lui Carmen Doreal, ordinarul devine extraordinar şi de aceea, bucuria ei de viaţă este de neuitat!

Vise la zid

ideograme
cu picioarele goale
cuibăresc îngeri
în măduva cuvintelor
fără emfază
şi cu multă timiditate
magia promisiunilor
cu dinţi ascuţiţi
muşcă în tăcere
inimi cu aripi înalte

Transcendenţă

dragoste fără griji
versuri albe pentru poeţi trişti
publicaţi în manuale şcolare
nu am uitat numele voastre
Eminescu, Arghezi, Sorescu
Nichita Stănescu

orice atelier de creaţie este pustiu
recit prea rar
rime romantice provocatoare

nu mai ştiu unde creşte
visul meu românesc de sub pernă
îmi amintesc doar
că voiam să impresionez
Dali, Elytis, Claude Debussy,
agăţau oglinzi suprarealiste înalte
deasupra patului meu la Paris

tulburată de umbre lumina mă împarte
poem însufleţit în valuri de culoare
pasăre cu inima de foc
reiterez din propria-mi cenuşă
aripi de neuitare
dragostea ucide
dar inima nu moare
pe malul fluviului Saint-Laurent
divaghez în vers alb cu Pierre Morency
dansez fericită pe frunze de arţar cu picioarele
goale
reinventez iubirea sub Poartă de Sărut
Coloană Nesfârşită în nopţi incendiare
la Masa Tăcerii albastre
mă aşteaptă la cenaclu umbre stelare
cu dragoste fără griji
Eminescu, Arghezi, Sorescu,
Nichita Stănescu şi Pierre Morency

Trăiesc la prezent

Trăiesc la prezent viaţa mea nouă,
indefinită
bucurându-mă şi lipsindu-mă
totodată de cei dragi
îmi fac zilnic promisiunea
că voi face pace cu mine însămi
şi mă voi reapuca de scris
ca de un fel de fitness intelectual
visez un roman despre arta reuşitei în
exil
dedublată de senzaţia placută
că aş calatori încă în ţara mea
acolo unde mă puteam dezbrăca de mine
însămi
în cel mai simplu şi curat limbaj românesc
noua mea identitate mă sarută
cu un aer realistic pe frunte

îmi asează pe buze cuvinte bilinguale
mă imbrăţişează cu sentimentul larg
că sunt la mine acasă în noua mea ţară
asimilată de visul alb
ce împarte cu dărnicie la toţi conflicte
interioare
îmi place din ce in ce mai mult să călătoresc
să aflu lucruri noi, să le experimentez
gătesc romaneşte, vorbesc amestecat
adaptată la o lume de sporturi extreme
mă războiesc duios cu cei ce m-au uitat
copiii mei au devenit canadieni înainte de
vreme
uneori privesc cerul şi mă întreb
oare mama mea şi bunicile mele
sunt mulţumite de mine?
poezia mi-a dăruit piramidă cu scară la stele
patria mea este în aer
în vis rădăcinile meie

Tu nu îmi poţi atinge cuvintele

trebuie doar să înveţi
cum să îmi citeşti sufletul

imaginează-ţi întâlnirea ta proprie
bărbatul ideal cu femeia ideală
trăind amorul bumerang
în a vieţii spirală
surprinşi pe o canava iluzorie

de fiecare dată
când dorinţa ta converge
in jurul visului meu
eu reîncep poemul

despre această noapte
pe care tu o mângâi
doar cu vârful degetelor

uitând de mine însămi
în braţele tale

în timp ce luna
înoată cu noi
în ape virtuale

Siddhartha

vânzătorul de iluzii
dedublat de un vis secret
îmi invadează sângele
cu telepatii halucinante

o altă toamnă
mă regăseşte înveşmântată
în rochița de funze
colorată rebel
de un pictor expresionist

dansez pe o fantezie de Bach
simulând vântul îndrăgostit
de imaginea mea din oglindă
complicitate şi foşnet
harpha serii mă desfrunzeşte

deasupra unui lac necunoscut
mă dezbracă de aşteptări
cu opusuri transparente
insinuate profund
în arşiţa cărţii

în numele iubirii semnez
aerul pictat cu fluturi adictivi
când aprind noaptea
sub ceruri străine

fluxul pur sânge contează
nu desfrunzirea de sine
to be or not to be
pată vie cu-vân-tul
decantează mirajul luminii
cu mine
C*D*

Refined Attic Greek:
Hallmark of the Emerging Phanariot Nobility

Jacques Bouchard

Centre d'études néo-helléniques, Université de Montréal

jacques.bouchard@umontreal.ca

Abstract: In the 17th century, in the field of language and culture, Romanian boyars and intellectuals gradually replaced Slavonic by Greek. Spoken modern Greek became a vernacular language in the Danubian principalities. To dissociate themselves from the local society, the emerging Phanariot élite used refined Attic Greek as a symbol of their ascendency and prestige.

Introduction

The historical period proposed here is quite exact: it begins in 1641, the date of birth of Alexander Mavrocordatos the Exaporit and continues until 1730, death year of his son Nicholas then prince of Wallachia; it marks also the end of the "Tulip Era", when Sultan Ahmet III was deposed.

I will focus my presentation on the first members of the Mavrocordatos family, particularly on prince Nicholas who reigned from 1709 to 1730, a period that I have described in other papers as that of "Reasoned Absolutism", a prelude to the Enlightened Despotism.

I intend to examine the linguistic levels used during that period that preceded the rise of the Phanariot caste, by the élite, princes, scholars and clergymen. Indeed, C.Th. Dimaras exposed the subject with acuity and pertinence in his foreword to my edition of the *Leisures of Philotheos,* more than twenty-five years ago.

Ancient vs Modern Greek

Firstly, we can empirically recognize at a glance what texts were written in ancient Greek: they have no particles θα, no conjunctions να or ας, no prepositions με or σε, and do not use the negative δεν. Of course even if the vocabulary and morphology are ancient, some texts belong to the *Kunstprosa,* while others are simply written in "tyronic Greek". Among the other documents one can distinguish texts in vernacular Greek (εις απλήν γλώσσαν), and those composed in elevated style, like the documents redacted by ecclesiastical authorities.

A second remark concerns the type of document: letters, essays, sermons, literary fiction, philosophical treatises, scientific memoirs, poetry must all be written in a language appropriate to the addressee: a sermon pronounced before poorly-educated congregation cannot be composed in a learned language. On the other hand, an erudite person can address a letter or an essay in ancient Greek to a scholar since he knows that it is easier to understand ancient Greek reading rather than listening.

Texts to be analyzed are found in the collection *Hurmuzaki,* the Βασική Βιβλιοθήκη, Legrand's editions, the *Bibliografia românească veche* (BRV) by Bianu and Hodoş, and some works of authors like Mavrocordatos, Cantemir, Notaras, Voulgaris, etc.

Of course, the Greek language existed from Antiquity on the site of Byzantium, but its presence was also attested during Antiquity on the Black Sea littoral, nowadays Dobrogea. Closer to the period under study, the Greek language was found in both principalities from the 14th century onwards, in spite of the fact that the Orthodox Church had even then imposed the slavonic language in the liturgy and civil administration[1].

Vernacular Greek

During the 17th and 18th centuries most texts with ecclesial or pastoral content destined to edify believers were composed in a mixed language, where a syntax with a dusting of ancient Greek imitates vernacular Greek in order to be understood by the audience. The best examples are the sermons of preachers and hagiographers, like Frangiskos Skoufos

(1644-1697), who published the "Panegyrical discourse on the Baptist's birth" in Venice in 1670, or Ilias Miniatis (1669-1714); subsequent preachers, like Cosmas Aitolos (1714-1779) and Nikiforos Theotokis (1731-1800) would do the same. They were aware that, to convince their flock, they had to use an understandable language. In his *Art of Rhetoric* published in Venice in 1681, Skoufos described his aim as follows: «να ωφελήσω το Γένος μου... αποφάσισα, ω Αναγνώστα μου φιλάνθρωπε, να συνθέσω το Βιβλίον τούτο, και δια να γενή κοινή εις όλους απλώς η ωφέλεια, ηθέλησα να ομιλήσω και με κοινήν γλώσσαν, επιθυμώντας να το δεχθούν όχιμόνον οι αγκάλες των σοφών και εναρέτων, αμή και εκείνες των απλουστέρων ανθρώπων»[2]. [To be useful to my Nation... I decided, o benevolent Reader, to compose this book, and with the purpose to beuseful to all men, I wanted to speak the common tongue, wishing to be accepted not only by scholars, but also by the most simple men].

Vernacular Greek in the Romanian countries

In the domain of sciences and learning, popularisers adopted a simple language; so did Chrysanthos Notaras in his learned work *Introduction to Geography and the Sphere*, printed in Paris in 1716, and reprinted in Venice in 1718. He exposed, in vernacular language, the heliocentric theories of Copernic, Kepler, Galileo and Descartes, even if in the end he preferred to follow the opinion of the Church concerning the geocentric system. Chrysanthos had also written an essay on excommunication in simple language at the request of Voivode Constantine Brâncoveanu, himself under threat of excommunication[3].

If we examine the bulk of letters written by Moldavian and Wallachian nobles and published by Émile Legrand and the editors of the *Hurmuzaki* collection, we notice that most of them knew Greek well and used to write it correctly. Among the authors of Greek letters there are the Voivode Vasile Lupu and his daughter Ruxandra, the stolnic Constantine and the spatharius Mihail Cantacuzino, the Brâncoveanu family, Constantine, Ştefan and Radu, the princes Mihai Racoviţă, Antioh

Cantemir, Gheorghe Creţulescu, the boyars Nicolae and Ioan Ruset, Grigore Filipescu, Radu Dudescu, Dumitrachi Hirosculaiu and many other letter writers from the Romanian principalities. All of them, without exception, wrote vernacular Greek with or without a touch of ancient Greek, under the influence of the Patriarchate[4].

We can conclude that knowledge of Modern Greek existed among the political and ecclesial élite, in civil society and the world of commerce. The influence of Greek culture was further encouraged by the establishment of two princely academies, one in Bucharest founded in 1689 by Şerban Cantacuzino, and the other in Jassy by Antioh Cantemir in 1707[5]. These institutions attracted Greek professors from many parts of the Ottoman Empire and encouraged an interest in Greek letters. In 1691 Constantine Brâncoveanu subsidized the edition published in Bucharest (έν τῇ περιφήμῳ πόλει Μπουκουρέστῃ) of *The Sixty six admonitory chapters* by Basil the Macedon, translated "εις την απλήν των ρωμαίων γλώσσαν" by Chrysanthos Notaras[6]. The Georgian scholar Antim, the Patriarch of Jerusalem Dositheos, his nephew Chrysanthos Notaras and Jeremy Cacavelas helped make of Jassy and Bucharest, andeven Târgovişte and Snagov, centers of Greek book diffusion. When the Moldavian Dimitrie Cantemir published his *Divanul* in Jassy in 1698, he juxtaposed with it the Modern Greek translation of his tutor Jeremy Cacavelas. On the other hand, Antim printed Plutarch's *Parallel Lives* in Bucharest in 1704, translated into Modern Greek and Romanian by Constantine, prince Constantine Brâncoveanu's son. In 1713, the same Antim printed *PhilosophicalMaxims*[7], in Târgovişte, translated from Italian to modern Greek. In 1715, Antim printed *Νουθεσίαι Χριστιανοπολιτικαί* (Christian Political Admonitions) in Bucharest with a dedication to Voivode Ştefan Cantacuzino as follows: "Ιξεύρωντας (ευσεβέστατε και φιλόχριστε Αυθέντα) την θερμοτάτην αγάπην, και τον ένθεον πόθον, οπού έχει η υμετέρα Υψηλότης εις το να κυβερνήση αμέμπτως και θεάρεστα το Υπήκοον [...] εσύναξα από τα Γνωμικά των παλαιών σοφών Διδασκάλων τα πλέον εξαίρετα και αρμοδιώτερα...»[8]. [Most pious and Christian

Prince, knowing your very fervent love and the divine desire of Your Excellency to govern your People impeccably and according to God [...] I have collected from the aphorisms of the ancient Masters' the most exquisite and pertinent ones.]

The Phanariots in the Danubian Principalities

Further historical events are well known: since Sultan Ahmet III had lost confidence in Prince Mihai Racoviță, on November 6th 1709 he appointed Nicholas Mavrocordatos to the throne of Moldavia. The Phanariot kept his throne until November 27th 1710. After the short reign of Dimitrie Cantemir, Nicholas recovered his principality on September 26th 1712. In the meanwhile, Nicholas learnt Romanian well enough to read Romanian chronicles[9]. Without losing time, he supported the Romanian edition of a *Synopsis* in Jassy in 1714, an edition decorated with the coat of arms of Moldavia[10]. One year later, a *Liturghie* [Liturgy] was printed under the Voivode Nicholas' high patronage, in Slavonic with explanations and prayers in Romanian[11], and in September, he published in Greek John Damascene's *Profession of* faith [12]. We see very clearly in this last example that the propagation of religious books served both the voivode's political propaganda and personal glorification: one can even decipher flanking the emblem of Moldavia the acronyms Ε Θ Ι Ν Α Β Α Η Π Μ: "By God's grace John Nicholas, son of Alexander, Voivode, Prince, Sovereign of all Moldavia". The Romanian slavist Emil Vîrtosu has demonstrated that the enigmatic "ΙΩ" before Romanian princes' names is a cryptograph that comes from Ιωάννης: a name whose Hebraic etymology means "Dei gratia - by the grace of God"[13].

Refined Attic Greek
1. Alexander Mavrocordatos the Exaporit

On December 25th 1715, Nicholas was appointed Voivode of Wallachia; he arrived in Bucharest on January 30th 1716. Immediately after his enthronement, Nicholas, in an act of exemplary filial devotion, hastened

to edit a *Holy History* written in koinè by his father Alexander the Exaporit. With its magnificent quality and the precision of its typography the volume is indeed imposing. The book was published in August in Bucharest at Voivode Nicholas' expense and was offered gracefully to the public for its spiritual edification. The double intention of self-promotion and propaganda is obvious as can be seen in the emblem especially executed for Nicholas: it unifies for the first time the arms of Moldavia and of Wallachia side by side, the aurochs head and the crow holding a Latin cross in his beak both coats of arms surmounted by one princely crown. In order to personalize his design, the engraver had inscribed the initials Ι Ν Α Β, meaning Ιω Νικόλαος Αλεξάνδρου Βοεβόδας. Nicholas' arms are followed by a twelve-verse poem composed by the Postelnic Ioannis that describes the coat of arms. The next three pages consist of a pious homage to the Virgin Mary and a profession of faith in the Holy Trinity. Nicholas also offers fulsome praise of his father: "Σοῦ δὲ πατέρων Φιλοστοργότατε, Ἡρώων εὐκλεέστατε, τῶν ἐν λόγοις, καὶ σοφίᾳ θαυμαζομένων Σοφώτατε, πῶς ἂν ἐπιλήσμων γένοιμι τοῦ φῦντος;"[14] [O you, the most affectionate father, the most glorious Hero, the Wisest among admired men for their speech and wisdom, how could I forget my genitor?] Next comes an epigram composed by Hierotheos Metro- politan of Drystra, who lauded Alexander and his sons' merits. Finally, Iacovos of Argos, former tutor of Nicholas, extols the merits of Alexander and his family, a ceremonial speech enhanced with metaphors and classical clichés that praise Prince Nicholas' virtues and administrative aptitudes.

Alexander Mavrocordatos' volume comprises five chapters and a table of contents; composed in refined Attic Greek, it has an austere and elevated style about which we could quote Dionysius of Halicarnassus' expression: "τὸ ἀξίωμα καὶ ἡ σεμνότης τῶν ὀνομάτων εὔμορφον πεποίηκε τὴν φράσιν"[15].[The dignity and grandeur of the words has given the style its pleasing form.] A solid knowledge of Attic Greek is necessary to read this Biblical History, even if the content is well known. It constitutes an example *par excellence* of what

Dionysius calls an "αὐστηρά ἁρμονία"[16] [austere harmony]. But the intention is without doubt to construct for the autochthonous noble families' reputation a monument to the grandeur and majesty to the prince who had inherited his prestige from his father, the founder of the dynasty, and close counsellor to the Sultan, and from the capital with its aura of imperial and ecumenical power, namely Constantinople, "τῶν πόλεων ἡ Βασιλίς, ἡ τῆς οἰκουμένης μητρόπολις"[17] [the queen of the cities, the metropolis of the world], and from the Phanar, the beacon that illuminates the world.

2. Voivode Nicholas Mavrocordatos

On November 14ᵗʰ of the same year 1716, Nicholas was abducted by the Austrians who had invaded Bucharest with the complicity of some boyars, and perhaps with Metropolitan Antim's assent. During his detention in Transylvania, Nicholas composed a *Treatise on Duties*, and a novel, the *Leisures of Philotheos,* two major works written in refined Attic Greek, known in French as "grec littéral". Nicholas recovered his throne in Bucharest after the signature of the Treaty of Passarowitz on July 10-21, 1718.

The 12ᵗʰ article of the treaty concerned the liberation of Nicholas. Sultan Ahmet III reconfirmed Nicholas on the throne of Wallachia on March 2ⁿᵈ 1719. The Voivode entered Bucharest on April 26ᵗʰ with great pomp and ceremony[18] where he resumed the publishing activities that had been interrupted by his abduction three years earlier. This time he was to assert his personal authority over the Wallachian nobility, and on religious and civil society by displaying the splendour of "noblesse oblige". Though Modern Greek, both vernacular and learned, was daily used in court and in the city, Nicholas would display his intellectual superiority as a Phanariot born in the capital of Hellenism, by publishing volumes written in "grec littéral", a form of purist and elegant ancient Greek. In December 1719 he brought out his *Treaty on Duties*. The book is illustrated with the Voivode Nicholas' coat of arms, finely worked, where one can see the emblems of both principalities, unified under only one crown and flanked by the

acronyms even more explicit than in the *Holy History* edition of 1716: IΩ NI AΛ BO. Here the Voivode's armorial shield brings together the scepter and the sword under the crown, flanked by two trumpets of fame blown by two cherubs.

The main text of the *Treaty on Duties* is preceded by seven eulogies, poems or letters that vaunt the prince's merits composed by George Trapezountios, the Metropolitan of Drystra Hierotheos Komninos and by Dimitrios Georgoulis Notaras. Concerning these flatterers the Amsterdam erudite Jean Le Clerc made, in his *Bibliothèque Ancienne et Moderne* (BAM), the following comment: "Quoi que ces Messiers entendent le Grec *litéral,* comme on parle dans le Levant, il s'en faut bien, qu'ils aprochent des Anciens, dans leurs vers. La Prose du Livre du Vaivode est d'un stile, beaucoup meilleur, que le leur"[19]. Much as Alexander the Exaporit's prose is stamped with an austere coldness and magnificence, so his son Nicholas' style stands out by virtue of its limpidity, sobriety, facility and elegance. The author of a commentary published in the learned review *Acta Eruditorum* of Leipzig in 1720 summarized the general opinion saying that: «Oratio hic est elegans, pura, perspicua, ingenua, non ad servilem imitationem composita, non Atticismos aut alia obsoleta temere affectans, non dialectos diversas imperite miscens, non calamistrata, nec sophistico tumore inflata»[20]. [The text in question is elegant, pure, clear, frank, not composed with servile imitation, without affecting here and there Atticisms or archaisms, without mixing awkwardly different dialects, not fussy, not inflated with sophistries].

The volume contains 19 chapters and constitutes a masterpiece of refined Attic Greek, or "grec littéral", as it was called in French. It helped to consolidate Voivode Nicholas' reputation as a wise philosopher-prince, and even as an exemplary Christian, but also as an author who could handle the koinè, a language with a high coefficient of refinement, with elegance and facility. Nicholas combines, with virtuosity, appropriateness - το πρέπον - and elegance - ἡ γλαφυρὰ σύνθεσις - according to the literary critic Dionysius of Halicarnassus [21].

As a code of deontology the work drew the attention of the scholarly public in Europe thanks to its publication by Thomas Fritsch in Leipzig in 1722, accompanied by Ştefan Bergler's Latin translation. This edition featured a renewed presentation: in addition to its refined and artistic typography, it presented an exquisite portrait of Voivode Nicholas, executed in 1721 by the engraver of the Prussian court, Johann G. Wolfgang [22]. Henceforth, in addition to the coat of arms, the Republic of Letters could now attach a countenance to his name and appreciate this fundamental text of neo-hellenic moral philosophy thanks to Bergler's Latin translation. The Leipzig edition of 1722 had such a great success that it was re-issued in London and Amsterdam in 1724 by Samuel Palmer and Gysbert Dommer. Unfortunately the portrait of Nicholas by Wolfgang was clumsily reversed by Joh. Georg. Schniebes[23].

The reputation of the prince-author now extended both to Europe and in the Romanian principalities, in spite of the fact that his other works, also composed in koinè took a much longer time to be published, as did his novel *The Leisures of Philotheos* (1800). But the reputation of the emerging Phanariot nobility was firmly established as it asserted itself in «un langage prestigieux: c'était la langue de Platon, de Thucydide, d'Isocrate», as Constantin Dimaras put it[24].

Two examples will be sufficient to demonstrate the evocative power and elegance of Nicholas Mavrocordatos' style. First, here is a lapidary nominal period extracted from the chapter XIII of his *Treatise on Duties:* «Δικαιοσύνης δὲ οὐ μόνον τὸ μὴ ἀδικεῖν, ἀλλὰ καὶ τὸ μὴ ἀδικεῖσθαι»[25]. Translation cannot easily render the density of the apophthegm: "Justice is not only not to commit injustice, but also to avoid being a victim of it". This sentence illustrates very well Demetrios' definition of the apophthegm in his Περὶ Ἑρμηνείας [On Style]: «ἔστι γὰρ καὶ ἀποφθεγματικὸν ἡ βραχύτης καὶ γνωμολογικόν, καὶ σοφώτερον τὸ ἐν ὀλίγῳ πολλὴν διάνοιαν ἠθροῖσθαι»[26] [For brevity characterizes proverbs and maxims; and a compression of a lot of meaning into a small space shows more skill]. Finally, let us quote a short extract from the chapter XI, "Περὶ ἀνδρείας": it corresponds to Demetrios' definition of the period: "ἔστι γὰρ ἡ περίοδος σύστημα ἐκ κώλων ἢ κομμάτων εὐκαταστρόφως πρὸς τὴν διάνοιαν τὴν ὑποκειμένην ἀπηρτισμένον"[27] [The period is a combination of clauses and phrases arranged to conclude the underlying thought with a well-turned ending.]; it is a masterpiece of rythmical balance, formed by a varied construction of κῶλα (clauses) and of two cascades of homeoteleutons[28]: «Ἔστι δὲ ἀληθὴς ἀνδρεία ἡ πάντοτε πραγματευο-μένη τὰ Θεῷ ἀρέσκοντα, τὰ εἰς δόξαν αὐτοῦ τείνοντα, τὰ κοινῇ συνοίσοντα. Οἱ δὲ τὰς ἰδίας ὠφελείας μαστεύοντες, καὶ δι'αὐτὰς κινδυνεύοντες, τῶν δὲ ὄντως καλῶν ὀλιγώρως ἔχοντες, ὤνιοί τινές εἰσι, καὶ βάναυσοι, τυφλώττοντες περὶ τὴν κρίσιν τοῦ ἀληθῶς καὶ ὄντως καλοῦ»[29]. Jean Le Clerc translates as follows: «La véritable force de l'esprit est celle, qui s'attache aux choses, qui plaisent à Dieu, qui tendent à sa gloire, qui sont utiles au Public. Mais ceux qui recherchent leurs avantages particuliers, sont des ames vénales, viles & aveugles, lors qu'il s'agit de juger de ce qui est réellement & véritablement bon»[30.] Le Clerc concludes: «Il faut avouër qu'il [le Vaivode] avoit bien du génie & de la connoissance de la Langue Greque»[31].

Conclusion

By way of conclusion, we note that no other contemporary author could equal Alexander and Nicholas Mavrocordatos' quality of style. Not only did they promote the "grec littéral"— refined Attic Greek—to the rank of the "hallmark of the emerging Phanariot nobility", but they also achieved a superior harmony of sonorities, images and metaphors, that reduced vernacular and ecclesiastical Greek to the level of common idioms.

Indeed, the Church used a simple language in order to be accessible to its congregations, or else the traditional idiom of the patriarchal chancellery, like the one in which Dositheos, patriarch of Jerusalem, composed the history of his patriarchate in 1247 pages, Ιστορία περί των εν Ιεροσολύμοις Πατριαρχευσάντων, printed in Bucharest by his nephew Chrysanthos Notaras in 1715[32]. A few decades later, when some ecclesiastics began archaizing, they did it so to excess, tastelessly, without any relation with Mavrocordatos' elegant koinè. For example, the famous Eugen Voulgaris composed his manual of logic *Η Λογική εκ Παλαιών τε και Νεωτέρων* in a language so archaic that the professor of logic Iosipos Moisiodax judged it incomprehensible[33]. In his leisure time the same Voulgaris translated Virgil's *Eneid* into Homeric verses much more abstruse than the original[34].

Once the Phanariot nobility's supremacy was recognized by the autochthonous nobility, prince Nicholas' successors did not deem it necessary to retain the same loftiness of style. Voivode Constantin Mavrocordatos, Nicholas' son, favoured instead the use of Romanian and vernacular Greek. Later, in 1780, strictly speaking at the beginning of the Greek and Romanian Enlightenment, Prince Alexander Hypsilantis published his legal code *Σύνταγμάτιον Νομικόν / Pravilniceasca condică*, in Romanian and vernacular Greek[35].

Finally, Nicholas, just like Dimitrie Cantemir, relied upon Latin to promote their written works among the savants of the Republic of Letters. Cantemir wrote his *Description of Moldavia* and his *History of the Ottoman Empire* in Latin: but, one is forced to recognize that Cantemir, in his haste to inform the reading public of the day, did not care to be as much a stylist in Latin, as Nicholas was in koinè. Certainly, Nicholas had at his disposal a language whose vocabulary and syntax had been refined by centuries, whereas Cantemir devised up a rough style, a language "rău scrisă"[36] both in Latin and in Moldavian.

Alexander the Exaporit and his son Nicholas enjoyed a posthumous glory as writers: their other works were published and widely diffused during the 18th and 19th centuries. After a period of latency corresponding to the decline of classical studies on a world-wide scale, we now notice a real revival of interest in Phanariot philology. But some quite significant texts are still waiting for translators and commentators if they are to take once again the place they deserve in the history of the Republic of Letters in the Early Enlightenment.

I would like to conclude with a symbolic comparison, that between two coats of arms:

First, one can see in a double-headed eagle, the arms of Ştefan Cantacuzino, hospodar of Hongro- Wallachia in 1714-1716. Thus the prince intends to claim his Byzantine heritage, pretending to belong to an imperial family settled in the Romanian countries. His armorial shield was published in the *Christian political Admonitions*, Bucharest 1715, by Antim, the metropolitan of Bucharest. It is followed by ten traditional political verses, written in vernacular Greek, with rhymes.

Io Ştefan Cantacuzino Voevod
Domn Oblăduitoriu Ţării Rumâneşti

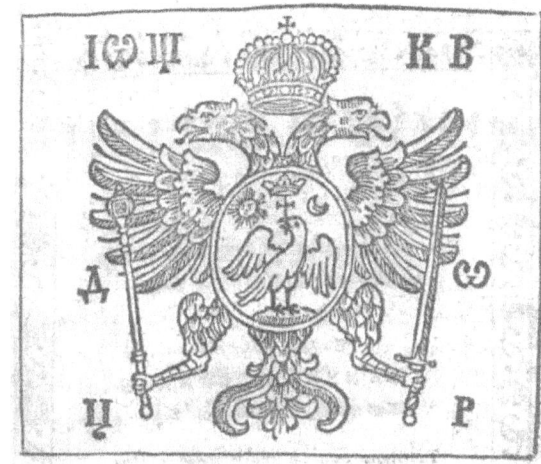

The second blazon represents the coat of arms of Nicholas Mavrocordatos, published in his father's *Holy History,* Bucharest 1716. Under the united arms of the two principalities, the description of Nicholas' shield, composed in archaic Greek, marks the difference with the preceding hospodars: ancient Greek is thence the hallmark of this emerging dynasty.

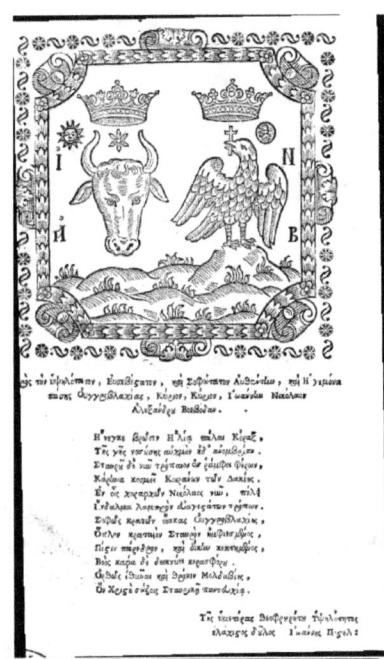

Notes:

[1]Stelian Brezeanu *et al., Relaţiile româno-elene,* Bucharest, Omonia, 2003. Cf. Paula Scalcău, *Hellenism in Romania,* Bucharest, Omonia, 2007.

[2]*Βασική Βιβλιοθήκη,* 8, p. 50.

[3]Παναγιώτις Δ. Μιχαηλάρης, *Η πραγματεία του Χρύσανθου Ιεροσολύμων "Περί Αφορισμού",* Athens, Poreia, 2002.

[4]Émile Legrand, *Ελληνικόν Επιστολάριον - Épistolaire grec,* Paris, Maisonneuve, 1888, passim. Cf. Athanasios E. Karathanassis, *Οι Έλληνες λόγιοι στη Βλαχία (1670-1714),* Thessaloniki, Ed. Bros. Kyriakidis, 2000.

[5]Ariadna Camariano-Cioran, *Les académies princières de Bucarest et de Jassy et leurs professeurs,* Thessaloniki, Institute for Balkan Studies, 1974.

[6]BRV, 1, p. 324-325. Legrand, *Bibliographie 17e,* 3, p. 5.

[7]Dan Râpă-Buicliu, BRV, Add. I, Galaţi, 2000, p. 226. Al. Duţu, *Coordonate ale culturii româneşti în sec. XVIII,* Buc., 1966, p. 47-49.

[8]BRV, 1, p. 499.

[9]É. Legrand, *Épistolaire, op.cit.,* p. 84.

[10]BRV, 1, p. 494. Cf. Ana Andreescu, *Cartea românească în veacul al XVIII-lea,* Bucharest, Editura Vremea XXI, 2004, p. 43.

[11]BRV, 1, p. 497-498.

[12]*Ibid.,* p. 501.

[13]Emil Vîrtosu, *Titulatura domnilor şi asocierea la domnie în Ţara Romînească şi Moldova (pînă în secolul al XVI- lea),* Bucharest, Editura Academiei Republicii PopulareRomîne, 1960, p. 84-86.

[14]Α. Μαυροκορδάτος, *Ιστορία Ιερά,* Bucharest,1716,p. (4).

[15]*Περί συνθέσεως ονομάτων,* VI, 3, 16. Translation by Stephen Usher, in Dionysius of Halicarnassus, *The Critical Essays in Two Volumes II,* Cambridge Mass., Harvard University Press, 1985, p. 33.

[16]*Ibid.,* VI, 22, 1.

[17]Α. Μαυροκορδάτος, *Ιστορία Ιερά,*Bucharest,1716,p. 11.

[18] Mihai Ţipău, *Domnii fanarioţi în Ţările Române (1711-1821),* Bucharest, Omonia, 22008, p. 133.

[19] BAM, Amsterdam, tome XIV, 1720, p. 116. Cf. BAM, tome XIV, 1720, p. 114-131. And, *ibid.,* tome XV, 1721, p. 84-95.

[20]*Acta Eruditorum,* Leipzig, 1720, p. 38.

[21]*Περί συνθέσεως ονομάτων,* VI, 20, 1 et VI, 23,1.

[22]Dan Râpă-Buicliu,*Bibliografia românească veche, Additamental,1536-1830,*Galaţi,Editura Alma,2000, p232

[23]*Ibid.,* p. 234.

[24]C.Th.Dimaras, Avant-propos, in Nicolas Mavrocordatos, *Les Loisirs de Philothée,* texte établi, traduit et commenté par Jacques Bouchard, Athènes-Montréal, Association pour l'étude des Lumières en Grèce - Les Presses de l'Université de Montréal, 1989, p. 9.

[25]*Περὶ Καθηκόντων Βίβλος,* Leipzig, Thomas Fritsch, 1722, p. 89. L. Kamperidis translates: "La justice consiste non seulement à ne pas commettre l'injustice, mais aussi à ne pas la subir", Nicolas Mavrocordatos, *Traité des Devoirs,* Athens, Fondation culturelle de la Banque Nationale de Grèce, 2014, p. 115. Cf. Plato, *Gorgias,* 474b- c.

[2] *Περί Ερμηνείας,* 9.

[27]*Ibid.,* 10.

[28]Aristotle, *Rhetoric,* 3, 9. 1410b1.

[29]*Περὶ Καθηκόντων Βίβλος, op.cit.,* p. 46.

[30]BAM, 1720, p. 125.

[31]*Ibid.,* p. 126.

[32]Μαγδαληνή Παρχαρίδου, Η γλώσσα του Οικουμενικού Πατριαρχείου, in *Ιστορία της Ελληνικής γλώσσας,* Αθήνα, Ελληνικό Λογοτεχνικό και Ιστορικό Αρχείο, 2000, σ. 208-209.

[33] Ι. Μοισιόδαξ, *Απολογία,* επιμ. Άλκης Αγγέλου, Αθήνα, Ερμής, 1976, p. 3.

[34]Μανόλης Α. Τριανταφυλλίδης, *Νεοελληνική Γραμματική, Ιστορική εισαγωγή,* Αθήνα, 1938, p. 322.

[35]*Suntagmavtion Nomikovn Alexavndrou Iwavnnou Uyhlavnth Boebovda hgemovno~ pavsh~ Ouggroblaciva~ 1780,* ekdidovmenon metæeisagwghv~ kai istorikhv~anaskophvsew~ twn en autwv qesmwvn ed. Panagiotes I. Zepos, Athens, Académie d'Athènes, 1936. Cf. *Pravilniceasca condica`, 1780,* edi†ie critica` Andrei Ra`dulescu, Bucharest, Editura Academiei Republicii Populare Romîne, 1957.

[36]N. Iorga, *Istoria literaturii româneşti în sec. al XVIII-lea,* p. 275. Cf. Virgil Cândea, "Stilul", *in* Dimitrie Cantemir, *Divanul,* ed. Virgil Cândea, Bucharest, Editura Academiei Republicii Socialiste România, p. 59-64.

From the Indo-European Trickster to the Romanian Păcală; a comparative study

Ana R. Chelariu, ARA member
achelariu@verizon.net
http://anarchelariu.wordpress.com

Abstract: The Trickster, as he is referred to in all the studies of mythology and folklore, is a fascinating character, whose function in the Indo-European social-cultural context is yet to be clarified. His presence is strongly visible from classic mythology to the recent folklore.

In mythology the Trickster is known as Hermes in the Greek heritage, Mercury in Roman culture, Loki in German sagas, Bricriu in the Irish songs, Pekulis/Patullos in the Baltic heritage, or Varuna in Hinduism.

In the European folklore the Trickster is found in each cultural complex: he is well known as Tyl Ulenspiegel in German countries, or Peik in Scandinavia, Pooca/Puca in Ireland, Velnius in the Baltic region, or Păcală in the Romanian tradition.

This paper will attempt to outline the Indo-European Trickster's traits as preserved in folklore particularly in the Romanian heritage.

1. Trickster in Mythology

According to Homer Hermes is the son of Zeus and Maia. Only a few hours old he steals Apollo's herds, and performs his first tricks: he wears huge sandals to hide his footprints, and makes all the cows walk backwards so that no one could tell where they are hidden. Then Hermes performs a significant ritual action: he sacrifices two cows, dividing the meat into twelve portions in honor of the twelve gods, but as the rules of sacrificial ritual require, he, as one of them, must not eat the flesh. He makes a fire by rubbing two laurel twigs, and cooks the meat. Although the aroma is appealing to him, he must restrain from eating. This episode explains why perhaps Hermes is often regarded as the god who taught people how to make fire. As the messenger of gods, master of the ancient mysteries of initiation, he is the instrument of transmitting the divine instructions to people. Returning home, he is no longer seen by humans, and dogs don't bark at him, signifying perhaps that his sacrificial act transformed him into the god that flies and can become invisible at wish [1].

Apollo's gratification for all his troubles is a musical instrument: a lyre, which makes beautiful sounds by the vibration of the air. In turn Apollo entrusts him with protecting his herds, thus confirming Hermes' position as the protector of shepherds, their flocks and the thieves.

Herodotus tells us that Hermes was regarded as a Pelasgian divinity of Thracian origin [2]. He was born in Arcadia, and particularly honored by the Arcadian shepherds as the protector of their flocks and huts. A rudimentary image of him was often found by the shepherds' doors as a symbol of veneration. This image was a bearded head pillar, and realistically modeled erect phallus. The phallus shaped stones or painted images are considered aggressive symbols of boundaries and fecundity. These 'hermai' are to be found in front of houses, in market places, but also at crossroads and at frontiers, signifying his role as the ethereal guide of the soul into the world of the dead.

The Lithuanian Pekulis, or the Prussian Patollus, are regarded as chthonic deities, agrarian helpers, who live near bodies of water. In the ancient times Pekulis was feared as the god of death. In the Christian era he was replaced by Velnias, the flying evil spirit, thus keeping the element of air as a medium and an attribute. According to Gimbutas [3], Velnias is a pastoral god, who helps the poor. He exhibits a perpetual resistance to Perkunas, the sovereign god As Hermes' symbol was a phallus, so Pekulis/Velnias often shows his enormous phallus to women. Velnias goes to weddings and frightens women who will not dance with him. He is known as the one who punishes the unfaithful. In a Prussian triptych Patullos is represented in the form of a horse's skull; his sacred objects are "the skulls of a man, a horse, and a cow" [4]. Velnias' parallel in the Old Russian is Veles or Volos, a god of horned cattle [5].

In Northern mythology Loki appears more as a companion of gods, and sometimes as a giant. He can change his sex and shape at will: in his female form he gives birth to monsters, thus causing considerable troubles for his companions. He is the 'mother' of Hell, the giantess who rules the realm of death. Loki scares the gods with old age and dying. He is an arch-thief stealing, or helping others to steal gods' treasures, and then helping the gods to recover them. In the poem "Lokasenna" [6] the gods try to keep him out of the hall of Aegir, where they are having a feast, but he manages to enter anyway. Once inside he betrays shameful secrets about the gods' cowardice and the goddesses' infidelities.

The Irish god Bricriu, whose nickname was Nemthenga, or Poison-Tongue, builds a splendid hall, preparing a feast to which he invites all the gods of Ulster. The gods refuse his invitation, but he threatens to make them kill each other, and to turn daughter against mother. After arriving at the party they try to keep him out of the hall, but he, as in the German version, manages to get in, and to incite the gods against one another. In the end, the hall is damaged, and he is covered with grime [7].

In the Indian mythology the pair Varuna and Mitra, have as their essential function to maintain the universal order. While Mitra looks after friendships and ratifies contracts, Varuna is the guardian of the oaths, upholds *ṛtá* the cosmic moral norm, thus he is the lord of moral order watching sinful behavior. He has ferocious, chaotic impulses if the rules are broken, punishing the guilty, actions that relate him to the Trickster character. Varuna is present at every gathering, witnesses every action. Varuna sees or 'shines' at night thus he is linked to the Moon, the place of the dead. He shares with Yama the title of King of the Dead. He regulates the motions of the heavens and the circulation of waters [8].

2. Trickster in folklore: Romanian Păcală

The Trickster from the folklore does not display major changes versus the mythological character. For comparison we will take the Romanian folk story of Păcală, and observe the corresponding motifs, as well as the differences.

In most of the Romanian folk tales anthologies the story starts with Păcală, the Trickster (Rom *a păcăli* 'to trick') who goes on top of a mountain, and builds a big fire, adding plenty of resin. A thick smoke rises up to God, who happens to have a cold. The smoke cures God's cold, who graciously offers Păcală anything he wants. The Trickster asks for a flute, a musical instrument very common among shepherds. The flute has magic powers; it makes anyone dance constantly until the music stops, as Pan's flute was making the satyrs and the fairies dance. Other mythical examples of this motif are Amphion, Orpheus, Oberon and Hameln, who are all creating magic with their musical instrument. The magic flute episode is found in Neo-Greek, Irish and German folklore [9].

Next Păcală inherits a cow, or a bull, and goes to the market to sell it. On his way he stops in the forest and sells his cow to a tree. Păcală gets upset because the tree would not answer him, or give him his money. Angrily, he hits the tree, and finds a treasure in its hollow. In the Western folklore the tree is replaced by a statue of a god or saint. In Aesop's fables one story relates how a man who asked Mercury's statue for money gets upset because the god does not seem to hear him, he breaks it and finds a treasure in the statue's head. In Cosquin's French collection Cadet Cruchon sells a piece of linen to a saint's statue, and because the saint does not want to pay him, he breaks the statue and finds a treasure. In the Neapolitan version Vardiello does the same thing, and Giufa from Sicily dye his linen green and sells it to a lizard; the treasure is then found in the lizard's house [10]. Remarkably, in the East European folklore this motif includes the cow/bull and the tree elements, whereas in the West European these elements are a saint statue and linen.

As the story goes on Păcală is hired by a priest to take care of his cattle (in some collections the story begins with this motif). The Trickster enters into a contract with his master: they agree to obey their deal until one of them gets angry, at which point the one who breaks the contract ought to be punished in various ways, most commonly, to have a piece of skin cut off his back, or to have his nose cut off. The Trickster starts immediately to do his malicious pranks, trying to force the priest to lose his patience. Some of his pranks are: making the priest and his wife dance until they are completely exhausted; while grazing the herds in the meadow he cuts a cow and eats it on the spot because the priest's wife doesn't give him any food; the next day he cuts a pig, than a lamb, and so on. Being told to clean the child, he does this is by peeling off the child's skin. When the wife tells him to cook a soup, and to add parsley to it, he also cooks the dog named Parsley.

In stories from Corsica, Picardia, Serbo-Croatia the priest from the Romanian folk story is replaced by a rich man or a king, who endure almost the same types of pranks. In the end the rich man or the king loses his patience, and gets punished [11].

In the Irish folklore the Trickster Pooka is considered essentially an animal spirit, whose name 'poc' means 'he-goat.' But he takes many shapes, horse, ass, bull, goat, etc. He could make a fine horse, but he has to be kept away from the sight of water, because he will plunge in with his rider and kill him. In an Irish folk story a piper meets with Pooka on his way to sing at a party; after a few spiteful pranks the Trickster gives him a pipe, and he becomes a famous piper [12].

The Scandinavian Peik gets horses and cattle from the bottom of the sea. Both, the Irish Pooka and the Scandinavian Peik have the magic flute that makes people dance until exhaustion. Peik is famous as a Trickster, whom the king always attempts to outsmart, but with no success [13].

The Trickster's name appears to be related to the Indo-European root * pek-. [14] which stands for livestock, domestic animal, cattle. It is possible that the Indo-European root is at the base of all these names: Pekulis, Patulos, Pooka, Pooc, Peik, Puc, Păcală, proving once more the incredible vitality of this character.

3. Similarities and differences

In comparing Hermes' actions with those of the Romanian Păcală, we can easily signal a number of conclusive similarities: the use of fire as a way please the divinity, the cows/animals sacrifice, the musical instrument, the fact that both take care of herds as shepherds, Hermes entrusted by Apollo with protecting his herds – Păcală hired as a shepherd. Both are supernatural characters tricking humans into trials. If Hermes' essential function is that of messenger of gods, in charge with bringing down the divine instruction, the Trickster may accomplish this task through his trial-pranks that actually may hide encoded messages. Playing the part of a catalyst the mythological character impacts the fortunes and behavior of the other gods. In folk stories the Trickster always puts an unsuspected fellow on trial with his tricks, bringing forward a moral message.

Besides similarities, there are some rather important differences between the two: Păcală doesn't steal the herd, while Hermes is entrusted with Apollo's herds after the boy steals (hides?) them from him. In the folk story the Trickster enters into a contract with the authority, and then starts his malicious pranks. Another difference to be observed is that Hermes as a god doesn't eat the meat he has sacrificed, whereas Păcală eats it. The significance of these differences may rest with the fact that ritualistic practices suffered modifications over time from myth to folklore.

4. Tricksters position in the Indo-European Pantheon

If we regard the main qualifications of the Trickster as a shepherd of flocks, or the association with sexual symbols, such as Hermes' phallic 'hermai' or Pekulis/Velnias showing his enormous phallus to women, the Trickster according to the Dumezilian Indo-European "class" system, represents the third class that governs wealth and fertility. Therefore the folk Tricksters, Păcală, Pooka, Peik, and ultimately the Lithuanian god Pekulis, belong to the agrarian pantheon, the third class function.

On the societal level the Trickster's pranks are addressed to the divine authority, Apollo or the other gods. In the West European folklore this divine authority is represented by the king of the land, or by the landlord. In the Eastern Europe the priest takes the function of divine authority. Parallel to the Trickster of ancient mythology who plays tricks on the other gods, the folk character plays tricks on the god's representative on Earth, the authority represented by a priest or a king. When the Trickster enters into a contract with the priest and forces him to break it, he maintains the judicial order, entering into the first class attributes as gods' messenger. Păcală always punishes the dishonest priest, the unfaithful wife, the thrifty fellow, and the perverts. He is in a permanent conflict with the bad spirits and the devil. He keeps the ethical order within the community. He seems to be closer in his vigilance to Varuna, the god who keeps a very sharp eye on every contract, and punishes bad behavior.

Why is this character playing tricks on the other gods? Is it because, as suggested by Gimbutas and others, he may be part of an older pantheon, a pre-Indo-European god? Due to lack of documentation such argument remains a speculation. Applying the Dumezilian classification we could argue that the Trickster's malicious pranks may be in some cases the result of his third function position. When he exposes the greediness and the stupidity of the king or the priest and their reaction through his pranks, the story enhances moral values within the social group, and the ethical and moral principles are transmitted from generation to generation.

References

1) *Homeric Hymn 4 to Hermes.* H. G. Evelyn-White, ed.,
http://www.perseus.tufts.edu/hopper/text?doc=Perseus%3Atext%3A1999.01.0138%3Ahymn%3D4

2) *Herodotus, Histories 2. 51.* translation by A. D. Godley. Harvard University Press, Cambridge, 1920.

3) Gimbutas Marjia. *The Balts.* Frederick A. Praeger, New York: 1963

4) Puhvel, Jaan. *Comparative Mythology.* Johns Hopkins Univ. Press, Baltimore, London, 1987, 224.

5) Eliade, Mircea. *Istoria Credintelor si Ideilor Religioase,* Universitas, Chişinău, 1994.

6) *Poetic Edda* translated by Lee M. Hollander, University of Texas Press, Austin, 2nd ed., 1986.

7) *Early Irish Myths and Sagas:* translated with an introduction and notes by Jeffrey Gantz,Penguin, 1981.

8) Dumezil, Georges. *Mitra-Varuna; an Essay on Two Indo-European Representations of Sovereignty,* Zone Books, New York, 1988.

9) Şăineanu, Lazăr. *Basmele Române în comparaţiune cu legendele antice clasice.* Minerva, Bucureşti, 1978.

10) Şăineanu, Lazăr. *Ibidem.*

11) Şăineanu, Lazăr. *Ibidem.*

12) *Treasury of Irish Folklore.* Ed. Padraic Colum, New York, 1954.

13) *Scandinavian Folk and Fairy Tales.* Edited by Claire Booss, Avenel Books, New York, 1984.

14) Pokorny, Julius. *Indogermanisches etymologiscches worterbuch,* Francke Verlag, Bern, 1959, 797.

LES PROBLEMES IDENTITAIRES EN REPUBLIQUE DE MOLDAVIE – APPROCHE LINGUISTIQUE

Ana Guţu
docteur, professeur universitaire
Membre honoraire à vie de l'Assemblée Parlementaire de l'Europe
Université Libre Internationale de Moldavie

Dans une Europe qui se diversifie progressivement le bilinguisme semble ne plus satisfaire les besoins de communication sociétale autant au niveau officiel qu'au niveau de la locution courante à travers l'espace communautaire. Le slogan européen « unité dans la diversité » s'appuie, au premier chef, sur la diversité linguistique, et, par conséquent, culturelle, de l'Europe.

Langue, pluri-(bi)-linguisme – approche doctrinaire. « *Nous sommes tous des polyglottes…ou presque, ou nous pouvons du moins le devenir* » (Walter, 1997 : p.9). Dans une Europe qui se diversifie progressivement le bilinguisme semble ne plus satisfaire les besoins de communication sociétale autant au niveau officiel qu'au niveau de la locution courante à travers l'espace communautaire. Jamais l'esprit babélien n'aurait connu une telle profusion linguistique et n'aurait atteint un degré tellement haut de transversalité dialogique, si aujourd'hui ce n'était pas l'époque de grandes délocalisations culturelles, et, par conséquent, langagières.

Le slogan européen « unité dans la diversité » et celui francophone– « vivre dans la diversité » s'appuie, au premier chef, sur la diversité linguistique, et, par conséquent, culturelle, de l'Europe. L'aventure du concept est beaucoup plus ancienne que cela ne paraît.

Dans la période où apparaissent les premiers volumes de l'Encyclopédie, l'abbé Pluche, dans *La Mécanique des langues et l'Art de les enseigner* (1751), avait rappelé qu'une première différenciation de la langue, sinon dans le lexique, au moins dans la variété d'inflexions entre une famille et l'autre, avait déjà commencé à l'époque de Noé. Pluche va plus loin : la multiplication (qui n'est pas la confusion) des langues apparaît comme un phénomène, à la fois naturel, et socialement positif. « *La confusio linguarum devient la condition historique de la stabilisation de certaines valeurs de l'Etat. En paraphrasant Louis XIV,* Pluche est en train d'affirmer que " L'Etat c'est la langue „ »* (cité d'après Eco, 1997 : p.383).

Face au cloisonnement médiéval, à l'élitisme de la Renaissance, à la rigueur classiciste, au cosmopolitisme conservateur, l'explosion révolutionnaire des mobilités humaines transocéaniques et transcontinentales fait naître sous nos yeux une nouvelle société: omnisciente, interculturelle, cosmopolite et polyglotte. A la recherche d'une théorie réconciliatrice entre le structuralisme classique et le pragmatisme langagier, les savants continuent leurs débats sur les concepts de langue, parole, discours, texte etc. Une terminologie abondante, parfois difficile à gérer du point de vue logique – hiérarchique, circule à l'intérieur de différentes écoles doctrinaires.

Notre préoccupation dans cet article sera la problématique des langues et du schéma de son fonctionnement du point de vue de son cadre social.

Nous proposons une définition fort générale de la langue, et notamment : *la langue en tant qu'outil de la communication c'est ce qu'une élite scientifique dans une société donnée à une époque donnée considère langue.* Malgré toutes les atomisations possibles du concept (saussurienne, greimassienne, peircienne, jakobsonienne etc), la langue ne peut fonctionner que sur des segments socio-historiques donnés, en stricte concordance avec la culture, les traditions et la mentalité de l'époque.

A part d'être un instrument de communication, *la langue est un pouvoir du point de vue politique et social.*

Alors, être bi- ou multilingue - c'est quoi, précisément? L.Bloomfield définit le bilinguisme par la "*maîtrise de deux langues comme si elles étaient toutes deux la langue maternelle"*. Cette position absolutiste définit de fait les "bilingues parfaits" ou "vrais bilingues" ou encore les "ambilingues". E.Weinreich définit le bilinguisme de façon moins absolue : "*Est bilingue celui qui possède au moins une des quatre capacités (parler, comprendre, lire, écrire) dans une langue autre que sa langue maternelle.*" E.Haugen se place résolument dans les compétences de production : "*Le bilinguisme commence lorsque l'individu peut produire des énoncés ayant un sens dans une langue autre que sa langue maternelle.*" C.Hagège considère une personne comme étant bilingue lorsque ses compétences linguistiques sont comparables dans les deux langues (cité d'après Claude Stoll <http://averreman.free.fr/aplv/num54-bilinguisme.htm>). Georges Mounin considère: « *bilinguisme - le fait pour un individu de parler indifféremment deux langues. Egalement - coexistence de deux langues dans la même communauté, pourvu que la majorité des locuteurs soit effectivement bilingue: on peut parler du bilinguisme espagnol-catalan pour la Catalogne espagnole. Certains sociolinguistes américains réservent le terme bilinguisme à la première définition seulement, et utilise diglossia (diglossie) pour le bilinguisme des collectivités* » (Mounin, 2004: p.54).

La question de savoir laquelle des langues peut être considérée pour une personne langue maternelle (*«langue de la mère, par abus de langage, langue première d'un sujet donné, même si ce n'est pas la langue de sa mère* » - Mounin, 2004 : p.198) a eu plusieurs réponses dans les études sociolinguistiques. Certains sont d'avis qu'une fois que la personne **pense dans une langue**, celle-ci peut être considérée sa langue maternelle. Par exemple, je me surprends souvent à penser (à part le roumain) en français, en espagnol, en russe. Des fragments de raisonnements m'arrivent aussi en anglais, langue que je n'ai jamais apprise, mais qui s'est emparée de mon esprit en vertu de son utilisation dans les médias. Du point de vue scientifique on pourrait rajouter à cette caractéristique de la **pensée** les quatre composantes de la connaissance approfondie d'une langue afin d'exercer d'une manière plénipotentiaire l'acte de la communication – **expression écrite, expression orale, compréhension écrite, compréhension orale**. Nous considérons que pour compléter la définition des caractéristiques de la langue maternelle, il faut y rajouter une, fort importante : **la création poétique**. Autrement dit, si la personne fait des vers, de la poésie, dans une langue sans difficulté et empêchement, cette dite langue est pour elle maternelle.

Sans doute, une personne polyglotte est dans la plupart des cas une personne érudite. Schleiermacher écrivait très éloquemment à propos des polyglottes : «*ces maîtres admirables qui se meuvent avec une égale aisance dans plusieurs langues, pour lesquels une langue apprise parvient à devenir plus maternelle que la langue maternelle.* » (Schleiermacher, 199 : p.63). La connaissance de plusieurs langues implique indubitablement l'activation (le déclic) de plusieurs centres neuronaux qui réfère à des réalités extralinguistiques multiples: aimer en français, penser à des choses philosophiques en roumain, chanter en espagnol, jurer en russe ou en anglais. Cette fonction civilisatrice de la langue, réfère-t-elle à des identités multiples ou pas? L'identité multiple serait-ce une fiction ?

Le représentant d'un trilinguisme en exercice, George Steiner est embarrassé de s'autoidentifier Anglais, Français ou Allemand. Descendant d'une famille mixte, Steiner a acquis les trois langues dans son enfance précoce. Steiner affirme qu'il lui est difficile de dire avec précision quelle a été la langue qu'il a commencé à parler la première: «*Je n'ai pas le moindre souvenir d'une première langue. Autant que je puisse m'en rendre compte, je suis aussi à l'aise en anglais qu'en français ou en allemand. Les autres langues que je possède, qu'il s'agisse de les parler, de les lire ou de les écrire, sont venues par la suite et sont marquées par cet apprentissage conscient* » (Steiner, 1998 : p. 173). Le cas de Steiner est une solution heureuse pour un polyglotte – s'autoidentifier de manière multiple, une autoidentification qui va jusqu'à un cosmopolitisme acceptable, autant qu'il promeut l'enrichissement interculturel, la tolérance et la libre circulation des valeurs.

Politiques linguistiques. Les politiques linguistiques menées par l'état vise, d'un coté, **l'aménagement linguistiques au niveau sociétal dans un espace multilingue** (l'exemple du Canada, de la Belgique, de la Suisse, de la République de Moldavie), et d'autre coté, **les politiques**

éducationnelles dans le domaine de l'enseignement des langues. Les deux axes ne sont exempts de conflits linguistiques, dégénérant parfois dans des conflits politiques et interethniques. L'élaboration et l'application des politiques linguistiques correctes constituent le garant de la démocratie dans une société. «*L'exercice de la démocratie et l'intégration sociale dépendent des politiques linguistiques éducatives: la capacité et les occasions d'utiliser toute la richesse de son répertoire linguistique sont essentielles pour participer aux processus démocratique et social et, en conséquence, aux politiques d'intégration sociale.* » (Le plurilinguisme, la citoyenneté démocratique en Europe et le rôle de l'anglais, in http://www.coe.int/t/dg4/linguistic/Source/BreidbachFR.pdf, consulté le 24 septembre 2012).

La question la plus difficile réside dans le fait comment trouver l'équilibre entre une langue majoritaire, par exemple, qui ne s'est pas encore affirmée en tant que langue assurant la communication entre les locuteurs des langues minoritaires et comment avancer sur la voie de l'intégration sociétale des locuteurs des langues minoritaires?

Situation en République de Moldavie. Mais, comment définir l'autoidentification dans l'espace post-communiste ? Dans la République de Moldavie la coercition de la langue russe a généré un bi- ou multilinguisme acquis par naissance (en vertu des mariages mixtes juifs, roumains, russe, gagaouzes, bulgares) qui réfère à une seule identité – l'identité russe. Plus que cela, il y a pas mal de cas où les parents sont de différentes nationalités, mais les enfants ne connaissent ni la langue de la mère, ni la langue du père, sinon seulement le russe, la langue de la formation à l'école ou à l'université. Ce phénomène de la *réduction linguistique* ne devrait pas se propager en Europe, qui se propose un apprentissage de deux langues étrangères à part la langue maternelle.

La coercition de la langue russe. A part d'être un instrument de communication, la langue est un pouvoir du point de vue politique et social. Or, selon nous, le bon exercice de la langue, aussi bien dans la variante écrite que dans celle orale, mène inévitablement à la coercition de la langue (Gutu, 2014, p. 15-33). Nous proposons ce terme juridique pour justifier toute une série de phénomènes sociaux, ayant des racines linguistiques. L'histoire

des sociétés modernes nous démontre largement les manifestations coercitives de la langue qui a été mise au service des pouvoirs politiques. A partir de 1812 – après l'annexion de la Bessarabie par la Russie – démarre un processus accru de dénationalisation et de russification par le biais d'abord de l'Église, ensuite par l'évincement du roumain comme langue de l'éducation. C'est là que commence la confusion identitaire sur le territoire actuel de la République de Moldova. La coercition du russe s'est manifestée dans son statut de langue de communication dans une fédération de 15 républiques, au sein de laquelle le russe est devenu la langue officielle de l'Union, sans que ce principe n'ait jamais été reconnu dans la Constitution soviétique. Le russe est devenu la langue de l'empire de 285 millions de personnes, comprenant quelque 130 langues nationales. Durant 70 ans la langue russe a exporté dans les 15 républiques attitudes et comportements, idéologie et réactions. Cette exportation coercitive de la langue et avec elle de l'idéologie, continue de multiplier le brouillage impressionnant des essences identitaires.

Brouillages identitaires. Qui étions-nous, les habitants de la République de Moldavie dans l'ex-URSS ? Je suis née dans un village aux bords du fleuve Prout, le fleuve qui sépare la République de Moldavie de la Roumanie. Dès mon enfance j'ai toujours écouté et regardé la radio et la télévision roumaines sans aucun problème (la proximité frontalière des antennes le permettait largement). Il y avait des livres roumains dans la bibliothèque de la famille, mes parents étant professeurs de langue et littérature roumaines (ainsi dite langue moldave). J'éprouvais une confusion que je sensibilisais dans mon for intérieur – pourquoi écrire en roumain tout en utilisant l'alphabet cyrillique ? La phobie envers l'écriture en langue roumaine en alphabet cyrillique a mené à un autre phénomène: celui de l'appropriation d'une autre langue et de sa culture avec elle. J'ai lu et j'ai connu la majorité des chefs–d'œuvre de la littérature universelle en russe. Cette littérature de traduction a laissé une empreinte colossale dans mon esprit de futur linguiste et professeur. C'est à cette époque, pareillement à d'autres compatriotes miens, que je suis devenue une bilingue parfaite, le roumain étant ma langue maternelle, le russe étant une sorte de langue maternelle seconde par adoption. Le russe était conçu par à l'époque comme un fétiche. Cette langue d'un grand peuple a fait naître un sentiment identitaire qui, pensait-on souvent, ennoblissait spirituellement la personne. A l'époque soviétique il convenait mieux de s'autoidentifier russe que d'une autre nationalité. Les Moldaves, les Géorgiens, les Asiatiques, étaient implicitement des gens de catégorie seconde. Dans les villes on préférait s'identifier comme des Russes. Les parents préféraient envoyer leurs enfants dans des écoles russes, pour avoir, disait-on, « un avenir sûr ». Les hommes moldaves, surtout les hauts fonctionnaires du parti communiste, préféraient se marier avec des femmes russes, car c'était « de bon ton », c'était bien vu par le parti et cela assurait une carrière professionnelle brillante dans la nomenklatura communiste.

Les Russes qui sont venus habiter en Moldavie après 1944 ont réussi à s'adapter dans un milieu linguistique non-slave, sans apprendre la langue roumaine. A présent le bilinguisme est souvent pratiqué par les habitants autochtones et non pas par leurs concitoyens d'autres ethnies. Le bilinguisme moldave est une véritable bigamie linguistique due à la jouissance quotidienne. Mais comme dans tout couple, pour la solidité et le bonheur durable, cette jouissance doit être réciproque. Sinon, le phénomène dérape et donne naissance aux conflits linguistiques. Le pire est que la langue russe était porteuse d'une idéologie. C'était la malédiction du phénomène ; d'un autre côté, la langue russe est la langue d'un grand peuple, qui a donné à l'universalité de grands écrivains, poètes, philosophes et penseurs et de ce point de vue, l'apprentissage du russe a servi à l'ouverture des esprits intellectuels.

Antinomie roumain-moldave. Il faut préciser que la langue roumaine fait partie de la famille des langues romanes, elle est d'origine latine et use de l'alphabet latin. Selon l'opinion des linguistes, y compris russes, la langue moldave n'existe pas (c'est un glottonyme). Le parler moldave, à côté de celui munténien a servi de base pour la constitution de la langue roumaine moderne. En 1924 a été créée la République Autonome Soviétique Socialiste Moldave (l'actuel territoire de la Transnistrie), où le dialecte, qui y était parlé, a été élevé au rang de langue officielle utilisant l'alphabet cyrillique. Entre les années 1918 et 1940 et 1941 et 1944 le territoire de la Bessarabie a été réuni avec la Roumanie. En 1944 avec l'arrivée de l'armée russe/soviétique, la langue roumaine a été complètement évincée de l'éducation, et la langue moldave écrite en alphabet cyrillique a été utilisée jusqu'en 1989 en tant que langue de la République Soviétique Socialiste de Moldavie.

En 1989 le parlement moldave a voté le passage de l'alphabet cyrillique à l'alphabet latin. La même année la loi sur le fonctionnement des langues sur le territoire de la République de Moldavie fut adoptée. Cette loi permettait aux minorités ethniques de développer et préserver leurs langues – le russe, le gagaouze, le bulgare, l'ukrainien - en les utilisant dans le système d'éducation. En 1991 a été adoptée la Déclaration de souveraineté, dans laquelle le roumain figure comme langue de l'État souverain créé. Malheureusement, dans la Constitution de la République de Moldavie, adoptée par une majorité du parti politique agraire (héritier de l'ancien parti communiste), l'article 13 stipule que la « langue d'État de la RM est le moldave écrit en alphabet latin ». Cette réalité a permis le déclanchement de disputes politiques et sociales, dont l'essence se réduisait à un conflit entre les nostalgiques communistes et la vague démocratique.

La problématique de la dénomination correcte de la langue officielle en République de Moldavie est bien connue par son contexte politique, ayant une histoire mouvementée, descendant dans la nuit des siècles. La langue officielle parlée en République de Moldavie est le roumain – la même langue que celle parlée en Roumanie, langue officielle de l'Union Européenne. C'est une réalité correspondant également à une vérité scientifique, car, les linguistes le savent bien, il n'existe pas de langue moldave. Pourtant, à cause des événements historiques liés à l'occupation tsariste russe et ensuite à celle soviétique, le politonyme « langue moldave » est devenu courant avec le temps sur le territoire de la Bessarabie, qui est couvert aujourd'hui partiellement par la République de Moldavie, état indépendant depuis 1991. Dans la Déclaration d'Indépendance de 1991, le nom de la langue officielle de la République de Moldavie est introduit correctement – la langue roumaine. Mais, dans la constitution moldave, adoptée en 1994 par un parlement majoritairement nomenklaturiste, très loin intellectuellement des vérités scientifiques, l'article 13 mentionne que la langue officielle de la République de Moldavie est « *le moldave fonctionnant en utilisant l'alphabet latin* ». Cette dualité de dénomination de la langue officielle semble être anodine à première vue, mais elle a causé et continue de causer beaucoup de scandales dans les débats publics de nature politique, sociale et économique. La dispute autour de la dénomination correcte de la langue officielle en République de Moldavie donne naissance à des argumentations et contre-argumentations spéculatives, qui n'ont rien à voir avec les sciences du langage, mais plutôt alimentent la nostalgie soviétique dans une partie de la population. Il faut mentionner que le binôme **langue roumaine-langue moldave** est une réalité en République de Moldavie, car la moitié de la population appelle la langue qu'elle parle **le roumain**, une autre moitié de la population l'appelle **langue moldave**. Une question surgit : que disent les intellectuels ? La réponse est univoque – les intellectuels, les écrivains, les philosophes du pays sont du côté de la science, bien sûr – le nom correct de la langue est le roumain.

Pourtant, le « maudit » article 13 de la Constitution est toujours là, impossible à changer, car il n'y aura pas de sitôt au parlement des académiciens qui fassent une majorité constitutionnelle afin de modifier l'article 13 en faveur de la vérité scientifique. Cette situation génère des « lâchetés » bureaucratiques – dans les textes de lois, les documents normatifs, les fonctionnaires évitent d'utiliser le syntagme « *langue roumaine* », en le paraphrasant par « *langue d'état* » (une autre erreur, cette fois d'abord de traduction et terminologique, car le terme approprié est « *langue officielle* », « *langue d'état* » étant une transposition du russe « *gosudarstveniy iazyk* »).

Avant d'entrer dans la politique, à coté de mes collègues linguistes et littéraires, nous étions un peu « réfugiés » dans notre monde académique, où tous parlent le même langage, celui de la raison scientifique. Une fois placée dans un contexte politique explicite, j'ai pu sentir « sur ma propre peau » tous les inconvénients de la dualité allant jusqu'à la rivalité du binôme « langue roumaine-langue moldave ». Subir des offenses de la part d'une opposition communiste extrémiste à cause du simple fait que tu es un scientifique et que tu appelles les choses par leurs propres noms – voilà une frustration qui est inimaginable dans un milieu académique. L'argument de base pour toutes les spéculations en faveur de « la langue moldave » reste toujours et encore l'article 13 de la constitution moldave. Les militants d'extrême gauche, eux-mêmes plaidant pour l'orientation vers l'Est de la politique extérieure de la République de Moldavie, se positionnent en tant que juri(dici)stes et affirment

« A, donc C » : dans l'article 13 de la constitution il est écrit « langue moldave », donc la langue officielle de la RM est la langue moldave. Je suis d'avis que cette structure argumentative proactive est très simplifiée, Ducrot vient encore appuyer de manière théorique, plus générale, notre perception « *Dans cet enchaînement argumentatif le sens de l'argument A contient en lui-même l'indication qu'il doit être complété par la conclusion... Il n'y a donc pas à proprement parler passage de A à C, il n'y a pas de justification de C par un énoncé... Par conséquent, il n'y a pas de transport de vérité, transport d'acceptabilité, depuis A jusqu'à C... »* (Ducrot, p. 22).

Les scientifiques disent : « la langue parlée en République de Moldavie est le roumain, car, d'abord, il existe une identité claire et nette entre la langue littéraire de la Roumanie et celle de la République de Moldavie, c'est-à-dire qu'il n'y a pas de différences entre la norme littéraire de la langue parlée en Roumanie et celle parlée en République de Moldavie... Cette structure argumentative rétroactive semble convaincre davantage, elle est plus profonde, plus raisonnée. Les arguments complémentaires s'enchaîneront, ils seront nombreux, complexifiés, très pertinents du point de vue scientifique, mais, hélas, très longs ! C'est un sérieux inconvénient quand on essaie d'être persuasif devant un public moins élevé, moins informé, plus « simple d'esprit ». Ainsi, en République de Moldavie nous assistons à un schisme de l'imaginaire linguistique collectif, une partie de la population, consciente et bien formée, connaît la vérité scientifique, et l'autre se complaît dans une sorte d'égocentrisme régional, adorant le qualificatif « moldave », l'utilisant à tort et à travers, y compris par rapport au nom de la langue, même s'ils se rendent très bien compte de l'absurdité de leurs convictions, car aucune source lexicographique ou scientifique n'atteste la langue moldave.

En tant que linguiste œuvrant dans le domaine politique, j'ai depuis toujours été préoccupée par la nécessité de faire émerger la vérité scientifique visant le nom correct de la langue officielle de la République de Moldavie, en dépit des normes juridiques qui ont élevé au rang de loi constitutionnelle une formule fausse. L'idée m'est venue de saisir la Cour Constitutionnelle en vu d'interpréter l'article 13 de la constitution moldave par rapport au texte de la Déclaration d'Indépendance de 1991, où est utilisé le syntagme

« langue roumaine ». Après avoir eu un échange d'opinions avec le linguiste Bernard Cerquiglini, qui m'a fait part de l'expérience de la France en matière de constitutionnalisation de certains principes de la Déclaration des Droits de l'Homme et du Citoyen du 26 août 1789, j'ai pensé à une possible acceptation du modèle français. Depuis 1971 cette Déclaration fait partie du bloc de constitutionnalité en vertu de la référence faite à la Déclaration dans le Préambule de la Constitution française de 1958 [1]. Cette même année 1971, le Conseil Constitutionnel de France a constitutionnalisé le principe de la liberté de réunion. *« Le Conseil constitutionnel a ainsi attribué valeur constitutionnelle aux textes auxquels la Constitution du 4 octobre 1958 fait référence dans son Préambule et qui ont été approuvés comme tels par le peuple français. Appartiennent donc désormais au « bloc de constitutionnalité » non seulement les articles de la Constitution proprement dite, mais aussi la Déclaration de 1789, les principes politiques économiques et sociaux particulièrement nécessaires à notre temps contenus dans le Préambule de la Constitution de 1946, les principes fondamentaux reconnus par les lois de la République, auxquels ce Préambule se réfère, ainsi que, depuis la révision constitutionnelle du 1er mars 2005, la Charte de l'environnement de 2004. »* [2] La saisine que j'ai faite à l'adresse de la Cour Constitutionnelle le 26 mars 2013 comportait une question [3] : peut-on établir le signe d'égalité, de manière formelle, entre le syntagme « la langue moldave fonctionnant en utilisant l'alphabet latin » et le syntagme « la langue roumaine » ? Dans l'argumentation je faisais référence au texte de la Déclaration d'Indépendance de 1991, à quelques

articles de la Convention Européenne des Droits de l'Homme et à la Déclaration Universelle des Droits de l'Homme, ainsi qu'à l'opinion scientifique nationale et internationale à propos de la langue roumaine. Par cette question je voulais obtenir une réponse positive afin de permettre aux bureaucrates de l'État d'utiliser sans empêchement dans les textes normatifs le nom correct de la langue – la langue roumaine. Dans la deuxième question de ma saisine, déposée le 19 septembre 2013, je faisais référence directe à l'expérience du Conseil Constitutionnel de la France, en sollicitant la constitutionnalisation du texte de la Déclaration d'Indépendance de la République de Moldavie de 1991.

Le 5 septembre la Cour Constitutionnelle a examiné ma saisine et a émis une décision historique [4] : la Déclaration d'Indépendance fait bloc constitutionnel commun avec la constitution de la République de Moldavie et en cas de prévision divergente le texte de la Déclaration prévaut sur celui de la Constitution. La divergence la plus saillante entre la Déclaration d'Indépendance et la Constitution moldave c'est le binôme langue roumaine-langue moldave, qui au cours des siècles s'est transformé en une véritable antinomie, parfois irréconciliable, divisant la société, rendant difficile la popularisation des arguments scientifiques, servant de fondement pour la manipulation politique.

Il est aussi certain que la seule décision de la CC ne peut pas garantir la mise en place immédiate de l'utilisation seule acceptable et correcte du glottonyme « langue roumaine ». L'expérience nous montre qu'on atteste une résistance obstinée à la mise en place de la décision de la CC, les bureaucrates ayant l'habitude de se débarrasser difficilement de leurs usages ambigus et non-engageants (une peur et une lâcheté provenant de l'époque soviétique et néo-communiste des années 2001-2009).

Selon l'opinion de la majorité des intellectuels de Moldova, la République de Moldova est un « morceau » d'espace géopolitique, un « morceau » de nation, un « morceau » de culture qui est roumain. En République de Moldavie le

---

[1]   Décisions n° 71-44 DC [archive] « Liberté d'association » du 16 juillet 1971 et n° 73-51 DC [archive] du 27 décembre 1973, qui l'ont intégrée dans le bloc de constitutionnalité en raison de la référence faite à la Déclaration dans le préambule de la Constitution de 1958. << http://fr.wikipedia.org/wiki/D%C3%A9claration_des_dr oits_de_l'homme_et_du_citoyen_de_1789#cite_note-2>> Consulté le 4 mars 2014.

[2] Droit pénal et droit constitutionnel <<http://www.conseil-constitutionnel.fr/conseil-constitutionnel/root/bank_mm/pdf/Conseil/penalconsti t.pdf>> Consulté le 4 mars 2014.

[3] Saisine Nr.8b du 26 mars 2013. <<http://www.constcourt.md/public/files/file/Sesizari/2 013/08b_26.03.2013.pdf>> Consulté le 4 mars 2014.

[4] <<http://www.constcourt.md/libview.php?l=ro&idc=7 &id=512&t=/Prezentare-generala/Serviciul-de-presa/Noutati/Textul-Declaratiei-de-Independenta-prevaleaza-in-raport-cu-textul-Constitutiei-Sesizarile-nr-8b2013-si-41b2013/>> Consulté le 4 mars 2014.

discours identitaire est loin de correspondre à la conception d'une monade nationale, il est plutôt triadique : roumain, russe et moldave. C'est un discours qui détermine même l'essence des partis politiques. Nous sommes à la veille des élections parlementaires, peu importe la doctrine politique, c'est le discours identitaire qui compte : aller vers la Roumanie (Union Européenne), s'orienter vers la Russie, ou plaider pour le renforcement de l'État moldave. Ceux qui s'autoidentifient comme Roumains rendent tribut à la vérité scientifique linguistique et historique, dans le contexte intégrationniste avec la Roumanie. Ceux qui s'autoidentifient comme Russes – en dépit du fait qu'ils sont des citoyens de la République de Moldavie – sont soit des Russes de génération en génération, soit des descendants de familles mixtes (Ukrainiens plus Russes, Gagaouzes plus Russes, Moldaves plus Russes etc). Ceux qui s'autoidentifient comme Moldaves sont les représentants de la génération des nostalgiques, qui ont été profondément marqués par l'idéologie communiste.

Selon les données statistiques du dernier recensement (2004) en République de Moldavie (excepté la Transnistrie) il y a environ 77% de Moldaves, 8,3 d'Ukrainiens, 5,9% de Russes, 4,4 de Gagaouzes et 2,2% d'autres ethnies. Le gouvernement communiste, qui se maintient au pouvoir depuis 2001, est accusé d'avoir truqué les données, surtout du point de vue de l'autoidentification. Car, selon ces données, seulement 2,2 % de la population recensée se sont identifiés comme des Roumains, ce qui semble être une falsification. Le nombre des intellectuels faisant partie d'une organisation non-gouvernementale, le Forum Démocratique Roumain de la République de Moldavie, dépasse 250.000 (6,4% de la population recensée en 2004). C'est une ONG pro-roumaine, qui plaide pour la réunification de la République de Moldova avec la Roumanie.

Solutions. La République de Moldavie actuellement doit brûler l'étape de la formation d'un État-nation, il n'en est pas question dans les conditions des échecs consécutifs de 25 ans d'indépendance. Selon nous, la République de Moldavie ne sera jamais un État-nation. L'identité moldave, à la rigueur, celle civique, peut être difficilement construite durant plusieurs décennies à savoir centaines d'années. La seule solution viable pour l'instant qui sauvegarderait la langue roumaine

parlée en République de Moldavie, mais aussi les citoyens, de plus en plus nombreux quittant le pays, c'est la réunification de la République de Moldavie avec la Roumanie. La nation roumaine est la seule nation européenne constituée historiquement qui reste divisée après la seconde guerre mondiale. Il est temps de réparer cette injustice historique et politique.

En matière de politiques linguistiques la situation peut changer surtout dans la perspective d'une réforme profonde dans l'éducation. En République de Moldavie cette question n'a pas été encore résolue de manière judicieuse. C'est aux facteurs de décision politique de prévoir dans leurs programmes électoraux de manière explicite une stratégie à part visant les politiques linguistiques sur le territoire actuel de la République de Moldavie, des politiques basées sur le respect de la population autochtone et sa langue, mais aussi sur des instruments efficaces de stimulation de l'inclusion sociale des minorités ethniques.

« *La vérification de la connaissance de la langue du pays d'accueil est en train d'être placée au centre des politiques d'immigration dans la plupart des pays européens, du moins, dans ceux de l'UE. L'évolution sur cette question est très rapide* » (Truchaut, 2008, p.67). Malheureusement, en République de Moldavie, cette exigence fonctionne uniquement dans le cas des immigrations récentes, et non pas dans le cas des minorités nationales et ethniques qui habitent le pays depuis des décennies. La non-acceptation de l'obligation de connaître la langue du pays constitue une véritable source du conflit linguistique roumain-russe, qui dure à la longue des années.

La solution acceptée par la population moldave est celle du cosmopolitisme : tous parlent roumain, mais certains l'appellent le roumain (niveau de formation supérieur), d'autres, le moldave. Les jeunes citoyens de la République de Moldova parlent facilement les langues : roumain, russe, anglais, français ou autres. Ils s'auto-identifient comme des Roumains. Et c'est l'éducation qui y contribue largement.

L'avenir linguistique de l'Europe réside dans le développement de la communication interculturelle, qui, à son tour, est inconcevable en dehors d'un héritage polyglotte. Les langues ouvrent les frontières et les horizons.

La République de Moldavie doit résoudre immédiatement ses problèmes linguistiques et instituer des politiques linguistiques correctes. Et ce sont les linguistes qui doivent apporter leur contribution au sujet, non pas seulement les politiciens. L'action des linguistes doit être concertée, fondée scientifiquement et orientée pragmatiquement vers la vie réelle. Les linguistes doivent quitter leurs bureaux et les salles des bibliothèques pour annoncer aux politiciens qu'ils existent, qu'ils sont là, nombreux et décidés à changer les choses, et que c'est à eux de mettre le point final sur le I dans le problème de la langue roumaine en RM, qui doit être indépendante de tout pouvoir politique, afin d'exclure à l'avenir toute tentative d'exploitation de la langue au profit des intérêts politiques. *« C'est pourquoi l'action humaine sur les langues, si elle veut être un peu plus qu'un fantasme de maître, doit se développer en indépendance de tout pouvoir. La part que prend le linguiste, lorsqu'une situation le rend légitime, dans le travail de planification et de réforme, est, à côté de l'enseignement des langues, de la traduction et de la réponse au défi informatique, une des grandes voies d'application qui peuvent donner à son activité une prise réelle sur le cours des choses. »* (Hagège, 2002, pp. 270-271). Nous assistons actuellement à un changement des générations, qui sont au moins bilingues (roumain-anglais), ou tri-quadrilingue (roumain-anglais-français-russe). Ce n'est pas par hasard que j'ai mis le russe en quatrième position. L'intérêt pour cette langue a sensiblement diminué avec les changements socio-économiques de la transition et avec l'exode en masse des citoyens de Moldavie à l'étranger. La facilité pour les langues de la population de notre république est un atout incontestable, mais la tragédie de nos compatriotes bilingues ou polyglottes consiste encore dans le problème de la confusion identitaire qui règne dans les mentalités sociales.

Références bibliographqiues :

1) Barthes R. *Le bruissement de la langue. Essais critiques IV*. Paris, Editions du Seuil, 2002.
2) Eco U. *La recherche de la langue parfaite dans la culture européenne*. Traduit de l'italien par J.-P.Manganaro. Paris, Editions du Seuil, 1997.
3) Gutu A. *Les pouvoir de la langue*. Chişinău, ULIM, 2014.
4) Hagège C. *L'homme de paroles. Contributions linguistiques aux sciences humaines*. Paris, Fayard, 2002.
5) *Les langues du monde*. Paris, Pour la Science, 1999. Mounin G. *Dictionnaire de la linguistique*. Paris, Quadrige/Puf , 2004.
6) Modreanu S. « Identité (s) fuyante (s) ». In : *La francophonie et la nouvelle identité européenne*. Iasi, Éditions Universitaires « Alexandru Ioan Cuza », 2008. P. 333-339.
7) Schleiermacher F . *Des différentes méthodes du traduire*. Paris, 1999.
8) Steiner G.(1998) : *Après Babel*. Paris, Albin Michel, 1998.
9) Truchaud C. Europe : enjeu linguistique. La documentation française, Paris, 2008.
10) Walter H. L'aventure des mots français venus d'ailleurs. Paris, Laffont, 1997.
11) Décisions n° 71-44 DC [archive] « Liberté d'association » du 16 juillet 1971 et n° 73-51 DC [archive] du 27 décembre 1973, qui l'ont intégrée dans le bloc de constitutionnalité en raison de la référence faite à la Déclaration dans le préambule de la Constitution de 1958. << http://fr.wikipedia.org/wiki/D%C3%A9claration_des_droits_de_l'homme_et_du_citoyen_de_1789#cite_note-2>> Consulté le 4 mars 2014.
12) Droit pénal et droit constitutionnel <<http://www.conseil-constitutionnel.fr/conseil-constitutionnel/root/bank_mm/pdf/Conseil/penalconstit.pdf>> Consulté le 4 mars 2014.
13) Saisine Nr.8b du 26 mars 2013. <<http://www.constcourt.md/public/files/file/Sesizari/2013/08b_26.03.2013.pdf>> Consulté le 4 mars 2014.
14) Déclaration d'indépendance de la République de Moldavie <<http://www.constcourt.md/libview.php?l=ro&idc=7&id=512&t=/Prezentare-generala/Serviciul-de-presa/Noutati/Textul-Declaratiei-de-Independenta-prevaleaza-in-raport-cu-textul-Constitutiei-Sesizarile-nr-8b2013-si-41b2013/>> Consulté le 4 mars 2014.

ONLY SHOUTS AND WHISPERS ~ NUMAI STRIGĂTE ŞI ŞOAPTE

Dr. Dinu Leonte, ARA MEMBER
ESS Global, Hayward, California U.S.A.
dinu.leonte@yahoo.com

Abstract: This paper is an extension of the ideas initiated eight years ago at the ARA 32nd Congress in Boston, USA [1], many of these poems have been written when the main topic of eschatology was initiated, being faithfully illustrated later within my papers published by the ARA 33rd Congress in Sibiu and the ARA 34th Congress in Bucharest, Romania. Preserving the initial format of those papers, new poems from my book "Amorfe si Cristaline" [2] and [3] were added, as reflected at the ARA 35th in Timisoara, where, for the first time here the topic of TRANSDISCIPLINARITY [4] has been approached. This last subject was enthusiastically adopted by the attendees of the ARA 36th Congres in Bari, Italy, within a Round-Table [5]. The last chapter of [3] has Romanian lyrics for some international hits I wrote, or English lyrics well suited for unforgettable Romanian evergreens. In addition, in this paper is also continued the topic of LIMERICKS. But as always, beyond any shadow of doubt, most of these poems express my gratitude to God, to whom alone should be given all the praise and glory for our accomplishments in this life.

Keywords: Eschatology, Transdisciplinarity, Education, Limericks.

1. INTRODUCTION

I would love to present with boldness and humility alike two of my poems belonging to the Christian topic. Some of these poems are not published yet. Some others were already printed in [3], the new revised and extended edition of [1]. In the next paragraphs, fix-form poems, epic poems, philosophical poems and lyrics poems will follow. Finally, in Romanian, a new kind of fix-form poetry, the limerick - is written.

2. CHRISTIAN POETRY

According to the Word of God, the Bible, the most important thing for each one of us is to not disregard the awesome warning about the imminent End of Time, when it will be too late to seek God's redemption…

SECOND PETER THRE

The Holy Bible teaches bold
In Second Peter, Chapter Three
The way the prophets were foretold
Of when the End of Time will be.

For GOD designed two Judgment Day
At seven thousand years apart,
Because the mankind went astray
In lawlessness of mind and heart.

In Second Peter Three, verse Six
GOD gives a warning: seven days!
Of Noah's Flood destruction speaks
And mercy in the Ark of Grace!

We learn what GOD has meant to say
In Second Peter Three, Verse Eight
About a thousand years - one day…
Therefore, GOD gave us here a date!!

Just seven days 'till Flood begun
Had Noah's family to spend;
So - seven thousand years are gone
With selfsame day for this world end!

8 February 2009, Hayward, CA

TREI FELURI DE OAMENI

Trei feluri de oameni în lume trăiesc
Şi-n vremea din urmă asemenea sînt
De cum îşi înclină balansul lumesc
Spre tot ce se spune de ei în Cuvânt…

O parte-s aceia ce nu au habar
Că lumea-i creată de Domnul Iisus.

Ei nu știu nimica de-al dragostei har
Ce-aduce -ntre oameni Lumina de Sus.

Sînt alții ce știu de Scripturi oarecum
Și-ncearcă să schimbe la bunul lor plac
A Domnului Lege-n a vieții lor drum,
Cătând împăcare-ntre înger și drac...

Și-n urmă sînt unii, ce-i drept mai puțini,
Ce viața cu grijă în rugă și-o cern.
Acei care-ascultă Cuvântul divin
Dorindu-și salvarea în Raiul etern.

30 Mai 2008, Miami, FL

3. FIX-FORM POETRY

There are many species within the fix-form poetry, some of them more usual such as the sonnets, and the rondels, other ones less usual such as the pantoums, the gazels, the tertines. Because of the space constraints, I selected here only a rondel and a sonnet.

DE VINĂ-S GROPIȚELE TALE
-rondel -

De vină-s gropițele tale
Nu-i nimeni să ți le priceapă
Că-n ele-ar putea să încapă
Duzini de palate regale!...

Privindu-le nimeni nu scapă
Și moare în cântec de jale.
De vină-s gropițele tale
Nu-i nimeni să ți le priceapă...

Și eu întâlnindu-te-n cale
Calc azi tot din groapă în groapă.
Colind străzi în noapte, agale
Cu genele pline de apă...
De vină-s gropițele tale.

15 Noiembrie 1960, Ploiești

PĂCAT...
-sonnet -

Păcat de ochii tăi de îngeraș
Că-i umple-adesea râs de drăcușor
Să nu-i mai pot nicicând uita ușor
Când îmi aduc minciunile răvaș...

Păcat și de-ai obrajilor bujori...
Prea nu știu cum de fragezi, prea gingași!
Din ce grădină oare îi furași
Cu-atâta farmec în petala lor?

Păcat de-al buzelor lichior amar
Un strop din ele doar, m-a îmbătat
Când l-am sorbit din zâmbetul fugar!

Păcat că țes și eu un vis ciudat
Și-l zvârlu-n calea ta ca pe un zar...
Păcat de mine și de toți păcat!...

20 Decembrie 1959, Ploiești

4. EPIC POETRY

This category of poetry in [3] is found within many poems written even from my teenager years. For instance, here I choose *La Moartea lui Cuza*, which illustrates my answer in the last year of high school to the centennial celebration of the Union of the Romanian Principates at 24 January 1859.

LA MOARTEA LUI CUZA
-fragment-

Doinea pustiu un cânt de jale vântul.
Peste câmpii cădea în picuri seara.
Și-n orizont își ridica argintul
Un chip de lună gălbejit ca ceara.

Drumul de fier părea un trup de șarpe.
Zăcea bolnav pe mutele întinsuri,
Când se-auzi din frunze, ca din harpe,
Un șuier surd desprins din necuprinsuri...

Tren mortuar. Venea ca o omidă
Târându-și trupul obosit pe șine.
Și se trudea alene, să-și deschidă
Un drum prin câmpuri de tăcere pline.

Tren mortuar. Şireag de boabe negre.
Aluneca din sânul mort al serii.
Iar lângă drum şedeau salcâmi de veghe
Cântând în cor un cântec al durerii...

Şi trenul poposea din gară-n gară,
Din sat în sat, în fiecare haltă.
În urma lui curgea popor de ţară.
Şi se-auzea în piepturi cum tresaltă

Un cântec trist. Şi s-a deschis deodată
O uşă grea la pat-vagon, în laturi...
Se adunau sătenii să-l mai vadă
Şi tot plângând, făceau moşnegii sfaturi...

Acolo sus zăcea-ntre flori sicriul.
Şi era lună. Razele de ghiaţă
Din chipul ei cioplit cu bisturiul
Înseninau făptura fără viaţă.

Şi el dormea. Ce dulce-i era somnul
Pe chipul blând, mai blând ca niciodată!
El era Cuza Vodă. Era domnul.
Şi faţa lui zâmbea ca şi-altădată...

Femei, flăcăi, copii de prin tot satul
Uitaseră de-a zilei sărbătoare.
Şi vestea alerga de-a lung, de-a latul...
Şi hora se rupea în drumul mare...

Tăceau cobzarii. Scripcile asameni.
Ţăranii-şi luau căciulile în mână
Şi nu era din cât amar de oameni
Să n-aibă unul inimă română...

Se-apropiau de tren, suiau într-însul.
Ştiau că domnul lor le-a dat o lege.
Şi jăluind şi podidindu-i plânsul
Îi sărutau lui Vodă fruntea rece...
....

Aşa a fost. Toţi fraţii lui Moş Roată
Au mas de cum a spus povestitorul...
Furat, pierzându-şi încropirea toată
S-a calicit mereu mai rău poporul...

19 Ianuarie 1959, Ploieşti

5. PHILOSOPHICAL POETRY

Now, after more than twenty five years from those so-called "first free democratic elections" in Romania, where in fact the second-line communists took over the power, I can still remember vividly how we sang full of hopes for the first time that night of December 21st 1989 in the University Plaza of Bucharest the refrain which soon will become the Romanian National Anthem "Wake Up, Romanian!" …

Later, one of those days of spring before the election, I climbed to the Balcony of Geography Faculty among others, and I recited to the crowd my poem, entitled "Dreptate", which ends with these verses :

Destul ai tot răbdat până acum
În beznă, foame, frig şi mascaradă,
Cu noi speranţe, la-nceput de drum
Dreptate, ochii plânşi vor să te vadă !

6 Mai 1990, Bucureşti

Fifteen years later, as an immigrant from my native country, Romania, I wrote the English version of this poem entitled "Justice" , ending with these verses :

You bear enough in darkness still to stay!
You must break your red chains right now or never
Renewing hopes are shining now your way
O, Justice, weeping eyes call you forever!

SPERANŢĂ ŞI NOROC

Noian de amintiri ne împresoară
Când lunecăm cu gândul spre trecut
Ne-atinge-aripa unui dor de ţară
Pe care încă nu l-am cunoscut.

Se luptă-n noi Românul cu Pribeagul
Şi nu ştii cine-o fi învingător.
Pe-acest meleag ţi-ar fi mai mare dragul
Să te-mplineşti şi să nu-ţi fie dor!

Tu vezi cum azi norocul te răsfaţă
De-ai să mai dai prin lume un ocol
Deşi lăsaşi prieteni de o viaţă

Și mama-ți simte-acasă locul gol...

Când ai știut ce-nseamnă suferința
Trăind aceleași vremuri cu cei mulți,
Ce răzvrătire ți-a umplut ființa
De nu mai poți de păsul lor s-asculți?

Atunci când ți-a fost țara la răscruce
Și ai văzut murind atâți eroi
Ai înțeles că orișiund' te-ai duce
Nu-i pasă nimănui ce-a fost la noi...

Ai înțeles că s-au schimbat stăpânii
Și că în rest sunt toate cum au fost,
Deși ici colo latră liberi câinii
La caravana vieții fără rost...

Și-atunci ai ascultat de-un glas al firii
Și te-ai mutat sub soare în alt loc -
Aici, pe-un țărm al păcii și-al iubirii
Cu muguri de speranță și noroc!

Thanksgiving Day 1998, San Jose, CA

Să-ți aduci aminte de iubirea mea mereu...
Poate-or fi unii când te-or dansa
Să-ți dea vre-un fior vânzându-ți vorbe mari!
Tu să știi să nu-ți dai inima
Pentru vraja lor c-ai să regreți amar!

Refren: Dar nu uita că mâine-oi fi tot eu
Acela care te-o chema...
Să-ți aduci aminte de iubirea mea mereu...

Dar mai mult aș vrea să fii a mea
Cel din urmă dans, te rog!
Și-atunci timpu-aș vrea să stea în loc,
Altceva nimic n-aș vrea...

Hai, vino, să știi că inima mea
Nu uită ușor că mi-ai luat-o tu.
Dac-o fi un altcineva să te conducă-n zori
Te rog răspunde-i NU!

Refren: Și nu uita că mâine-oi fi tot eu
Acela care te-o chema...
Să-ți aduci aminte de iubirea mea mereu...

Să-ți aduci aminte de iubirea mea mereu...

31 Martie 1962, București

6. LYRICS

From the high school years, when the direct access to the western music was a rare privilege in my native country, I found a way of singing beautiful songs whose lyrics I did not know, by writing my own Romanian texts for them. I continued to write these kinds of lyrics for many years to come. The book [3] has an entire chapter, which brings together these lyrics. Among them, an emotional connotation is tied to the original French version *La dernière Danse* in the interpretation of Dalida. Finally, my own English translation in 1977 for the song *Rugă pentru Părinți*.

CEL DIN URMĂ DANS

Poți dansa câte dansuri dorești
Și cu toți acei pe care-i vei plăcea!
Draga mea, nu te-opresc să zâmbești
Nici celui care-l vrei să strângă mâna ta!

Refren: Dar nu uita că mâine-oi fi tot eu
Acela care te-o chema...

DAY BY DAY
(Adrian Păunescu: Rugă pentru Părinți)

Day by day the truth is sad
Though we're trying to deny -
Among us grow weak and die
Our dearest Mom and Dad!

Among us grow weak and die
Our dearest Mom and Dad!

Lord, alleviate their pain
By Thine everlasting Word!
Bring them back to us, my Lord!
Make them strong and young again!

Bring them back to us, my Lord!
Make them strong and young again!

11 November 1977 , București

7. LIMERICKS POETRY

A **limerick** is a five-line poem in anapestic or amphibrachic meter with a strict rhyme scheme (aabba), which intends to be witty or humorous. It may have its roots in the 18th-century Maigue Poets of Ireland, although the form can be found in England in the early years of the century in literature, arts, science and many other human fields. The following example is a limerick of unknown origin.

> *The limerick packs laughs anatomical*
> *In a space that is quite economical,*
> *But the good ones I've seen*
> *So seldom are clean,*
> *And the clean ones so seldom are comical.*

PENTRU CE?

Pentru ce lăsăm lumea cuvântului
Ca s-o bată hachițele vântului?
 Sînt atâtea de spus
Prin poruncă de Sus...
Să nu facem doar umbră...Pământului!

19 Decembrie 2014, Truckee, CA

QUANTUM [6]

O dezbatere-n termeni de etică
Pentru QUANTUM nu-i una cosmetică-
Între Einstein și Bohr
N-a fost lucru ușor
Să-și dispute esența estetică...

29 Februarie 2016, Bahamas

REFERENCES:

1) *PROCEEDINGS ARA Congress XXXII*, July 22-26, 2008, Wentworth Institute of Technology, Boston, MA, USA pp 337.

2) Leonte, D. I., *Amorfe și Cristaline,* Casa Cărții de Știință, Cluj-Napoca, 2006, 309 pages.

3) Leonte, D. I., *Amorfe și Cristaline,* Casa Cărții de Știință, Cluj-Napoca, 2009, 307 pages – ISBN 973-686-823-0.

4) *** Transdisciplinarity Theory and Practice – Edited by **Basarab Nicolescu**, (Advances in Systems Theory, complexity, and the human sciences)   Hampton Press, Inc Cresskill, NJ 07676

5) *PROCEEDINGS ARA Congress XXXVI* May 29-June 3, 2012 Italy-PROLEGOMENE IN TRANSDISCIPLINARITY Round Table.

6) Manjit Kumar, QUANTUM EINSTEIN, BOHR, and the great debate about the nature of Reality.

Tectonics of the Symbol

Dr. Livio Dimitriu, Professor of Architecture and Theory/History

Sponsored by Pratt Institute School of Architecture
200 Willoughby Avenue, Brooklyn, NY, USA, 11205
usainst@gmail.com

Abstract: There are two reoccurring representations of the Cross in Christian icononography. If they are approached tectonically instead of a symbolism disconnected from fabrication, the attention focuses on the crucified body acting as a structurally stabilizing element. This allows for a totally new approach to the meaning of the Crucifixion and its symbolism, this time literally rising from the reality of fabrication.

The two representations of the cross are fused with their respective meanings. The standard cross is simply symbolic, where the T-cross is symbolic but also tectonic. The result forms a "stable" image at many levels of discourse. The body of Christ nailed to the Cross provides the triangulation required for a stable structure. It is not the Cross but the body of the Savior that provides stability and thus salvation.

The *Colonel Uricariu Monument* built by the architect Livio Dimitriu in the Ghencea Military Heroes Cemetery in Bucharest, Romania (1998), exemplifies the plastic potential of this new palimpsest-like logic as related to the representation of the cross over time. The development of the project argues for a process of design which does not follow the standard obsession with the linear development of the image and frozen early canon, but allows materials, methods, environmental factors, societal concerns and a multitude of other real factors, to take the design into constantly surprising directions until reaching the moment of building.

Essay

This project for *Colonel Uricariu Monument* was developed and concluded in a rapid succession of sketches all the way through its detailed execution drawings during a five-hour period. A mere thought and cultural references were used only as a catalyst for the design process. The transformation of the thought was due to a series of factors all involving reality checks in addition to formal concerns. The process was governed by an understanding and respect for the spirit of the early Christian faith as embodied in its thought and in the Orthodox faith. Above all, there was a constant reference to the simplicity of execution, particularly given the local material and technology available in Romania, and other assorted pragmatic concerns.

At every step of the design process the proposal changed under self-imposed and welcomed pressures and limitations offered by working with marble. The development of the final form appears and is non-linear. The only linearity present in the process is that of a concern with ease of fabrication given the limited machinery and craft available in Romania, the physical characteristics of the local Ruschita marble and its reaction to weathering in a climate that features rather drastic seasonal changes of temperature, resistance to vandalism, and cost control.

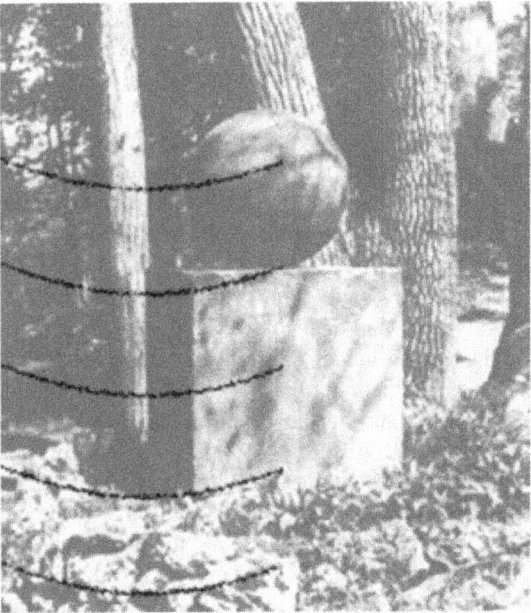

Figure 1. Johann Wolfgang von Goethe, *The Altar of Good Fortune / Monument for Charlotte*, Weimar, Germany (1777)

The point of departure was marked by a relatively little-known monument designed by Goethe, *The Altar of Good Fortune* (1777). This Illuminist homage to his lover Charlotte, placed in the poet's garden in Weimar, Germany (Figure 1). This beginning established the criteria for abstraction that remained present throughout the process and continued to dominate the final result. The motif of the cross was introduced immediately after, along with the lyrical theme of a perceptual illusion involving the image of a heart. Human perception and scale followed almost instantly.

Spherical volumes were abandoned rather early due to cost and material waste, and uncertainties regarding the craft and capability of the local stone cutters ability to turn out such a shape flawlessly. Solutions that involved protruding elements and/or assembly of standard thickness pre-cut slabs were soon eliminated because of concerns with the local seasonal variations of temperature as well as vandalism. (Figure 9)

The monument's measurements were determined by the height of the warrior, 1.80 meters, standing up-right in death as in life. The verticality is not only a metaphor. It refers to the tradition of many cultures around the world to have heroes buried vertically. The plan proportions ultimately became an anthropomorphic 30x50 cm due to limitations of weight and handling that could be accommodated by the type of lift-forks available locally and on-site. This limitation forced a welcomed frontality for the monument, with a perceptual rotation in space and around the monument achieved through the motif of the cross and its typological variations wrapping around this *stella*, and with the dynamic "axonometric heart" placed at one of its lower corners.

The outer surfaces of the marble block are slightly bush-hammered while the recessed surfaces of the carved-out cross motif are polished. This surface treatment virtually minimizes the blinding effect under sun-light that marble can have if seen as a large polished marble volume. In this monument, light is only reflected off the minimal surfaces receding in the mass of the block, and place in evidence the name of the deceased and dates. These factual concerns resonate deeply however with the early Christian Church. For St. Augustin, it is the Christian soul that shines bright, regardless of the outer body and worldly shape.

Galla Pacidia in Ravenna, like so many other early Christian religious buildings, echoes this view (Figure 2).

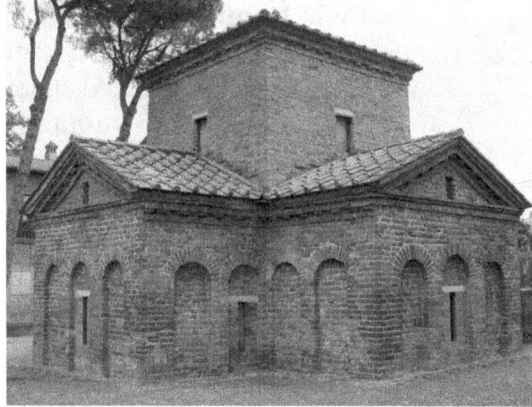

Figure 2. Galla Placidia Church, Ravenna, Italy, (425-450 AD).

Galla Placidia's simple, exposed and weathered brickwork façade highly contrasts with the rich soul of gold and polychrome mosaics on the inside. The monument incorporates the spirit of a tradition and religion. In the case of the *Monument for Colonel Uricariu*, the architecture and its contemporary language also alludes biographically to an imperfect body, bush-hammered by the war, in contrast with the human spirit that continued to shine bright in life as in afterlife.

The front of the monument carries the machine-cut incision of the standard four arms cross. The arms of the cross literally wrap around and embrace the body of marble and become, on the opposite face, the "true" three arms cross of martyrdom. The western painting tradition abounds in examples of both, and yet precious little has been analytically written on the subject. (Figure 3).

The four armed cross is a symbol. The three arms cross is the actual instrument of torture, as revealed by any effort at understanding efficient and simply detailed wood construction. It is the martyred body on the three arms cross that renders the structure stable by triangulation of the static forces at play. It is the crucified body, and thus the meaning of the sacrifice of the Saviour, that makes the symbol statically strong. (Figure 4). This is the tectonics of the symbol.

The wrapping of the two crosses joined in a continuous figure around the body of the monument is an invitation for the viewer to move around in space, despite the relative frontality of the work

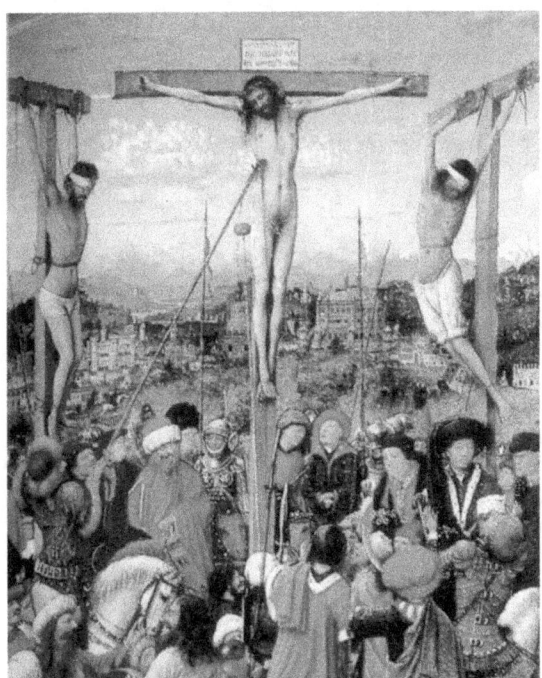

Figure 3. Hubert and/or Jan van Eyk. Detail, *The Crucifixion*, c. 1420-1425. Tempera and oil on canvass transferred from panel, The Metropolitan Museum of Art, New York, USA.

in a procession similar to the women in Christ's entourage that had witnessed the Crucifixion. Two matching semi-circular small incisions on two adjacent faces of the marble volume, and one of the lower corners, generate in anamorphic perspective the reading of a heart, modest and timid, but a reminder of an ever-present gratitude as well as an allusion to Christ's heart bleeding for us all. The "heart" is readable from eye level as one looks down while leaning against the monument. The carved inscription and the "heart" were gilded at a later date, so as to further underline the captive warm meaning frozen inside the coldness of the marble. (Figure 5, 6)

Doina Uricariu matched this essay in stone with a volume of poetry entitled *The Axonometric Heart*. The volume honours the memory of Colonel Uricariu, consistent with the family's centuries-long writing tradition traced back to Axinte Uricariul, the medieval chronicler of Romania.

While our position initially involves the preconceived abstraction contained in Goethe's monument only as a point of departure, it becomes eventually free of the dictatorship of this image.

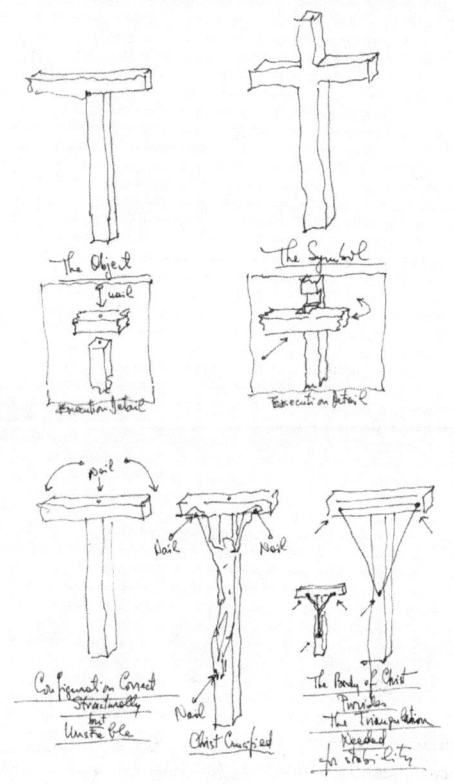

Figure 4. Livio Dimitriu, sketches. The tectonics of the Cross: actual feasible fabrication and the crucified and nailed body stabilizing an unstable structure.

The monument owes its abstraction to a rich and layered meaning involving the numerologically significant number 11, the total number of standard machine-generated cuts performed onto the marble block. The rotation of the stone block on the cutting bed during execution makes one recall the torture of the Saviour while the nails were driven into His body while He was still on the ground. (Figure 7, 8). The anamorphic heart is a carefully inserted rhetorical device generated by two additional half-round cuts that were obtained by simply lowering a circular blade of a selected diameter into the marble, and only half way its diameter.

Similarly to the volume and space of Galla Placidia, the spirit of this project attempts to address and to incorporate the continuity over millennia of a quote from Pindar, appropriated by St. Augustin in his *Confessions*, and brought into modernity by Paul Valéry in his poem *The Maritime Cemetery*: "The light foot-steps of a white

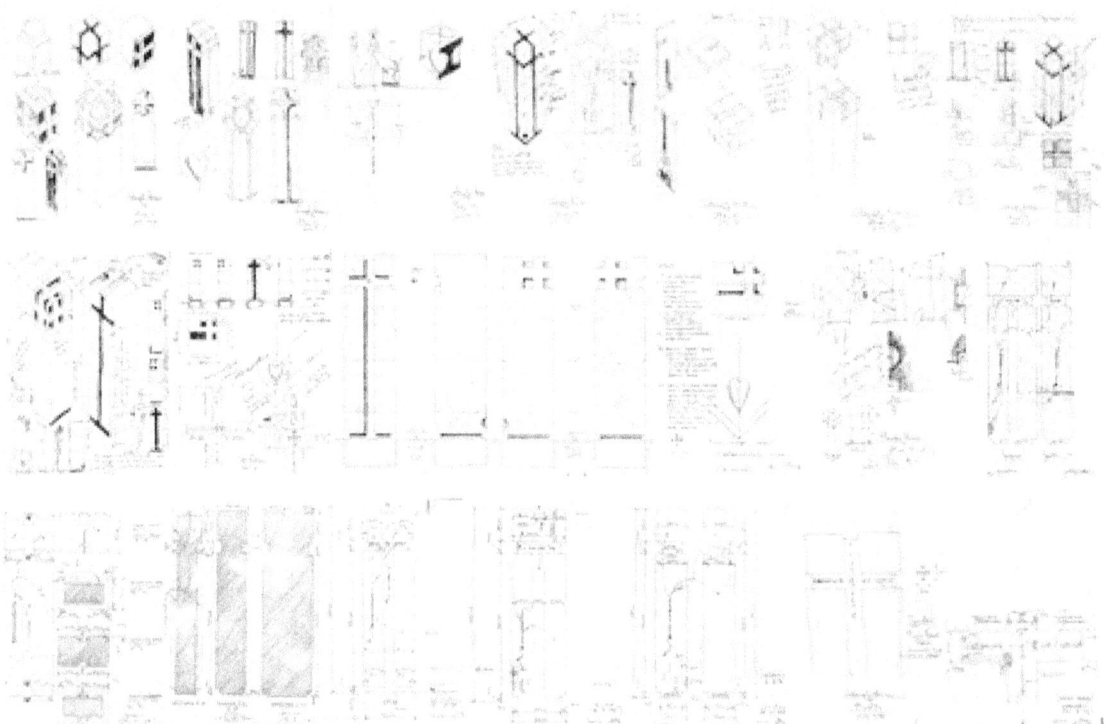

Figure 9. Livio Dimitriu, arch. Non-linear design development, first sketch to finished project.

dove on the roof of my soul". The project's conception constantly hovers between the perception as dictated by the physical eye and that of the axonometric governing the eye of the mind.

Bibliography:

Livio Dimitriu, *Monument for Colonel Gheorghe Uricariu*, in *Arhitext Magazine*, No. 5, May 2004, Bucharest, Romania.

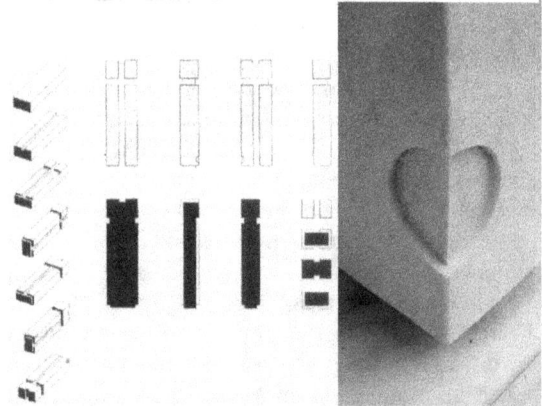

Figure 7, 8. Livio Dimitriu, arch. *Colonel Gheorghe Uricariu Monument*, instructions for numerologically significant eleven marble cuts; Detail, "the axonometric heart".

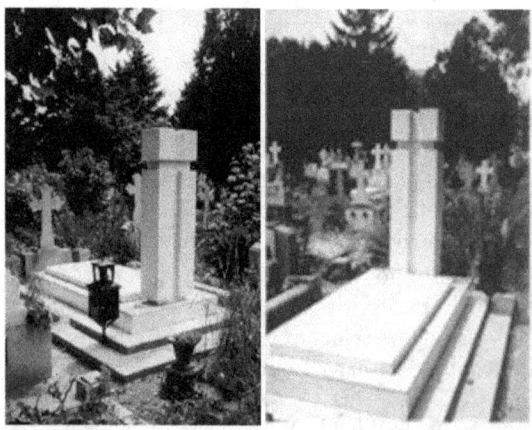

Figure 5,6,. Livio Dimitriu, arch. *Colonel Gheorghe Uricariu Monument*, 1998. Rear view with corner

Biographical Sketch:

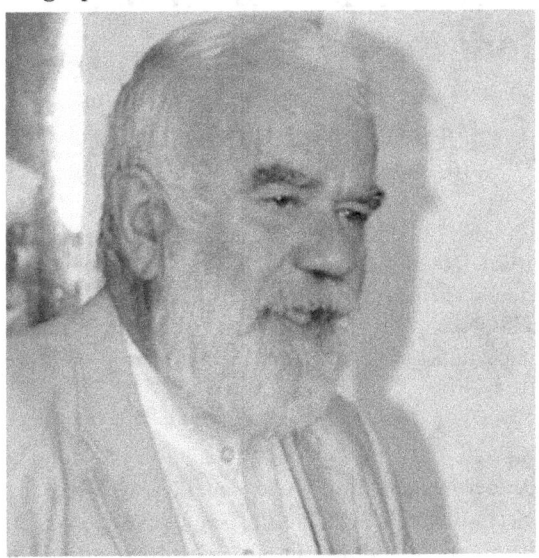

Dr. Livio Dimitriu is a Serbian-born American educator and architect in private practice in New York since 1978. His urban design and architecture projects received fifteen national and international awards in the Nord America, Europe, and Asia. His projects have been exhibited major museums and dozens of galleries and other institutions in twenty eight countries. Dr. Dimitriu has authored and/or contributed to thirty four volumes in ten countries and on three continents, along with articles and projects published in over one hundred magazines and periodical worldwide.

Dr. Dimitriu accepted a Ph.D. with the Highest Honors in Theory/History of Architecture from the "Ion Mincu" University for Architecture and Urbanism in Bucharest/Romania, where he was also awarded an Honorary Distinguished Professor *Bene Merenti*. He received an Honorary Master Architect in Stone from Antica Corporazione in Verona/Italy, and a B.Arch from the Cooper Union in New York. He is currently a tenured full professor at Pratt Institute in New York. Over 200 of his students received national and international awards in architecture and urban design over the past 37 years, including a US National First Prize for digital design in *FormZ*.

He founded and co-founded respectively two series of avant-garde publications in New York: *New York Architects* and *Pamphlet Architecture*. He is a founding member of the Islandic School of Architecture, an Olivetti Foundation Scientific Committee member, and a Senior International Editor with a cumulative 25 years of service with *Controspozio*, *Octogon*, and *Arhitext* magazines in Italy and Romania. Dr. Dimitriu was a Senior Fulbright Scholar to Europe, received a research grant from the National Endowment for the Arts/USA, and repeated research grants from the Romanian Government and Pratt Institute/USA.

Dr. Livio Dimitriu founded USA Institute in 1978, a public service design and research organization active in Nord America, Europe, and Asia, and is chief editor of USA Books He held academic appointments/directorships at twenty one universities/higher education institutions in Nord America, Europe, and Asia.

An emigration story, "Exercises in not-forgetting", and 3 ARA Congresses

Professor Ileana Costea, PhD
California State University (CSUN), Northridge, California, 91330, USA

Abstract: In this article Ileana Costea presents some highlights of her emigration, the two volumes of her book on "surprise-Romanian presence abroad" ("Exerciţii de Neuitare"), the articles she published about the two most recent ARA Congresses (Pasadena, California, 2014, and Frascati, Italy, 2015), and the ARA Congress 1992 whose general local organizer she was at CSUN, Northridge, California. Some details on the cultural events at the 1992 ARA Congress (exhibitions, book fair, a concert) and the two art exhibitions are also presented.

1. Introduction

This article is an accolade to Romanian creativity which for years and years I was passionate to discover through what I like to call "surprise-Romanian presence abroad".

2. My emigration

Like all of us, Romanian emigrants, each one has his/her own story. Mine happens to be a very romantic one. I left communist Romania in 1972, right after I graduated with a Master Degree in Architecture from the "Ion Mincu" Institute in Bucharest. I took off on a 7-day trip, to visit the castles on the Loire Valley in France, and those days were delayed until today. I left the group of "young people" (of which in fact the majority were way beyond the age of being members of the Union of the Working Youth, later called the Communist Youth – and this change in name gave me a lot of trouble with the immigration office in LA since they could not find this organization in their thick book!). With great difficulty I was finally able to obtain the approval to go on this trip, organized by the Tourist Office of Young People (BTT – "Biroul de Turism pentru Tineret"). Having the "burden" of a relative which was living in Chicago (my first degree cousin, the daughter of the sister of my mother, Alexandra Bellow, I was only allowed to travel the Soviet Block countries. After 5 years of filing for travelling to the Western side of the world, for the first time I was finally given the exit-visa. It was

through a mere luck: The Chief Visa-granting Officer happened to have a daughter, who like me was a student in architecture. Had I have been given the permission to travel to the West before, I would have returned to my country, since I was a child/a teenager. But now I was determined to leave my country, especially because I was in love with the man who will later become my life-long husband, Nicolas V. Costea, MD, professor of medicine. The two of us met briefly at a family dinner in Bucharest. He was sent by the American National Academy of Science to visit some hospitals in Romania, and he came to see his mother and brother who were left behind the iron curtain. It was in 1970. I was in the 5th year of my architecture studies. At a dinner given in the honor of Dr. Costea, I fell in love with him on the spot. It happened like in silent movies, when a bolt of lightning goes from the young girl's nose tip to his... It was only in that direction, and not from his to mine too. We saw each other again in Paris, when he was traveling on another US National Academy of Science trip to Romania in 1972. He was kind enough to bring me a piece of luggage with clothes from my mother. My luggage was confiscated by the Romanian consulate in Paris when I escaped from the hotel where the BTT group was hosted. My love story was really like in the movies, and my dream to marry Nic came true. I arrived to Los Angeles on the 14th of February (Valentine's Day) 1973, and we got married in April that year at the Santa Monica City Hall (at the same time and in the same hall where Dean Martin's 3rd marriage took place).

In July 1972 when I arrived in Paris, I did not know anybody, and I only had $10 in my pocket, and two packages of Snagov Romanian cigarettes and a doll in a national costume ("căluşar"). I never smoked but selected these items from my luggage just in case I will need to give a little gift to someone who would help me. I will never forget that early morning, when at 6AM I left the room so that I did not have to turn in my passport to our group guide who announced that he will take our documents and keep them during our trip. The streets of Paris were waking up, and I was telling myself "I will succeed. How? I do not know. But I am sure I will succeed." The lack of fear characteristic to young age. My story is long. My mother was called numerous times to the police and interrogated. They reproached her, a high-school teacher, that she did not know how to raise her daughter in the spirit of communist ethic. At that point I had two dreams: to find Nic, whose trace I lost, since he moved from Chicago, where he was a Professor at Illinois University to UCLA in California. When I left Romania I had the positive attitude of an optimist, rather than that of pessimist who always uses a negative. The joke goes: Pessimists said worst can't happen in our country. I, in an optimistic way told myself: Of course worst can happen, as it indeed did in the last years of Ceausescu's leadership. Mother was desperate. I was the only child, and both she and I thought we will never see each other "in this life". Like all Romanian emigrants I know I was wishing I could go back to Bucharest, to see the places I loved, and grew up in, my relatives and friends. Like everybody else I knew, I had the customary nightmare that I return, then they arrest me, and I cannot get out of the country anymore and ask myself why did I go back? The acute missing of my country ("dorul") kept growing and growing. But as the years went by it softened, especially with the arrival of my mother in LA to live with us. Through numerous letters to a senator we were able to get her out of the country. In 1992 it was because of her desire to visit Romania that I went back to visit.

3. "Exerciţii de Neuitare" – discovering surprise-Romanian presence abroad

In 1998 I started my "exercises in not-forgetting" choosing my self-imposed mission of discovering and writing about surprise-Romanian presences abroad. Since then I published over 40 articles on this subject in various Romanian magazines and newspapers of the Diaspora, as well as in Romania. In July 2015 I published my first book, Volume I of "Exerciţii de Neuitare"/"Exercises in not-forgetting" published by Reflection Publishing, and which can be found on Amazon.

3.1 Volume I – "Exerciţii de Neuitare"

My mission started based on my observation that Romanian culture and achievements are not well known abroad. Nevertheless, there are everywhere around the globe "well-kept secrets", such as a skilled hand, an innovative idea, an extraordinary brain, and all have a Romanian origin. The Volume I of my book makes known some of these secrets. It presents representatives of Romanian cultural life in different parts of the world, Romanians established abroad (like the musicians Liviu and Ovidiu Marinescu, the fashion designer Smaranda Schächtele, the writer Livia Medilanski Grama, the photo-journalist Emanuel Tânjală, the radio and TV show creator Benoni Todică, the architect Dino Tudor, the sculptor Patriciu Mateescu) or messengers from Romania (like the Eminescu specialist, writer, poet, and journalist from Botosani, Lucia Olaru Nenati, the poet and actress Lidia Lazu, from Bucharest). In his preface to my Volume I, Ion Lazu says: Ileana Costea "smells them" from any distance (I am referring in this case to inter-continent distance) not only when a Romanian cultural messenger arrives in Los Angeles, but also in New York, Paris, Venice, and Düsseldorf. In Palm Springs the author raises our attention on a building holding the signature of a famous Romanian architect, Haralamb Georgescu, and in the State of Kentucky she discovers the museum dedicated to the painter Dimitrie Berea. The author also spots very young Romanians, such as the anthropologist Mihai Anghel who successfully combines computer science with statistics and applied art. The author's articles also discuss classics of the Romanian art and culture, such as Eminescu, Blaga, Enescu and Brâncuşi. She also writes about creators who are not of Romanian origin, but who are attracted by or got inspiration from Romanian culture, such as the New York artist Jerry W. McDaniel or the opera specialist William Toutant... The author is not only very well informed, but also has a tomographic way of penetrating through the skin of things to their kernel and presenting them in very expressive, powerful, memorable ways."

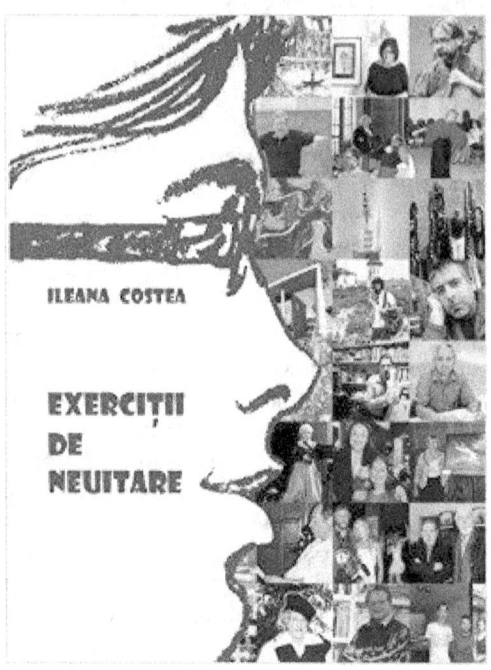

Figure 1. Ileana Costea, "Exerciţii de Neuitare", Volume I, Reflection Publishing, 2015. Cover design by Jerry W. McDaniel and Bogdan Alexandru Ungureanu.

3.2 Volume II– "Exerciţii de Neuitare"

I am now preparing the second volume of the articles on surprise-Romanian presence abroad. Here are a few of the articles which will be presented: Part of the story of my remaining in Paris as a political refugee in 1972, as it appeared as a chapter in the book "Romanians, from New York to Los Angeles" by Emanuel Tânjală and Dan Turturică; an interview taken of me by Ben Todică, during my visit in Melbourne, Australia in January 2011. It contains articles about writers (Bujor Nedelcovici, Ion Lazu, Lidia Lazu). I am posing a moment over the book "Refugiaţii"/"The Refugees", written by Mihai Vasilescu/Edgard Shelaru, former BBC and Free Europe newsman, with whom my family shared an apartment in Bucharest where later Saul Bellow stayed during his brief visit to Romania, accompanying my cousin Alexandra. An article about the Class of 66 at Spiru Haret High School, "Amintiri din Şcoală – Promoţia Spiru 66", a book written through online

correspondence by Mihai Vasilescu's colleagues, of the class one year younger than mine at the HS I too studied. Volume II will also contain articles in the field of art, music and movies. One article is about the Romanian-Cuban modern painter Sandu Darie, an interview with the heterogeneous American Artist Jerry W. McDaniel – painter with a Romanian flavor since he did illustrations for poems of Lucian Blaga (May 2008), and more recently (June 2014) for poems of Ana Blandiana. An article about the "last romantic in music", the composer Eugen Doga of Bessarabia. More general articles on culture, such as "The City of Los Angeles pulsates with Romanian Art", "Poetry, literature, history, and memory…", about the visit, to the US West Coast in the summer of 2014, of Ana Blandiana, Romulus Rusan and Doina Uricariu, an article about the art exhibition called Salon ARTIS 2010, organized by the architect-& graphic artist Marina Nicleaev in Bucharest, an article about the Contemporary Art Fair of Chatou, near Paris. An article about two women scientists, specialists in child psychology, Florica Nicolescu and Florica Bagdasar, as well as articles I created or contributed to on the Wikipedia (Pericle Papahagi, Ioan D. Caragiani, Mihail Magiari); articles about books such as "Veneticii" by Ion Lazu, "Refugiaţii" by Edgar Shelaru, "In two Worlds" ("In două Lumi") by Ben Todică, and a review made on Amazon on Andrei Codrescu's book "A hole in the Flag". Another article which will appear in Volume II is about a recent movie presented on the public television station PBS in January 2016 "Chuck Norris vs. communism and the entrepreneur Teodor Zamfir" – about the smuggling of videos with Western movies, especially American thrillers, during the harsh Ceausescu's time - an incredible story.

4. Three ARA Congresses

Both volumes of my book end with articles about ARA Congresses. The first volume ends with the article on the 38th ARA Congress which took place at Caltech, Pasadena, California, 2014. Volume II will conclude with two ARA articles: one on the most recent ARA Congress, the 39th, which took place at the National Institute of Nuclear Physics, Frascati, Rome (2015), and the Congress I organized in 1992 at the university where I teach, California State University, Northridge (CSUN).

The latter was the 17th ARA Congress and took place while Dr. Maria Manoliu Manea was the President of ARA, and when ARA Congresses were attended by a very large number of people. The 1992 Congress had about 400 attendees, and it presented two exhibitions of art (coordinated by the sculpture CSUN professor Robert (Bob) Bassler, and arranged by Dinu (Constantin) Rădulescu – a sculptor from Romania, and Ioana Sturdza, an artist living in California then, now long deceased. A Book Fair was also held– organized by Georgiana Fârnoaga Gălăţeanu who has taught Romanian at UCLA for several dozen years. Many well-known writers participated at that Conference, among which Augustin Buzura, Bujor. Nedelcovici, Petru Popescu. Several IREX, and SOROS scholars participated, among whom Gabriel Andreescu (Univ. of California, Irvine). A design/architecture exhibition had among the participants Romanian architects from California, Georgio Lupu and Dino Tudor, the Industrial Design UCLA Professor/Architect Nathan Shapira – deceased, and the Romanian architect from Târgu Mureş, Maria Dragotă. The CSUN newspaper wrote a praising article about the art exhibitions and placed a picture of a Hieronymus Bosch-like black-and-white drawing by gifted Romanian artist Marina Crainic, from Lausanne, Switzerland. There also was a beautiful piano concert where Lory Wallfish (Smith College, Northampton, Massachusetts - deceased), the soloist Julien Musafia (California State University, Long Beach), and the then very young pianist Virginia Munteanu performed.

4.1 ARA 1992 Art Exhibitions

A rich black-and-white Art Catalog was published on this occasion where Robert Bassler, well-known California sculptor and professor at CSUN wrote the introductory statement.

4.2 ART Catalog Introduction Statement – ARA 1992, CSUN

"I am pleased and honored to have been invited to host this unique assembly of accomplished artists from Romania. It is a wonderful opportunity to become personally acquainted with contemporary painters and sculptors who share the roots and history of a country and a culture so rich and diverse, yet have been subjected to severe deprivations of the rights and freedoms that we take for granted in the United States."

Their works are aesthetically various, addressing personal spiritual and political concerns, and are definitely within established modernist traditions. I applaud them all for their talent and their courage."

Prof. Robert Bassler
Dept. of Art, 3D Media

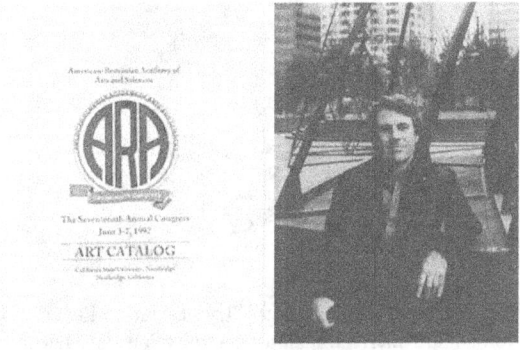

Figure 2 & 3. The black-and -white Art Catalog and sculptor Robert Bassler, coordinator of the art exhibitions at ARA 1992, CSUN.

The participants at the CSUN 1992 ARA Congress could admire on campus the beautiful modular sculpture of Patriciu Mateescu, known under the name "Love Flower" or "Carpathian Flower" donated to the University in 1985 by UCLA Professor Nicolas V, Costea, MD and CSUN Prof. Ileana Costea. The sculpture was later dedicated to the memory of Nicolas Costea, deceased in 2000.

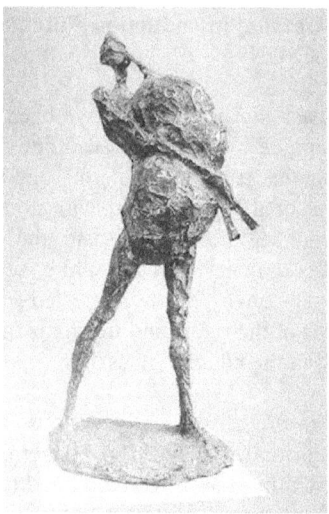

Figure 4. Dinu Rădulescu "Cimpoier" ("Bagpiper"), one of the numerous pieces in the art exhibitions at ARA 1992.

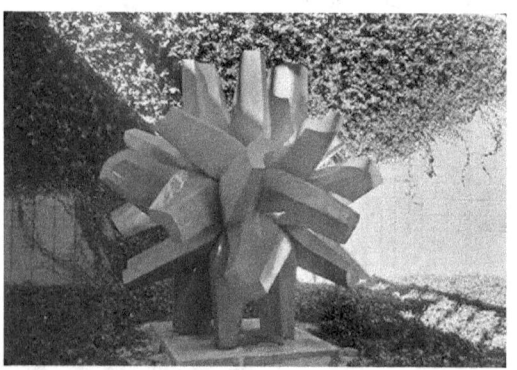

Figure 5. "Love Flower"/"Carpathian Flower" by Romanian-American sculptor ceramist Patriciu Mateescu, CSUN Student Union Courtyard, 1985. There are three sister sculptures of it, one at UCLA Sunset Recreation Center (yellow), and one at Cedar Sinai Hospital, Beverly Hills (white), and the most recent one installed in the main square in the Bistrita town in Norhtern Romania (red; launched in 2016).

5. Conclusion

The memory of the 1992 Congress still lingers upon me, since I danced "crazily" with my husband Nic and I had a quiet dance with the well-known Romanian artist Florin Piersic, who also attended that year's event. That ARA Congress really marked the beginning of my life-long activities of making Romanian creativity and culture known.

References:

Ileana Costea, Exerciții de Neuitare, Reflection Publishng, July 2015.
https://www.amazon.com/EXERCITII-NEUITARE-vol-I-romanesti-surpriza-strainatate/dp/1936629429).

Ileana Costea, Los Angeles-ul pulsează cu evenimente românești
http://melidonium.ro/2015/05/24/ileana-costea-los-angeles-ul-pulseaza-cu-evenimente-romanesti/

Ileana Costea, Românul Australian, Anul 24 Nr. 225, Melbourne, Australia, Martie-Aprilie, 2016
Comemorarea a 7 ani de lamoartea poetului Grigore Vieru - De vorbă cu scriitorii Ion Lazu și Lucia Olaru Nenati

Ileana Costea, Românul Australian, Anul 24 Nr. 226, Melbourne, Australian, Mai-Iunie, 2016
Chuck Norris vs. Comunism si Antreprenorul Theodor Zamfir

Alexandra Constanda, Mediafax, Documentarul "Chuck Norris vs Comunism", despre Irina-Margareta Nistor, printre filmele care trebuie vizionate la Sundance 2015
http://adevarul.ro/entertainment/film/documentarul-chuck-norris-vs-communism-despre-irina-margareta-nistor-printre-filmele-trebuievizionate-sundance-2015-1_54c2015a448e03c0fde0059b/index.html

Ileana Costea, Prezentări remarcabile de artă, cultură și știință românească la ARA 2015
http://acum.tv/articol/75146/

Ileana Costea, Al 38-lea Congres ARA, 23-26 Iulie 2014
Știință și artă românească minunat reprezentată la Caltech, Pasadena, California de Sud
http://www.agero-stuttgart.de/REVISTA-AGERO/JURNALISTICA/ARA38%20de%20Ileana%20Costea.htm

The 39th ARA Congress Frascati, Roma, July 28-31, 2015
http://www.americanromanianacademy.org/#!39th-congress-2015/c1vxd

The 38th ARA Congress, July 23-27, 2014, Pasadena, California
http://www.americanromanianacademy.org/#!38-congress-2014/cqg1

The 17th ARA Congress, 1992, CSUN, Northridge, California, General Local Organizer Professor Ileana Costea
Translation from Romanian by Ileana Orlich, of the article published in the Universul newspaper
www.ic-art-gallery.com/ARA/1992Congress.html

Congress Visions of Dobrogea and Balcic in Romanian Paintings

Dr. rer. nat. Carmen Sabau CSabau@comcast.net,

and Dr. Isabelle Sabau cyberedu22@netscape.net

Abstract: The area of Dobrogea with its tremendous biodiversity, beach resorts and fascinating vistas has inspired numerous artists for centuries. This presentation will discuss a number of visionary and artistic interpretations of this alluring landscape and its importance in the development of modern Romanian art.

1. Introduction

The Dobrogea region in Southeastern Romania, located between the Danube River and the Black Sea, is home to the UNESCO wildlife "Reservation of the Biosphere" of the Danube Delta and numerous historical ancient settlements going back to Neolithic times of which the famous "Thinker and his Wife" dates from the Hamagia period in the 6th Millennium BCE.

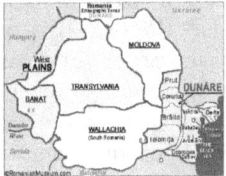

Figure 1. Map of Romania

The Mediterranean and Black sea basin (**Figure 1**) with its inviting beaches, therapeutic mud baths and ample sunshine has attracted numerous artists throughout the centuries and even more abundantly in the modern period from the 18th-20th centuries. During the rapid development of the modern world, momentous discoveries, inventions and dramatic changes in all domains of human activity from the social, political and technological realms to the artistic, cultural and philosophical areas, led to fundamental changes in the web of knowledge. As the Impressionist artists had discovered the Cote d'Azur and the minute variations of colors and light, and, as landscape painting itself gained popularity providing an escape from the bustle, noise, pollution and crowds of the modern city, seascapes and beachscapes beckoned the painters of the times with rich sources of inspiration facilitated by the introduction of oil paints in portable tubes. In Romania, while the prominent artists had traveled, studied and painted in France, the Romanian seaside of the abundant region of Dobrogea provided a similar source for inspiration as the French and Italian Rivieras. From the well-known Grigorescu, Tonitza, Luchian, Petrascu, to more contemporary artists such as David Croitor and George Stefanescu, the seaside, Dobrogea and the area of Balcic in present day Bulgaria, have inspired fascinating works of art to be discussed further in this presentation.

2. Romanian Marinescapes of Dobrogea

Nicolae Grigorescu (1838-1907), considered the greatest Romanian painter, was instrumental in the evolution of modern Romanian art and the artistic taste of the Romanian public. He was the foremost artist who expressed better than any other artist the nature and soul of Romania and its people as Nicolae Iorga commented [1], the stunning beauty of the Romanian land, the positive and poetic rural life of the peasants, succeeding to create his own unique style. While he studied and acknowledged the importance of the modern developments of Impressionism, the Barbizon School and various artistic techniques, he never became affiliated with any of the contemporary artistic movements, developing instead his own distinctive artistic sensibilities, which enabled him to represent with sensitivity and honesty the landscapes and portraits of peasants, presenting an optimistic aspect of rural life. His works, often idyllic and lyrical, display a vast array of techniques employed to evoke specific feelings in the viewer while concentrating on the atmosphere of the subject through a synthesis of ideas.

Diagonal streets, undulating houses and bright colors sparkling in the intense sun punctuated by broad, flat areas of color mark the work of Nicolae Tonitza (1886-1940). (**Figure 2**)

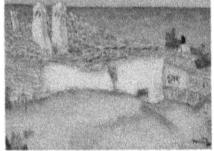

Figure 2. Nicolae Tonitza Houses Balcic
His rapid sketches of Turk and Tatar children reveal a deep sensitivity to the personality of the sitter, while the seascapes emphasize the vastness of the sea merging into the azure sky glimpsed through the arabesque shapes of the trees. He used to teach his students "The art of the great Colorists doesn't consist of the variety of colors he spreads on the canvas, but in the chromatic musicality that he knows to establish with the help of a minimum of colored substances" [2].

Stefan Luchian (1868-1916) approached the shore with more somber colors rendering the lyrical coloring in subtle nuances while summarily indicating its features. (**Figure 3**)

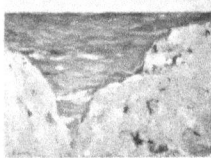

Figure 3. Stefan Luchian Black Sea at Tuzla
Luchian was a master in representing nature, especially Romanian landscapes and a staunch interpreter of the Romanian soul, with its tenderness, melancholy and confidence in the beauty of life [3]. In his landscapes, curved, calm lines dominate undisturbed by sharp angularities, emphasizing tall and slender trees, adorned by rich foliage, catching the sun sometimes with the bright sparkle of enamel and other times discreetly glimmering. He presents restricted views, showing the richness of a small number of trees, a steeple, a backyard, a well or a path in a graveyard, but his harmony of colours and the play of sun rays, brighten up the whole painting without stark contrasts, just shadows of different nuances [4].

Gheorghe Petrascu (1872-1949) highlighted the figures casually strolling on the shore contrasting their indistinct and sketchy features with the vibrant colours of the sea and the vastness of sand focusing the attention of the viewer to the infinite horizon somewhere in the center of the work. His works are characterized by a thick impasto of color applied with a palette knife often using a number of colors simultaneously [5]. His oeuvre shows a great preoccupation with color and light, as he mentioned in an interview in 1931, when asked how he creates such unique light in his works, smiling, he answered: "for more than 30 years, I have been asking myself that, too!"[6]. For Petrascu black is not shadow and white is not light, but they have the same role as every other single color [7].

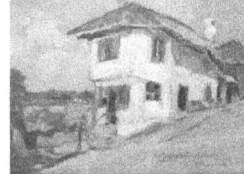

Figure 4. Gheorghe Petrascu Café in Turtucaia
"The Café in Turtucaia" (**Figure 4**) is typical for that town and occupies the center of the painting, on a downhill street. The white walls offer luminosity and a spotlight amidst the abundance of green hues that dominate the work. Spots of blue appear in the pond on the left side of the house as darker nuances fill the upper right corner. Between 1905 – 1932, Petrascu painted several windmills from Ialomita, Mangalia, Branesti, Balcic, and from Bretagne, as a romantic subject, repeated with different chromatic palettes [8]. In "Looking at the sea at Mangalia", the artist's wife and daughter break the saturated dominance of ultra-marine and the emerald-green. The contrast between the blue of the sea, the yellow of the sand and the red of the dress or between the sky-blue and sea blue and the beach yellow-green are striking and the shadows cast by the figures are apparent. Petrascu started painting marines, already as a student, discontinuing around 1940, because of health problems. Many of his marines are tranquil, serene, sometimes sad, sometimes brilliant, stressing the movement of light and often including his wife, Lucretia Petrascu, their daughter, Mariana, or Lila Marinescu, his wife's sister. In the year,

1930 Pertrascu painted numerous marines at Techirghiol. The foamy waves and the movement of the dresses exposed to the wind are concentrated in the centre of the composition dominated by grey, ultra-marine and emerald-green accentuated by the usual violet-red of the hats [9]. His chromatic harmonies intensified by the use of black, present a distinctive expression of a constructive vision of the beauty and nobility of matter clothed in light and shadow.

George Stefanescu (1914-2007) was born at Plainesti/Dumbraveni, a small village lying somewhere between Focsani and Ramnicul Sarat [10]. He studied at the Academy of Fine Arts in Bucharest with Nicolae Darascu, whom he accompanied to Balcic, the painters' paradise. (**Figure 5**)

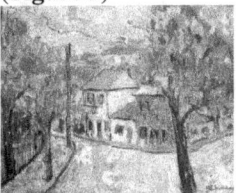

Figure 5. George Stefanescu Balcic street

Later he was a student of Lucian Grigorescu and during the period of 1958-1975 he worked as a set and costume designer for the Municipal Theater. Although better known for numerous costumes and set designs for the Theater, Stefanescu continued to paint and exhibit drawings and watercolors of the Romanian and Bulgarian seasides. In 1974 George Stefanescu retired from the Theater in order to devote himself exclusively to painting and in 1989 he moved to Germany, where he continued his artistic career punctuated by various personal and joint exhibits, culminating with a solo exhibit celebrating the artist's 90's birthday in 2004, organized by the city of Lüdinghausen and the Kunstverein KAKTus entitled "George Stefanescu - Licht und Farbe" [11]. Petru Comarnescu described the work of the artist: "The images of George Stefanescu appeal through their warmth of feelings, their bright harmonies and balanced tonalities…"[12].

Lucian Grigorescu (1894-1965), an original Romanian post-Impressionist, fused the perceptual vision of the impressionists with the structural analysis of the post-impressionists and unique Romanian motifs, creating a balance between tradition and modernity, infused with a personal sensibility and delight in nature and rendered through a vigorous construction of form bathed in colour and light. (**Figure 6**)

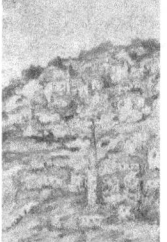

Figure 6. Lucian Grigorescu Balcic

3. Queen Marie and Balcic

Queen Marie (1875-1938), born Princess Marie Alexandra Victoria of Edinburgh, a granddaughter of Queen Victoria was married, in 1893 to Prince Ferdinand of Hohenzollern-Sigmaringen, who ascended the Romanian throne in 1914, after the death of his uncle King Carol I of Romania. Queen Marie was known as a great beauty, romantically inclined and idealistic, admired throughout the courts of Europe, who dedicated herself with strong resolve and courage to her adoptive country [13]. During WWI she devoted herself to charitable work serving tirelessly as a nurse, setting up her own hospital on the grounds of the Royal Palace, providing meals, blankets and comforts to the wounded and fearlessly touring typhoid stricken villages and foxholes on the front [14]. After WWI, she represented Romania at the Paris Peace Conference at Versailles, helping to smooth conflicts and bring favorable outcomes to Romania, with the promise of giving Romania back her lost territories and not partitioning the country [15].

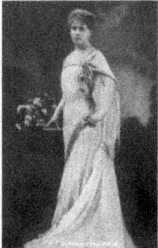

Figure 7. Queen Marie

In her Memoirs Queen Marie (**Figure 7**) described the Coronation Day and her love for Romania as the crowds cheered:" "*Regina Maria!*" And we faced each other then, my people and I. And that was my hour—mine—an hour it is not given to many to live; for at that moment it was not only an idea, not only a tradition or a symbol they were acclaiming, but a woman—a woman they loved. And at that hour I knew that I had won, that the stranger, the girl who had come from over the seas, was a stranger no more; I was theirs with every drop of my blood!" [16]. Queen Marie was also a great correspondent and writer of both, prose and poetry and in 1922 she wrote an "Ode to Roumania" in which she expresses her love of the country and its people: "...O Roumania, now are all thy children united, their chains broken, their captivity ended—and the mountains barriers no longer exist....And now my people, it behoves us to reconstruct. The foundations are laid, the great work is begun. Build, build! All differences set aside; let us draw close together in this sacred unity which we have bought with our blood and which will constitute our strength. Stone by stone build up the future; with courage and with confidence advance; but so that the edifice which thou buildest may be indestructible, forget not the hearts, the countless hearts on which thou hast placed its stones." [17]. Queen Marie was beloved by the Romanian people and hailed as "Mother Queen" for her utmost dedication and service to the country and its people.

The area of Balcic located on the Silver Coast of the Black Sea, a touristy, sea resort and spa today, in present-day Bulgaria belonged to Romania between 1913-1940 and was the Summer residence of the Queen. Here, on this ancient Greek colony, founded in the 6th century BCE by the Milesians and known as Dionysopolis in antiquity, Queen Marie completed her palace between 1926-1937, on a steep bluff overlooking the sea, inspired by the architectural style from the 16th century of Curtea de Arges in Romania and surrounded by a number of residential villas, a wine cellar, a chapel and monastery, extensive botanical gardens and vistas. (**Figure 8**)

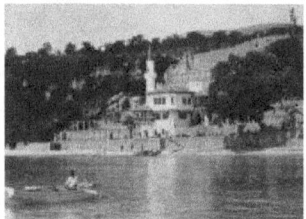

Figure 8. Queen Marie Palace Balcic

With the unfortunate partition of Romania in 1940, the palace was incorporated in Bulgaria and its numerous contents of paintings, ceramics,

sculptures and the Queen's archive of correspondences and historical photographs were transferred to Constanta. The Queen was buried at Curtea de Arges, but she had requested that her heart would be buried in the Stella Maris Chapel in her palace at Balcic. With the loss of Balcic, her heart was moved a number of times to eventually be placed at her beloved Pelisor Castle in southern Transylvania where it still resides today [18].

4.   Topalu treasures and Balcic artistic visions

On August 27, 1960 Dr. Gheorghe Vintila (1898-1978) one of the Chief and great surgeons of the 1930's to 1969 when he retired, donated 228 works of modern Romanian art, as a foundation for a museum, dedicated to commemorating and continuing the contribution of his parents to their community, Topalu, in Dobrogea. This museum was named "Dinu and Sevasta Vintila" in honor of his parents and is housed in his parent's home [19]. (**Figure 9**).

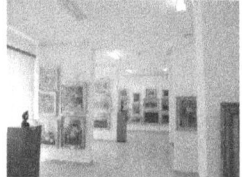

Figure 9. Topalu Museum interior view

The astonishing and remarkable collection includes a vast array of important Romanian artists like Grigorescu, Tonitza, Pallady, Petrascu, Ressu and many more – a surprising gem hidden in a small community off the beaten tourist track to the sea. Portraits, flowers, still lifes as well as seascapes and images of Balcic fill the rooms organized by period and grouped by artist, providing a venerable encyclopaedia of Romanian modern art. Upon its inauguration, the museum was hailed by Corneliu Baba 'a spiritual oasis in the desert' [20].

5.   Conclusion

This paper provided a short overview of the inspiration the Dobrogea region and the palace of Balcic offered to various prominent Romanian painters of the modern period. Each artist approached the subject in a special, unique and individual style revealing fresh perspectives and affording the viewer a glimpse at the past with nostalgic vistas and sun drenched villages, capturing a time prior to the commercialized and touristy frenzy of the resorts that have overtaken the sea shore in more recent times. Their paintings attest to the creative spirit of Romania on a par to other, more prominent artistic movements of the 19th and 20th centuries in Europe, whose lessons and experiments were synthesized in new and distinct ways by the greatly admired Romanian artists.

References

[1]. Pool, P. Impressionism Thames & Hudson 1991, p. 14-15.

[2]. Colection Genius Painters : *"Life and work of Tonitza"*. #11. The Truth. , Art Library. Truth Holding Press. 2009, ISBN 978-606-539-107-9, pp. 1-153.

[3]. Lassaigne Jacques, Stefan Luchian, Editions Meridiane, Bucharest, 1972, pp. 72-73, 110-113.

[4]. Jianu, Ionel, Comarnescu, Petru Stefan Luchian, Editura de Stat pentru Literatura si Arta, Bucuresti, 1956, pp. 133-147.

[5] Constantinescu Paula and Schobel Doina, *Exhibition of Painting, G. Petrascu, 1872 – 1972,* Museum of Art of the Romanian Socialist Republic Catalog, Bucharest, 1972, p. 66-70.

[6] Constantinescu Paula and Schobel Doina, *Exhibiton of Painting, G. Petrascu, 1872 – 1972,* Museum of Art of the Romanian Socialist Republic Catalog, Bucharest, 1972, p. 66-70.

[7]. Mesea Iulia, *Gallery of Romanian Art,* National Brukenthal Museum Guide, Sibiu, 2010, p. 98-102.

[8]. Constantinescu Paula and Schobel Doina, *Exhibition of Painting, G. Petrascu, 1872 – 1972,* Museum of Art of the Romanian Socialist Republic Catalog, Bucharest,1972, p. 55-57, 62, 76, 78, 80.

[9] Constantinescu Paula and Schobel Doina, *Exhibition of Painting, G. Petrascu, 1872 – 1972,* Museum of Art of the Romanian Socialist Republic Catalog, Bucharest,1972, p. 55-57, 62, 76, 78, 80.

[10]. George Stefanescu Biography, retrieved from http://www.georgestefanescu.ro/biografie.php

[11]. George Stefanescu Biography, retrieved from http://www.georgestefanescu.ro/biografie.php

[12]. George Stefanescu, retrieved from http://georgestefanescu.ro/category/articole/page/2/

[13]. Princess Marie of Edinburgh – Queen of Romania, retrieved from http://www.unofficialroyalty.com/princess-marie-of-edinburgh-queen-of-romania/

[14]. Brenda Ralph Lewis, Queen Marie, retrieved from http://www.tkinter.smig.net/queenmarie/mammaregina/index.htm

[15]. Queen Marie, retrieved from http://www.geh.org/link/SN/queen-marie.html

[16]. MY LIFE AS CROWN PRINCESS, Part 8, by Marie, Queen of Rumania, The Saturday Evening Post, 16 June 1934, retrieved from http://www.tkinter.smig.net/QueenMarie/SatEvePost/1934-06-16/index.htm

[17]. ODE TO ROMANIA by H. M. Queen Marie of Roumania, from Roumania - The Royal Edition, Marie Jonnesco, Paris, 1922, retrieved from http://www.tkinter.smig.net/QueenMarie/OdeToRoumania/index.htm

[18]. Elizabeth Jane Timms, Burying a Queen's Heart – Queen Marie of Romania, retrieved from http://royalcentral.co.uk/blogs/history/burying-a-queens-heart-queen-marie-of-romania-52050

[19]. Doina Pauleanu, Muzeul de Arta Dinu si Sevasta Vintila Topalu Catalog, Editura Arcade, Bucuresti 2012, p.5

[20]. Doina Pauleanu, Muzeul de Arta Dinu si Sevasta Vintila Topalu Catalog, Editura Arcade, Bucuresti 2012, p.5

The Legacy of Dada a Century Later

Dr. Doina Uricariu
111 South Elliott Place, Brooklyn, NY 11217, USA
doina.uricariu@gmail.com

Abstract:

A personal view of the art of our century, dealing with separate subjects that seem important for an understanding of Dadaism, modernism and post-modernism from the beginning of a sense of modernity in European Culture, that is roughly from 1880s to 1914-1916.

Charles Peguy considered that "the world has changed less since the time of Jesus Christ than it has in the last thirty years".

How long can this engine resist producing a *changement* of all conditions of Western, Eastern and American societies, from its sense of history, beliefs, pieties, modes of production, the idea of itself, its art, the refusal and rejection of social significance? Are the artists of our time still in the same state of great convulsion, social tumult, and cultural turmoil? Have we the sense of an accelerated rate of change in all areas of human discourse and languages, including art? The Dadaists were alive at the beginning of modernism along with the Futurist, Cubist, Surrealist, and Expressionist artists. Today, we live at the end of modernism.

Dadaism created the modernist laboratory, an arena of significant experiments. Now Dadaism is relegated to a period room in the museums, a series of exhibitions presented by Ian Dunlop in an important book entitled *The Shock of the New,* published in London and New York, the same year, in 1972.

In 2013, Robert Hughes published an oustanding volume with the same title, *The Shock of the New,* presenting the hundred-year History of Modern Art, its rise, dazzling achievement, and fall. What has our culture lost at the end of XXth century, in 1980, 1990 and 2016 that the avant-garde had in 1880, 1890, 1916 ? Idealism offered a belief that formed an immense territory to explore, new metaphors such as Tour Eiffel, the new sculpture such as Brancusi's oeuvre and its *Birds in Space*, the mechanical paradise of the ready-made, the *cadavre esquis*, the romance of technology, the evironment and the landscape, the city, and the metropolis. The view from the train/from the plane was neither the view of the eagle nor the view of the frog, mentioned by Friedrich Nietzsche. It was neither the view of the horse, nor of *la fourmillante cite* of Baudelaire, the machine-made environment, the city.

The Legacy of DADA a Century Later

I have forced myself to contradict myself in order to avoid conforming to my taste. I don't believe in art. I believe in artists. (Marcel Duchamp)

DADA after a Century and how might we understand the idea, the concept and the legacy of DADA? How can we do this today, in 2016? Thinking and working against banal and facile generalizations. DADA has been presented in terms of non-sense, anti-art, and absurdity. The Romanian and European Tristan Tzara asserted that the word DADA meant nothing—*ne signifie rien.*

DADA displayed skepticism with regards to accepted values

DADA's artefacts and performances project an image of group camaraderie and multi-signature manifestos vs. individualism

DADA, an abiding legacy for the century to come with programs, manifestoes and strategies that include ready-mades, performance, collage, assemblage, montage, media, chance, prankish mind and behavior, and different forms of automatization, photomechanical reproductions and printed ephemera

DADA's reconceptualization of artistic practice, as a form of tactics and strategies

DADA may have had the greatest influence on contemporary art of avant-garde movement

DADA's radicalism in concepts and strategies is foundational in modernism and postmodernism, representing today even a cliché, a déjà vu, a stereotype.

DADA's machine culture and anti-art

DADA in the movement's main centers: Zurich, Berlin, Cologne, Hannover, Paris, New York

DADA's "form of calculated irreverence" (Hugo Ball)

Dada's subversive reaction against the *status quo* of the system

DADA's "self-historicizing" (Tzara)

DADA's mischief-making

DADA in Eastern Europe

DADA's globalism and constellation of identities and ideas. Its aims were often supranational emerging amid the racially having a tinge of nationalistic discourse of W.W. I.

Leah Dickerman: "There had not been an artistic movement so self-consciously international".

DADA's promotion of a proto-globalized identity is evident in the stationary that Tristan Tzara produced for the movement Dada in Paris, by listing Dada branches below

DADA's global network of artists of diverse nationalities might be exemplified by Zurich and Cabaret Voltaire: Hans Arp (German-Swiss), Hugo Ball (German), Sophie Tauber, (Swiss-born), Marcel Ianco and Tristan Tzara (both Romanian-born), Walter Serner (Austrian of German-expression, born in Karlovy-Vary), Vikking Eggeling (Swedish), Emmy Hennings, (German-born and Hugo Ball's partner), the German Richard Huelsenbeck) and New York with John Covert (born in Pittsburg), Jean Crotti (born in Switzerland ,French-speaking), Marcel Duchamp, (French-born and naturalized American), the Brooklyn-born American Man Ray who spent most of his career in France, the French Francis Picabia, one of the early major figures of the Dada movement in United States, Italian-born Joseph Stella, who emigrated to New York, German-born Baroness Elsa von Freytag-Loringhoven, a central figure of Dada in the Greenwich Village, the Philadelphia-born American Morton Livingston Schamberg, one of the first American artists to explore the esthetic qualities of industrial subjects, Beatrice Wood, born in San Francisco and admirer of Marcel Duchamp, Great Potter called the "Mama of Dada" in a documentary film, written and directed by Thomas L. Neff.

DADA in Paris with Louis Aragon, Celine Arnauld, Hans Arp, Andre Bréton, Jean Crotti, Paul Dermée, Marcel Duchamp, Suzanne Duchamp, Paul Eluard, Max Ernst, Man Ray, Francis Picabia, Georges Ribemont-Dessaignes, Philippe Soupault, Tristan Tzara

DADA's antinationalism

DADA's compilation of materials, new technologies, and media promoted by iconoclastic credoes, manifestoes and commentaries.

DADA assumed like a generally SURREALISM. The first three exhibitions presented DADA in conjunction with surrealism:

-1936 *Fantastic Art, Dada, Surrealism, exhibition* presented at the Museum of Modern Art, New York

- 1968 *Dada, Surrealism and their Heritage,* Museum of Modern Art, New York

-1978 *Dada and Surrealism Reviewed,* Council of Great Britain

2005-2006

DADA, Centre Pompidou, Musée national d'art moderne, Paris, (Oct. 5, 2005- Jan. 9, 2006)

DADA, National Gallery of Art, Washington, DC, (Feb. 19, 2006-May 14, 2006)

DADA Museum of Modern Art, New York, (June 18-Sept.11, 2006)

-DADA and international art movements: Blaue Reiter, Cubism, Futurism, Expressionism, Constructivism.

There was a pressure exercised by the new experiences vs the demand for new forms to contain it, and to perform art as a work in progress.

A new perspective on the heroic and dynamic sense of cultural and artistic possibilitiy: Arthur Rimbaud's injunction *être absolument moderne*

NB William S. Rubin noted in exhibition catalogue *Dada, Surrealism and their Heritage:* "Dada was however the first programmatically international movement in the plastic arts".

DADA's supranational aspirations vs national cultural and artistic agendas

DADA's refusal / rejection of the transcendent and sublimated art as an illusionistic conjuring of imaginary worlds.

From the Apotheosis of machine to the apotheosis of computers and virtual artistic realities

DADA vs Minimalism.

Biographical Sketch:
Dr. Doina URICARIU

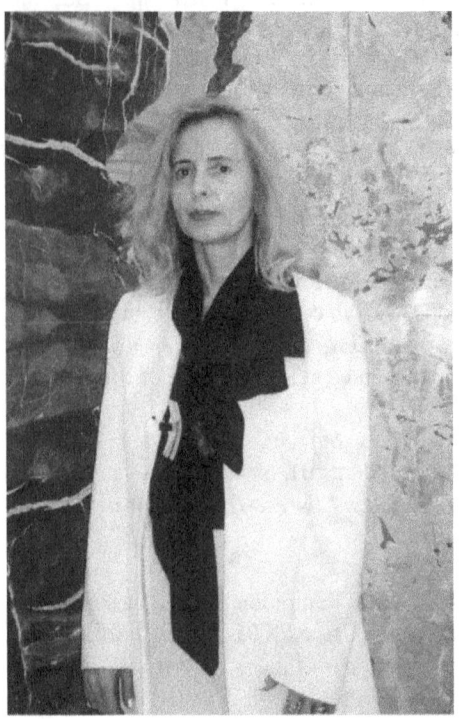

Doina Uricariu is an important contemporary Romanian writer with an unmistakable lyric voice born out of a generous sensuous universe, intersected by constant moral, philosophical, and political questioning. She started out in poetry with the volume *Healings*, published in 1976 after a seven-year wait in the hands of the Communist censorship. Eight other volumes of poetry followed: *The Heart Institute, Jugastru Sfiala, Happy Beings, The Hand-covered Face, Atrocious Eye, The Heart Institute, The Power of Leviathan, The Axonometric Heart,* one volume published by Khalit Lala National Academy in India, and three volumes of poetry published in Italy*, En plein air, Architettura scavata/Excavated Architecture, Inserzioni sovversive/Subversive Insertions, Abitare in una lente,* a bi-lingual German-Romanian volume *Das Herzinstitut/The Heart Institute.* Doina Uricariu published in 2015 a volume of new poems, *The Glass Book/Cartea de sticlă* in different editions, one in Romanian, two editions in English (2015, 2016). *The Glass Book* was recently translated in French.

Mrs. Uricariu's poetry has been translated in anthologies and individual volumes in English, German, French, Italian, Swedish, Norwegian, Russian, Serbian, Czech, Hebrew, Albanian and in twenty other languages. She published volumes of criticism and literary history, essays, and art criticism that are considered milestones on their respective subjects*: Apocrypha on Emil Botta, Ecorches, Nichita Stănescu-Paradoxical Lyricism, Søren Kierkegaard, Maria Pillat-Brates: Poetry and Reverie, Antonovici: Sculptor on Two Continents, Vlaicu Ionescu: The Artist.* Doina Uricariu prepared and edited reference editions on Emil Cioran, Dominic Stanca, Jeni Acterian, Emil Botta, and the *Podrom Filokalia.* She published four volumes of her memoirs in two books entitled *The Lower Jaw* and *The Lions Stair.*

Dr. Doina Uricariu is member of Romanian Writers Association, European Cultural Society, and the International PEN CLUB. She is a Knight of the Order of Faithful Service of Romania, distinction conferred by the President of the country, and Knight of the Order of the Romanian Crown bestowed by King Mihai I and Royal House of Romania. She was the Director of the Romanian Cultural Institute in New York (2013-2016).

The New Man and his films:
a glance at how contemporary Romanian cinema traces the recent past

Marina Vargau
Université de Montréal, Montréal, H3T 1N8, Canada
marinavargau@yahoo.ca

Abstract: Romanian filmmakers who have won prizes at international film festivals in the first decade of twenty-first century belong to the same generation: that of the New Man, a paradigm brought forth by Ceausescu's communist regime. My analysis will, in the first place, discuss the documentary *Nascuti la comanda. Decreteii* (*Children of the Decree*, 2004) by Florin Iepan. In this film we will follow the traces of the two traumatisms that marked this generation: the ban of abortion through decree by the State in 1966 - a moment which announces the birth of the New Man - and the anti-communist Romanian revolution of December 1989, in which most of the participants and victims belonged to the generation born under the order of the Party. The new generation of Romanian filmmakers have shown through their films that this New Man is actually rethinking the past and its traumas. The question of abortion and of its consequences will be discussed in relation to the film *4 luni, 3 saptamani si 2 zile* (*4 months, 3 weeks and 2 days*, 2007) by Cristian Mungiu and to the short *Visul lui Liviu* (*Liviu's Dream*, 2004) by Corneliu Porumboiu, while a polyphonic image of the Romanian revolution will be put into question in three films: *Cum am celebrat sfarsitul lumii* (*How I celebrated the end of the world*, 2006) by Catalin Mitulescu, *Hartia va fi albastra* (*The paper will be blu*, 2006) by Radu Muntean and *A fost sau n-a fost ?* (*12:08 East of Bucarest*, 2006) by Corneliu Porumboiu.

1 Introduction

Le documentaire *Le sang des bêtes*, réalisé en 1949 par Georges Franju, jette un regard cinématographique objectif sur le monde sanglant des abattoirs situés aux portes de Paris. Les images,

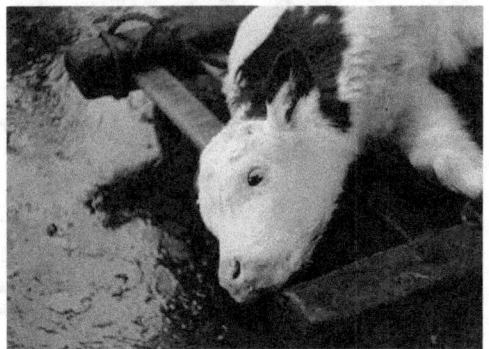

Figure 1. *Le sang des bêtes*, 1949, Georges Franju

d'une violence excessive, montrent en direct la tuerie des animaux. On voit l'agonie, le sang qui éclate, les derniers spasmes avant la mort, puis l'écorchement et le tranchement des corps. Siegfried Kracauer [1] relie ces images de violence contre les animaux aux images des corps humains torturés, montrées dans les films réalisés dans les champs de concentration nazis. Selon lui, grâce à ces images, on redonne à l'horreur, caché sous les voiles de la panique et de l'imagination, sa visibilité.

En 1985, en Roumanie, Copel Moscu réalise un film documentaire de 12 minutes qui présente une ferme avicole. Le documentaire, intitulé *Va veni o zi* («*Le jour viendra*»), est interdit (il sera présenté pour la première fois en 1992, après la chute du régime communiste) parce que Moscu a le courage de suggérer à travers des images brutales sur le processus de sélection des poussins ce qui se cache derrière un autre processus de sélection, celui de l'«Homme nouveau» : l'extermination des enfants avec des déficiences durant le régime communiste de Ceausescu. Pour reprendre le commentaire de

Kracauer, grâce à ce documentaire, Moscu donne la visibilité à une horrifiante réalité roumaine.

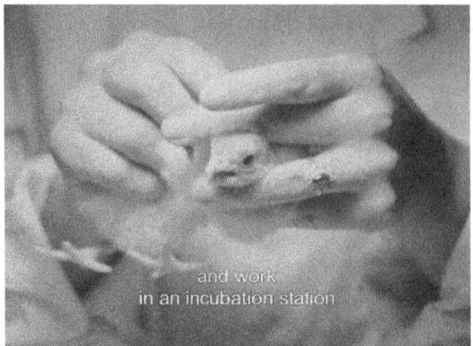

Figure 2. *Va veni o zi*, 1985, Copel Moscu

En 2005, Florin Iepan inclue quelques séquences de ce film dans son documentaire *Nascuti la comanda. Decreteii* («*Nés à la commande. Les enfants du décret*»), pour introduire des images absolument terrifiantes et choquantes, montrant les enfants trouvés en 1990, trois mois après la Révolution, dans l'orphelinat de Cighid (Roumanie), qui fonctionne comme centre d'extermination. Ces

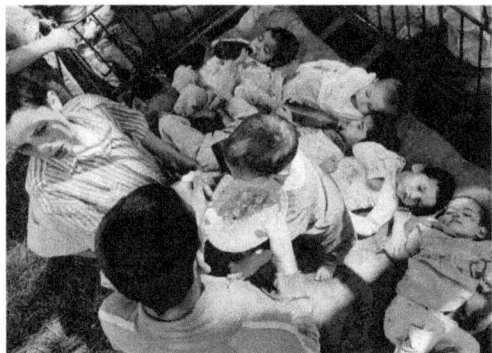

Figure3. *Nascuti la comanda. Decreteii,* 2003, Florin Iepan

enfants, pour la plupart des «conséquences» du décret 770 par lequel l'avortement est interdit, après avoir été «sélectionnés» par des commissions de médecins et trouvés inaptes pour la vie, sont abandonnés à l'orphelinat de Cighid, où, sans bénéficier de soins médicaux, d'aliments, de chauffage, d'électricité, d'un espace adéquat, sont laissés tout simplement mourir. Parfois, dans les

pièces lugubres, ces enfants maigres, malades, sous-développés, sont mangés vifs par les rats.

Les enfants de Cighid représentent le visage caché, monstrueux, d'une expérimentation sociale Étatique qui a duré 23 ans et qui avait comme but la création d'une nouvelle génération, endoctrinée politiquement et performante : la génération de l'«Homme nouveau».

2 Le décrèt 770 de 1966 et ses conséquences

Pour mieux comprendre cette réalité de la Roumanie communiste, il faut remonter au fil du temps, jusqu'en 1966, l'an qui marque l'interdiction de l'avortement et, implicitement, l'acte de naissance de l'«Homme nouveau». Le but de l'interdiction de l'avortement par le décret 770 de 1966 est l'augmentation de la population du pays pour des raisons économiques. L'État demande à chaque femme fertile de donner naissance à quatre ou cinq enfants, la reproduction est entièrement politisée et la maternité devient la responsabilité fondamentale des femmes roumaines envers l'État. Sans moyens de contraception, les femmes deviennent une proie pour l'État, qui les réduit à la fonction d'outils de reproduction. Un an après le décret, pris au dépourvu, sans contraceptifs et avec une vie sexuelle normale, les Roumains réussissent la «performance» d'augmenter deux fois le taux de natalité dans un seul an. Quand, en 1969, le Roumain no. 20000000 est né, Ceausescu est fier de son projet.

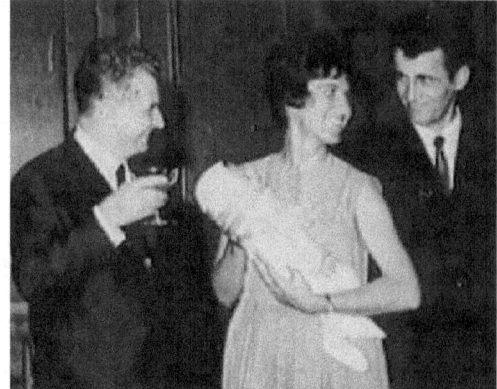

Figure 4: Ceausescu et les parents de l'enfant choisi pour être le 20 000 000 Roumain.

Les seules modalités de contrecarrer les doléances du Pouvoir sont l'abstinence comme moyen de contraception et l'avortement clandestin, fait dans des conditions impropres et dangereuses. Ceux qui, d'une façon ou d'une autre, sont impliqués dans l'avortement illégal sont poursuivis dans la justice et subissent diverses peines, parmi lesquelles la prison est la plus exemplaire.

Quand en 1973, pour la première fois depuis l'adoption du décret, la natalité commence à baisser, Ceausescu prend de nouvelles mesures coercitives. Les représentants de la Milice et de la Procurature occupent des positions dans les hôpitaux, même dans les salles d'opération, et ont le droit de vie et de mort sur les femmes qui se sont faites des avortements. Ces femmes, interrogées, battues, en absence de collaboration, sont laissées à mourir, les médecins n'ayant pas la permission d'intervenir. En 1984, le régime prend d'autres mesures. Pour surveiller encore mieux la vie intime des femmes, Ceausescu impose les contrôles gynécologiques périodiques mensuels, subis obligatoirement par les femmes à leurs places de travail, dans le but de découvrir la grossesse le plus tôt possible et de suivre son déroulement. Par cette mesure, les femmes n'ont aucune possibilité de s'échapper à leur obligation primordiale devant l'État : devenir mères. Mal nourris, dans le froid et le noir des appartements communistes, sans contraceptifs et aucune éducation sexuelle, les Roumains sont obligés de se reproduire pour augmenter la population du pays.

La chute du régime nationaliste et communiste de Ceausescu met fin à ce cauchemar, car le décret 770 est abrogé le 25 décembre 1989. À l'heure de son triste bilan, le projet de Ceausescu de créer l'«Homme Nouveau» est considéré aujourd'hui l'une des plus grandes expériences sociales de l'histoire de l'humanité : dans 23 ans, plus de 2 000 000 d'enfants non-désirés ont été nés et plus de 10 000 femmes sont mortes suite aux avortements clandestins. Paradoxalement, les «enfants du décret», nés à la commande et qui, théoriquement devraient servir les intérêts du régime politique de Ceausescu, ont été les premiers à sortir dans les rues en décembre 1989 et, toujours eux, ont constitué la plupart des victimes durant la Révolution anti-communiste et anti-Ceausescu.

3 L'Homme nouveau au cinéma

Les deux traumatismes qui ont marqué la naissance, le devenir et la mort symbolique de la génération de l'«Homme nouveau», notamment l'interdiction de l'avortement en 1966 et la Révolution roumaine de 1989, sont questionnées par la plupart des cinéastes qui font partie de ce qu'on appelle «la nouvelle vague roumaine» et qui sont les récipiendaires de nombreux prix dans les festivals internationaux à partir de 2005. Biologiquement, ils font partie de la génération de l'«Homme nouveau», (Mungiu et Iepan sont nés en 1968, Muntean en 1971, Mitulescu en 1972, Porumboiu en 1975), génération au-dessus de laquelle plane une question angoissante: est-ce que ces enfants nés durant cette période ont été désirés par leurs familles ou ils sont des *decretei*, des conséquences du décret 770 ?

Cette génération des cinéastes roumains montre dans ses films que l'«Homme nouveau» d'autrefois repense aujourd'hui le passé récent et ses traumas, notamment l'avortement clandestin et ses conséquences, comme dans le film *4 luni, 3 saptamani si 2 zile* (*4 mois, 3 semaines et 2 jours*) de Cristian Mungiu, gagnant du Prix Palme d'Or en 2007, et le court *Visul lui Liviu* (*Le rêve de Liviu*, 2004) de Corneliu Porumboiu, tandis que la révolution roumaine est questionnée dans trois films sortis en 2006 : *Cum mi-am petrecut sfarsitul lumii* (*Comment j'ai fêté la fin du monde*) de Catalin Mitulescu, *Hartia va fi albastra* (*Le papier sera bleu*) de Radu Muntean et *A fost sau n-a fost ?* (*12:08 A l'est de Bucarest*) de Corneliu Porumboiu. Ces films se constituent en autant de questions générationnelles et une recherche de repères.

Le film *4 mois, 3 semaines et 2 jours* de Cristian Mungiu raconte une journée de la vie de deux étudiantes à la Polytechnique. L'une d'entre elle, Găbita (Laura Vasiliu), enceinte, doit subir un avortement illégal le même jour. L'autre, Otilia (Anamaria Marinca), sa collègue de faculté et de chambre dans les résidences universitaires, fait tout pour l'aider. L'avortement illégal se passe dans une chambre d'hôtel et les jeunes filles doivent payer

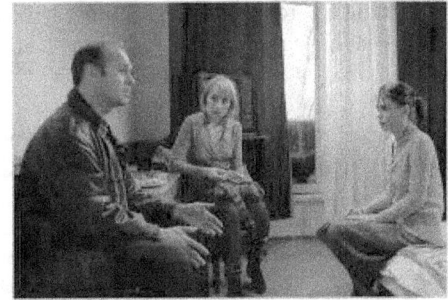

Figure 5. *4 luni, 3 saptamani si 2 zile*, 2007, Cristian Mungiu

d'avance le service de Monsieur Bebe (Vlad Ivanov), le faiseur d'anges, en acceptant d'être violées l'une après l'autre.

Lui-même enfant du décret, comme le déclare ouvertement, Cristian Mungiu a voulu raconter d'une manière neutre une histoire réelle passée dans les dernières années de dictature. Critiqué par certains pour avoir montré trop le fœtus avorté, le cinéaste se défend. Le traumatisme subi par sa génération des *decretei*, ces enfants non-désirés par leurs parents, est vif et douloureux et il doit être montré, devenir visible.

Un court-métrage réalisé par Corneliu Porumboiu, *Visul lui Liviu,* précède comme thématique le film de Mungiu, en montrant les conséquences de la loi contre l'avortement vues par ceux qui, d'une manière paradoxale, doivent être reconnaissants à Ceausescu pour leur existence. Liviu (Dragos Bucur), lui-même enfant du décret, fait un cauchemar récurrent qu'il est incapable de se rappeler. À la fin du film, il se souvient : chaque

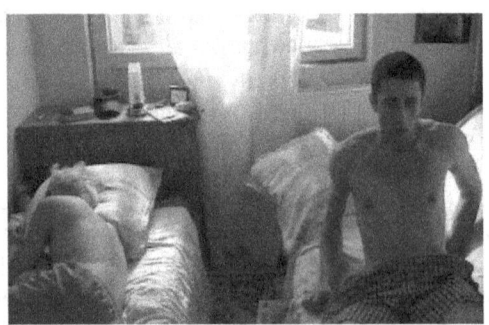

Figure 6. *Visul lui Liviu,* 2004, Corneliu Porumboiu

matin, il rêve de son frère jamais né qui, une chandelle dans la main, erre comme un fantôme dans le noir des souterrains, mais qui, finalement, conduira le protagoniste vers la lumière.

La Révolution roumaine est questionnée dans trois films sortis en 2006. Le film *Comment j'ai passé la fin du monde* de Catalin Mitulescu présente la vie d'une famille ordinaire qui vit dans une banlieue de Bucarest durant les deux dernières années de dictature de Ceausescu. La vie d'Eva (Dorotheea Petre) est bouleversée par un incident. L'écrasement accidentel d'un buste en plâtre de Ceausescu et le refus de l'adolescente de donner des explications entraînent son exclusion de l'organisation officielle des jeunes communistes (en

roumain, UTC) et du lycée. Sa vie change complètement et, à partir de ce moment-là, elle a un seul désir : de quitter son pays. Après des entraînements physiques pour pouvoir franchir le Danube en nageant, la fille renonce au dernier moment à son plan et retourne chez soi, à sa famille et son frère, l'émouvant personnage Lalalilu (Timotei Duma), et aussi au garçon qu'elle aime. La

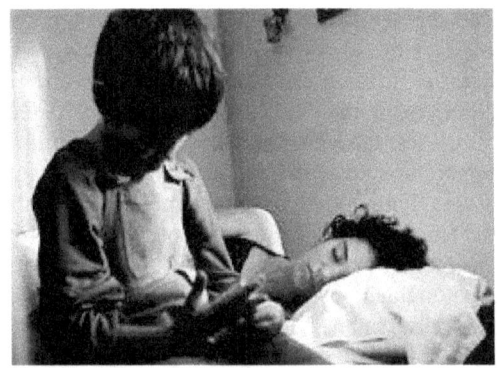

Figure 7. *Cum mi-am petrecut sfarsitul lumii,* 2006, Catalin Mitulescu

fin du film montre alternativement des images-document - transmissions directes de la télévision roumaine durant la Révolution de 1989 - et des séquences fictives montrant comment la petite communauté décrite dans le film a vécu ce moment historique. Enfant unique d'un père méprisé par la communauté à cause de son travail comme délateur pour la police secrète roumaine, le jeune Alexandru (Ionut Becheru) veut montrer à la fille aimée et à tous son courage et qu'il peut se battre pour la liberté. Mais il se fait tué durant le chaos qui règne dans les rues de Bucarest et devient l'une de nombreuses victimes innocentes de la Révolution.

Le film *Le papier sera bleu,* signé par Radu Muntean, est entièrement consacré à cet événement et surprend très bien la confusion générale de ces jours-là de décembre quand, dans les rues de Bucarest, il y avait des feux d'armes partout, quand on ne savait pas si l'armée roumaine et la milice étaient du côté des révolutionnaires ou de l'État communiste, qui donnait des ordres, qui tirait contre qui, qui étaient les terroristes. Le protagoniste collectif du film est une patrouille de milice qui, la nuit de 22 décembre 1989, erre dans un char de combat par un Bucarest chaotique, avec une mission ambiguë : défendre, surveiller, intervenir ou bien se cacher et fuir. Idéaliste, le jeune soldat qui fait son service militaire obligatoire déserte son poste et fuit pour défendre la Télévision, la place la plus

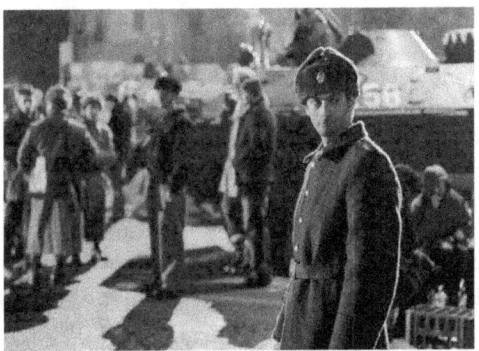

Figure 8 : *Hartia va fi albastra*, 2006, Radu Muntean

dangereuse à ce moment-là, mais sans y arriver à cause du chaos dans les rues. On voit des soldats qui tirent contre d'autres soldats, on voit comment le jeune homme et le tzigane rencontré par hasard sont pris pour des terroristes, et, finalement, quand le jour se lève et on a l'impression que le cauchemar nocturne a pris fin, on voit une scène absurde : la patrouille, faute d'avoir le bon mot de passe, est exterminée par d'autres jeunes soldats en service militaire obligatoire. Ce tragique imbroglio est dû au mal fonctionnement des transmissions d'émission-réception à cause des brouillages et des interférences. Pour que cet événement devienne plus absurde et incompréhensible, les images de la tuerie, les dernières dans la chronologie des évènements racontés, sont montrées au début du film, sans aucune introduction. Les spectateurs sont consternés: on a du mal à comprendre le pourquoi et comment on est arrivé dans une telle situation. Mais la réalité de ces jours et nuits-là de décembre 1989 est encore plus inexplicable et plus problématique, même regardée à une distance de plus de vingt ans.

Dans un registre parodique et absurde, le film *12h8. A l'est de Bucarest* de Corneliu Porumboiu joue sur la même confusion, restée irrésolue après des années. «Y a-t-il eu, oui ou non, une révolution dans notre ville?», se demande le réalisateur d'un débat télévisée seize ans après décembre 1989. Dans cette émission, transmise en direct sur le poste

local d'une petite ville moldave, «à l'est de Bucarest», Jderescu (Teo Corban), ex-ingénieur textile devenu propriétaire du poste local de télévision, présentateur et politologue, Manescu (Ioan Sapdaru), professeur d'histoire et ivrogne notoire dans la ville, et Piscoci (Mircea Andreescu), un vieux monsieur à la retraite qui toute sa vie a fait le Père Dugel (le substitut laïque du Papa Noel durant le communisme) essaient d'élucider l'énigme des événements de décembre 1989 dans leur ville.

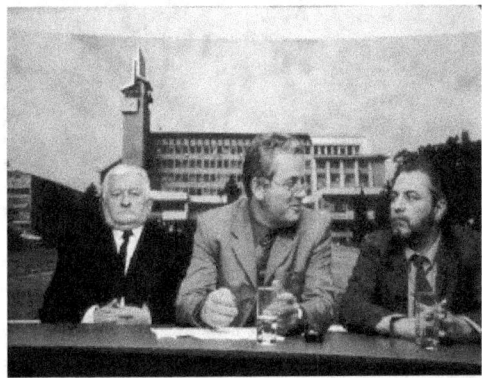

Figure 9. *A fost sau n-a fost ?*, 2006, Corneliu Porumboiu

Sous cette question réitérée avec insistance dans le film se cache le questionnement concernant les événements réels de décembre 1989 en Roumanie. Dans un article publié en 2007, Ruxandra Cesereanu [2] essaie de faire le point sur ces événements, en énumérant les trois interprétations «classiques» - «une révolution pure»; «un complot manipulé de l'extérieur»; «un coup d'État interne»-, et en avançant une autre hypothèse, selon laquelle «il s'est agi d'un événement hybride entre révolution spontané, révolution de palais et/ou complot de l'extérieur».

Si, aujourd'hui encore, le mystère et l'ambiguïté planent sur ces événements, il est certain que, depuis 1990, les centres de grandes villes de la Roumanie, parmi lesquels Bucarest, Timisoara, Brasov, sont devenus des lieux de mémoire, réaménagés pour abriter des cimetières où sont enterrés les héros de la Révolution, pour la plupart des jeunes qui sont descendus dans la rue en décembre 1989 et qui ont été tué dans le chaos et la confusion générales. Si les vrais coupables ne sont encore nulle part, les morts sont là, enterrés. Dans le documentaire *Né à la commande*, à la demande :

«Qui est-il l'«Homme nouveau»?» la réponse est une autre question : «Est-il un homme avec des qualités extraordinaires ou seulement une personne ordinaire qui a le courage de payer avec sa vie pour la liberté?»

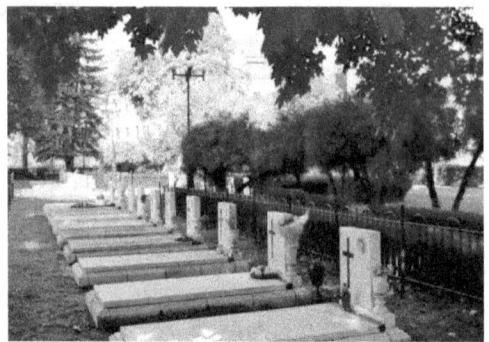

Figure 10. Brasov, centre-ville.

4 Conclusions

Dans des registres esthétiques divers, les cinéastes roumains, en réactualisant des événements traumatiques pour leur génération, ont proposé quelques interprétations et surtout ont soulevé des questions, qui restent ouvertes aujourd'hui. Leur mise en question de l'Histoire reflète la quête identitaire douloureuse de cette génération qui cherche encore de se redéfinir et de trouver des repères. En réponse à l'Histoire, avec ses déraillements et ses inconnues, et à un système totalitaire qui avait détruit systématiquement la vie quotidienne des gens, les cinéastes roumains ont présenté leur point de vue et leur vision sur les choses, à travers des histoires mineures, émouvantes, humaines. Au-delà des films discutés ici, le questionnement sur l'«Homme nouveau» se suit, car cette génération extraordinaire des cinéastes roumains continuent leurs interrogations et leur recherche des réponses dans d'autres films, plus récents. Par exemple, en changeant de registre esthétique, Cristian Mungiu continue sa trilogie annoncée, avec un film épisodique et collectif, dont il écrit également le scénario, et qui est une sorte de comédie amère portant sur quelques légendes urbaines durant le communisme, et où les autres réalisateurs sont Ioana Uricaru, Hanno Höfer, Razvan Marculescu et Constantin Popescu. Le titre,

Souvenirs de l'Époque d'or, pourrait devenir générique pour tous les films qu'on a vus ici, car il synthétise l'une des préoccupations primordiales de cette génération: le rapport problématique avec le passé récent et ses traumatismes et le devoir de se souvenir. Je crois que l'une des forces de ce cinéma réside justement dans cette prise de position devant l'Histoire. Dans une analyse politique de cette période de l'histoire roumaine, en réfléchissant sur la question de la mémoire, Vladimir Tismaneanu [3] affirme : «In the process of dealing with the communist past there is no surplus of memory. To forgive doesn't mean to forget, but to forget is simply impardonable.» Si en Roumanie, depuis les évènements de décembre 1989, tant la société politique que la société civile oscillent entre amnésie et hypermnésie - définie par Alain Besançon [4] comme un «déséquilibre entre la conscience collective d'un fait historique et celle d'autres faits contemporains qui, eux, sont l'objet d'amnésie collective» - avec ces films on voit clairement qu'oublier n'est pas une thérapie et que la génération de l'«Homme nouveau», obligée de se taire durant les années de dictature, fait de l'acte de se souvenir un devoir que les cinéastes assument pleinement.

References

1) Siegfried Kracauer, *Theory of film. The redemption of physical reality*, edited by Princeton University Press (Princeton, New Jersey, 1997), pages 305-6.

2) Ruxandra Cesereanu, "Timisoara 15-20 décembre 1989 et Bucarest 21-22 décembre 1989", in *Communisme*, no. 91/92 (2007), 161-172, page 161.

3) Vladimir Tismaneanu, "Coming to terms with a traumatic past: reflections on democracy, atonement, and memory", in *History of Communism in Europe, Politics of Memory in Post-communist Europe*, edited by Zeta books, vol. I. (2010), 15-20, page 18.

4) Alain Besançon, *Le malheur du siècle : sur le communisme, le nazisme et l'unicité de la Shoah*, edited by Fayard (Paris, 1998), page 165`

Dialogue and Critical Thinking Online

Dr. Isabelle Sabau, Dr. rer. nat. Carmen Sabau

e-mail cyberedu22@netscape.net

Abstract: Times New Roman 10 point,

1 Introduction

The exponential developments in all areas of knowledge that mark the modern and especially the contemporary world with its proliferation of digital and telecommunication innovations and advances, have led to both, the expansion in human understanding as well as the fragmentation of expertise and disciplines into multiple subject areas that often appear to be seemingly disconnected. Aristotle defined humanity as animals who talk, in the Greek sense of logos that is inquire, formulate opinions and arguments and communicate [1]. Today, the increase of mobile technologies in the form of smartphones and tablets promises unprecedented access and exchange of information and ideas along with heightened levels of continuous and constant communication. Yet, ironically, these very same developments may thwart the conversational uniqueness of humanity through the increase of shortened acronymic expressions in text messaging and various forms of social media such as Facebook and Twitter. The frenetic, fast pace of today's global world leaves increasingly less introspective space for contemplation and self-reflection, while the explosion of multimedia through the growth of images, music and videos widely distributed through the Internet, threatens to supplant the written word and diminish the enterprise of critical thinking. "Man is a social and political animal…the only animal with the gift of speech...[endowed] with a sense of good and evil, of just and unjust…" Aristotle explained in his Politics [2]. Thus conversation, and especially dialogue remains a most important aspect of human communication and expression in the ever present communal search for knowledge. The Delphic admonition of Apollo to 'Know Thyself' becomes a lifelong enterprise for Socrates, whose dialectic and unwavering search and love of wisdom entices all humans to follow his lead and develop their own freedom and power to think for themselves [3]. The paramount importance of dialogue and discussion has become a critical necessity in today's confusingly fractured digital environment and forms a pivotal center for online learning.

The tsunami of technological advances of the 21st century has brought a tremendous revolution in education spurred by the development of the computer and the Internet. E-learning, or online education has seen an exponential growth as the student body has expanded to include increasing numbers of non-traditional students returning to further their education, gain re-certification and even change careers. This increasingly varied population of 'consumer students' demands flexibility and learner control of their personal time management which the online environment amply supplies by fulfilling the promise of anytime, anywhere, anyplace opportunities for learning. These new delivery methods open possibilities for those who would otherwise be unable to participate in the more formal and rigid schedule of daily courses offered by most higher education institutions. Lifelong learning is emphasized based on globalization and the explosion of technology which evolves continuously at an accelerating pace. Online learning promises a democratic approach to the formation of virtual learning communities that stress the importance of critical thinking skills and self-reflection. The new paradigms of teaching and learning reflect different learning styles, independence and responsibility of the learner and relevance of the material in contemporary life. In addition, the online environment also requires new pedagogies which counter-balance the isolation that often accompanies the human-computer interface and the active interaction of the class [4]. Strategies for the use of dialogue and specifically the Socratic elenchus in the online teaching and learning environment will be further explored in this paper with reference to humanities and art history online courses, although the critical thinking skills developed through such interaction are applicable to all domains of learning.

2. Online Learning

According to the Instructional Technology Council's latest statistics, 5.5 million students in the U.S. were enrolled in at least one distance education course in 2013, an increase of 3.7% from the previous year [5]. Online learning is basically defined by the physical and sometimes temporal separation of learners, and their interactions being mediated through the computer and the Internet. The participation can be either synchronous or asynchronous or a combination of the two modalities. The addition of the Internet with its vast databases of materials greatly enhances research, interdisciplinary analyses and the creation of new ideas. Online learning which rests on the formation of virtual learning communities is based on mutual respect and collegiality in collaboration. Modern technologies enable the formation of an interactive learning environment separate from temporal and spatial constraints. Education must address the needs of the learners and provide quality programs which enable a basic understanding of the modern world coupled with critical thinking and critical scrutiny, based on logical analysis, research and synthesis skills. These critical thinking skills help promote the discovery of new approaches and solutions as well as new connections between domains. Since the beginning of the 20th century a number of new learning paradigms have been evolving, placing the learner at the center of the learning process. Learning which emphasizes the cognitive development in a variety of constructive approaches has emerged to accent critical thinking and logical argumentation coupled with increasing communication and dialogue between professor, learners and colleagues. Online pedagogies center on community building through collaboration and performance based assessment that employs the application of knowledge and discovery learning. The emphasis on learner freedom of choice and learner control of their own learning, places the responsibility for learning on the learner thus encouraging reflection on a lifelong basis.

3. Critical Thinking

According the Webster's Unabridged Dictionary, thinking is defined as a mental action, cogitation, judgment, while the act of thinking involves using the mind to arrive at conclusions, decisions and perform any mental operations [6], while "Critical thinking is thinking that proceeds on the basis of careful evaluation of premises and evidence and comes to conclusions as objectively as possible through the consideration of all pertinent factors and the use of valid procedures from logic" [7].

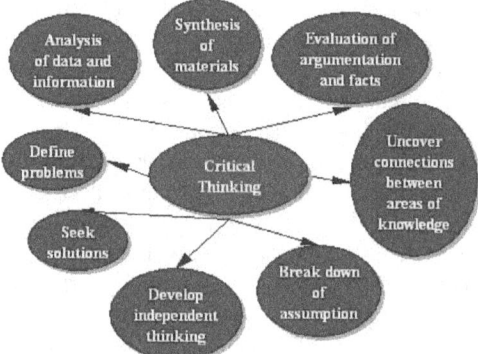

Figure 1. Critical Thinking

Critical thinking skills are paramount today to enable clear and informed decision making through the systematic breaking down of information, data and assumptions, analysis and synthesis of ideas and thorough evaluation of the situation and the logical arguments it involves. This systematic method aids in defining problems and uncovering various connections between subjects. The above diagram, **Figure 1** displays some of the important aspects of critical thinking. The first step in the analysis process requires identification of the components of the problem at hand and the breakdown of information and the examination of the data pertaining to a specific problem. Often the principles, theoretical underpinnings and concepts used require close scrutiny and the discussion usually moves from the general ideas to specific situations. The application of logical methods of argumentation enables accurate inferences, evaluations and predictions. The use of real-life examples and scenarios including case-base studies and ethical dilemmas promotes holistic approaches to decision making processes and problem solving and allow for connections with prior learning thereby increasing the relevance and continuity of learning and helps diminish the fragmentation of knowledge. Therefore, "Critical thinking is the use of those cognitive skills or strategies that increase the probability of a desirable outcome. It is used to describe thinking that is purposeful, reasoned and goal directed - the kind of thinking involved in solving problems, formulating meaningful inferences, calculating possible likelihoods, and making decisions when the thinker is using skills that are thoughtful and effective for the particular context and type of thinking task. Critical thinking also involves evaluating the thinking process - the reasoning that went into the conclusion we've arrived at, and the kinds of factors considered in making a decision. Critical thinking is sometimes called directed thinking because it focuses on a desired outcome" [8]. In the online education world, critical thinking serves multiple purposes: it enables quick analysis of information and issues, it fosters interactive participation of the learners via discussions and chats through a systematic examination of material, it stimulates ability to discover new ideas and approaches to problem solving, it helps enlarge horizons and eliminate bias, and it promotes synthesis of material and construction of new knowledge. Since the Internet has grown exponentially along with the accumulation and increase in human knowledge: "The sum total of humankind's knowledge doubled from 1750-1900. It doubled again from 1900-1950. Again from 1960-65. It has been estimated that the sum total of humankind's knowledge has doubled every five years since then....It has been further projected that by the year 2020, knowledge will double every 73 days"[9], it has become imperative for individuals to develop skills of timely and accurate information gathering and rapid evaluation of its relevance. Therefore, online courses must emphasize the development of strong critical thinking skills that enable the scrutiny of an ever expanding websites and other data, along with research skills based on constructivist learning methods [10]. Situated and distributive cognitive exercises and activities, which combine with the problem-based, collaborative construction of knowledge [11], are best suited for the electronic environment which can connect people and databases across the globe virtually instantly while at the same time accentuating meta-cognitive tasks of synthesis, reflection and evaluation. The practical application of acquired knowledge can be assessed through performance related tasks and collaborative projects. The most important aspect of e-learning is interaction, which leads to the creation of online learning communities that diminish isolation by bringing learners together in a common venture which increases the exchange of ideas, collegiality and networking. Cooperative and collaborative projects enable distributed cognition and conflict resolution skills by engaging learners in cooperative ventures [12]. Collaboration based on fairness and mutual respect is usually accomplished through the use of group projects that help learners negotiate the exchange of ideas and task oriented team work, enable greater diversification of content and ultimately prepare learners for a more active participation in both society and democracy [13].

3. Dialogue

Turning now to dialogue, we begin with a number of important questions - what is dialogue? - does dialogue require two or more interlocutors?- should dialogue be modelled by the Socratic elenchus? – does the Socratic dialogue require a test of character leading to the self-examination of the participants? What are the necessary components of dialogue? – does dialogue include virtue along with the search for truth? – what is the connection between dialogue and critical thinking as well as self-examination and what is the difference between dialogue and debate? How can dialogue enhance critical thinking especially in an online environment in which the participants communicate via the portal and thus never physically meet? Collaboration encourages constructive approaches to the creation of meaning through dialogue: "Making meaning is creating a shared perception of events that helps us all get more of what we want, when what we want is good for all of us. Dialogue plays a key role in making meaning and thinking together. The purpose of dialogue is 'seeking mutual understanding and harmony.' Dialogue is also seen as initiating team learning so the team members gain the ability to suspend assumptions and enter into genuine 'thinking together'"[14]. Through dialogue, the necessary components of critical thinking which require the application of a methodology that includes analysis of the problem at hand and the gathering of information, extracting the important components of relevant information from the sources, the synthesizing the material thus discovered and the evaluating the validity of the conclusion from the given premises, are jointly discussed as each interlocutor is respectfully encouraged to provide their own research and argumentation. This process is "holistic in content and integral in structure...the solution to any single existential issue necessarily involves the solution to all problems" [15]. In an environment of asynchronous discussion the questions used to initiate various applications of the critical thinking For example in the teaching of philosophy or ethics one could begin with general questions such as what is morality, what are virtues, is there a difference between ethics and morality, how do values relate to virtues, can one learn the virtues, or as Socrates asked, can one teach them. One method of tackling these kinds of questions would be to begin with the proverbial forest, the general concept, and then dissect it into its smaller components in order to arrive at the singular tree. An alternative could be to go in the opposite direction starting with the specific example and moving on to more general concepts of how that solution to the specific example might apply to similar situations. In the domain of the history of art it is useful to introduce the student to the development of the arts throughout the ages and its connection to political and socio-economic events. Within the broader context of antiquity, Middle Ages, the Renaissance and the modern periods one can begin to explore the artistic dialogue sculptors and painters have developed with their artistic predecessors. The visual arts reveal the real and imaginary aspirations of humanity through the ages including fantastic or ephemeral subjects as well as portraiture and sacred imagery. For example, one can see a progression from the idealism of the classical Greeks to the extravagant expressionism of the Hellenistic times to the veristic portrayal of the Romans and then under the influence of the growing Christian religion the imagery returns to the archaic, abstract style that attempted to de-materialize the physical body and focus the attention on the spiritual ideal. The rediscovery of antiquity during the Renaissance once again brought the attention of the artists to the human realm and its potential for knowledge. The dialogue between the artist and the viewer sometimes was mediated and motivated by the patron who commissioned the work and who often times provided specific instructions to the artist, not only about the subject matter, but even to the details of

the appropriate colours to be used. The biggest patrons of the arts for a considerable length of time in Europe were the church and nobility. In the pre-photographic world of the 19th century portraiture was a most important aspect of the arts to record specific individuals and their influence on particular events. And yet the stoic idealized and refined world of the Renaissance gave way to the exuberance of the Baroque with its swirling diagonals sometimes sombre, sometimes brightly coloured or extremely realistic. The rapid changes that propelled the modern world into the 20th century fostered an explosion of artistic styles and artistic personalities. The study of art history can provide a model for the development of critical thinking skills which connect specific artworks with events and innovations. Discussion of mythological works can refer the student back to ancient civilizations and lead the student to examine those myths in terms of their power of reviewing and understanding the human psyche. At the same time, those myths and stories can relate to contemporary events the artist may present in allegorical terms. With the advent of Realism and later Impressionism and Post-Impressionism the artistic dialogue expanded to incorporate the dialogue between the artist and the world of nature, atmosphere, and light. The production of oil painting in portable tubes freed the artist, liberating him from the confines of the studio to roam the countryside and explore the landscape and the world of nature, as well as their own perceptions of the changes that light and the weather can inflict on the physical world. Reluctantly, the public also was swept into this dialogue to reconsider its own perceptions and reactions to the environment. The experiments of the Impressionists opened the floodgates to a myriad of new developments and inventions, thereby initiating a proliferation of the variety of expressions the visual arts can produce which continues to expand. In the study of art one must follow the basic questions -why, what, where, when, how - in order to synthesize these various aspects of a work into a coherent understanding of the purpose and effects of a particular creation. During this process of analysis one follows the steps of critical thinking by gathering information that is available about the piece and then exploring the various tangents that might connect that work with its environment and time period. Such steps and methods can be applied throughout the study of the arts resulting in a deeper understanding of the interconnectedness of the different aspects of human behaviour and action. Successful development of such critical thinking skills can be transferred to other domains and subjects of study.

For example, upon first encountering a Bird in Space sculpture by Constantin Brancusi, one begins with the sensory and perceptual qualities of the piece. At first, one notes the streamlined shape, the high polished surface of the form which continuously reflects its surroundings, and the precarious balance of the piece on its slender leg. Upon further investigation one discovers that it is called a bird, yet it doesn't appear to indicate any of the expected features of a bird, no feathers, no wings, even no beak, perplexing and confusing the viewer and thereby encouraging further examination – now begins a process of analysis begins as one starts to research both the artist and the work. In today's online digital environment critical thinking and dialogue are beginning to merge, each strongly dependent on the other, to enable individuals to navigate the various features of the Internet and avoid the pitfalls of misinformation in order to discover the web of knowledge that is hidden within. The course management systems, like Blackboard, WebCt, Desire2Learn, Moodle and others, provide an elaborate template and depository for the course content and materials while at the same time it also allows for areas where participants can interact both synchronously via chats, and asynchronously in the discussion boards. Another feature may include email which is a more personal form of communication between participants which does not involve the whole class. To help foster active participation in the class, discussion forums are the most flexible and effective means because they allow participants to fully reflect and carefully prepare their responses as they log in at their convenience rather than at a set time. The discussion boards must be an integral part of the virtual classroom and are best utilized in conjunction with the course division into modules or units that create a specific structure for the course and may be used in assessment. Various strategies can be used to ensure successful moderation of the discussion forums and increase learner participation.

Frequent posts and log-ins sustain dialogue and increase participation – minimum requirements should be stated in the beginning of the course, the posing of engaging and motivating questions that encourage deeper analysis and prior knowledge connections promotes critical thinking and extend the relevance of the material through the discovery of relationships and implications while the stress on confidence building promotes mutual respect on the part of the participants [16]. The professor's role dramatically changes in the online environment as the professor becomes a participant alongside all other participants and adopts a role of guide, monitoring and facilitating the participation and dialogue of the class [17].

2 Conclusion

There are many creative ways of incorporating the most important skill of critical thinking into the online environment and thereby promote collaboration, analysis and synthesis of information. Probing, questioning and the Socratic dialectic can help encourage deeper reflection on the subject and new ways of integrating new learning into everyday life ensuring a successful virtual learning community development. Further study of the various tools and motivational strategies will continue to add valuable resources to the design and delivery of flexible anytime, anyplace educational opportunities to a vaster audience of learners.

References

[1]. Sinaiko Herman, Reclaiming the Canon Essays on Philosophy, Poetry and History, New Haven, Yale University Press, 1998, p.10

[2]. Aristotle, Politics I 2 1253a5-15, translated by Jowett B. in The Complete Works of Aristotle, volume 2, edited by Jonathan Barnes, New Jersey, Princeton University Press, 1984, p.1988

[3]. Sinaiko Herman, Reclaiming the Canon Essays on Philosophy, Poetry and History, New Haven, Yale University Press, 1998, p. 10-12

[4]. Portway, P.S., & Lane, C., 1992. Technical guide to teleconferencing & distance education. San Ramon, CA: Applied Business tele-communications.

[5]. Trends in eLearning: Tracking the Impact of eLearning at Community Colleges 10th Anniversary Edition April 2015, Instructional Technology Council retrieved from

http://www.itcnetwork.org/membership/itc-distance-education-survey-results.html

[6]. Webster's Unabridged Dictionary, 1983, DeLuxe Second Edition, NY: Simon and Schuster

[7]. Carter, C. V. Dictionary of education. New York: McGraw Hill, 1973

http://www.intime.uni.edu/model/democracy/crit.html

[8]. Halpern, Diane F., Thought and Knowledge: An Introduction to Critical Thinking. 1996.

[9]. Appleberry, James. (1992) "Changes in our Future: How will we cope?" Faculty speech presented at the California State University, Long Beach, California.

[10]. Jonassen David H., Peck L. Kyle and Wilson Brent G., 1999, Learning with Technology: A Constructivist Perspective, Prentice Hall Inc.

[11]. Oliver Ron and Herrington Jan, 2000, "Using Situated Cognition as a Design Strategy for Web-Based Learning", in Abbey Beverly, 2000, Instructional and Cognitive Impacts of Web-Based Education, Idea Group Publishing, London

[12]. Palloff Rena M. and Pratt Kevin, 2005, Collaborating Online: Learning Together in Community, Jossey-Bass A Wiley Imprint

[13]. Palloff Rena M. and Pratt Kevin, 2005, Collaborating Online: Learning Together in Community, Jossey-Bass A Wiley Imprint

[14]. Yankelovich, D. (1999). The magic of dialogue: Transforming conflict into cooperation. New York: Simon & Schuster.

[15]. Sinaiko Herman, Reclaiming the Canon Essays on Philosophy, Poetry and History, New Haven, Yale University Press, 1998, p. 331

[16]. Salmon Gill, E-Moderating: The Key to Teaching and Learning Online, second edition, RoutledgeFalmer – Taylor and Francis Books Ltd., 2003

[17]. Bonk Curtis J., Kirley Jamie, Hara Noriko and Dennen Vanessa Pez, 2001, "Finding the Instructor in post-secondary online learning: pedagogical, social, managerial and technological locations", in Stephenson John (ed.), 2001, Teaching and Learning Online, Kogan Page Limited, London Webster's Unabridged Dictionary. 1983. DeLuxe Second Edition, NY: Simon and Schuster

Vers une approche préventive en ethnothérapie

par **Angela Stoica**

Chargée de cours à l'Université de Montréal, Département des sciences de l'éducation

Abstract: Nous présenterons l'évolution d'une famille maghrébine suivie à la Clinique transculturelle de l'hôpital Jean-Talon en regard du dysfonctionnement scolaire de l'aîné de 12 ans, de ses comportements impulsifs et violents et de la crainte d'une escalade pouvant mener au placement du jeune sous la Loi de la protection de la Jeunesse. Nous démontrerons l'impact de ce dispositif thérapeutique groupal suivi de services dans le milieu de vie de l'enfant et comment toutes cette collaboration a permis l'apaisement de la souffrance de la famille. Le concept de ''famille autarcique'' a orienté la description clinique de cette famille.

Contexte de la référence :

La famille a été référée à la clinique par une psychologue qui rencontrait le couple en privé et se questionnait sur la nécessité de signaler l'aîné à la Direction de la protection de la jeunesse (DPJ). Cette psychologue avait été contactée pour ce cas par une intervenante maghrébine d'une organisation communautaire qui travaille avec des femmes immigrantes.

La psychologue référente se sentait dans une impasse, pas seulement à cause du manque de progrès, mais surtout parce qu'elle observait la détérioration des comportements du jeune Adam, l'aîné de la famille.

Suite à la prise en charge par la clinique, la famille a été rencontrée à 6 reprises sur une période de 18 mois. D'autres services ont étés utilisés par la famille : une éducatrice en milieu et une coach de vie (femme musulmane).

L'alliance thérapeutique a pu se former rapidement grâce à la présence d'un thérapeute principal d'origine maghrébine et de la présence de nombreux co-thérapeutes issus de l'immigration. Vers la fin de la thérapie, l'équipe a pu observer que le couple parental est beaucoup plus solide et présente moins de conflits avec la société d'accueil. Chacun des membres de la famille a repris sa place et son rôle. La famille est plus soudée et les comportements du jeune, beaucoup moins intenses et moins fréquents.

Histoire des familles maghrébines au Québec :

La cohorte arrivée à la fin des années 90 et le début de l'an 2000 est scolarisée (14 ans et plus de scolarité). Les familles appartenant à cette cohorte quittent un Maghreb tiraillé par des crises sociales et politiques, parsemé des attaques terroristes (en Algérie).

Au pays d'accueil, le conteste sociopolitique après le 11 septembre 2001 (attaques terroristes à New-York) engendre l'islamophobie. Des débats sociaux s'ensuivent autour du port du voile et la laïcité des services publics. Les ressortissants des pays musulmans, pourtant francophones et hautement scolarisés, subissent l'impact massif de la déqualification professionnelle.

Types de problématiques traitées

Le modèle de l'ethnothérapie est utilisé souvent avec de bons résultats avec des familles maghrébines qui n'évoluent pas avec des suivis conventionnels en place.

Hypothèses thérapeutiques :

Dans l'analyse de ce cas thérapeutique, nous nous sommes inspirées par le modèle d'analyse proposé par Marie Rose-Moro, ethnopsychanalyste française de prestige qui étudie depuis des décennies le drame de la parentalité en immigration.

L'hypothèse principale est la suivante : H1 :*l'interaction parent-enfant n'existe pas en* dehors d'un système interactif généralisé, le système culturel d'appartenance des parents.[1]

Cette hypothèse est ensuite décomposée en deux hypothèses, dont la première (H2) est *clinique*, alors que la deuxième (H3) est *technique*, elle rend compte du travail thérapeutique.

H2 : en dehors de la langue, les éléments fonctionnels spécifiques dans le système culturel des parents sont *: les représentations ontologiques* concernant l'origine et la nature de l'enfant; les *théories étiologiques* utilisées pour rendre compte des maladies de la mère, de l'enfant et de tout dysfonctionnement de la relation parents-enfants; *les logiques des thérapies traditionnelles* mises en œuvres par le groupe lors de tels dysfonctionnements au pays.

H3 : Pour modifier des interactions disharmonieuses parents-enfants en situation migratoire, il est nécessaire d'agir d'abord sur l'interaction entre les parents et leur système culturel d'appartenance pour modifier l'interaction observable parents-enfant.

Quant à lui, le concept de ''famille autarcique'' a ouvert de nouveaux angles à notre analyse. En effet, il désigne une ''famille fermée vers l'extérieur, comme si les fonctions paternelles de **passeur** (=père) ne s'exerçaient pas. Cette fermeture maintient les enfants dans un état de dépendance infantile''. [2]

Retombées pour les parents :

a) Mythe de la pensée commune, la non-différenciation parents/enfants qui amène l'absence de règles parentales
   Rien n'a à être demandé explicitement, tout doit être compris de manière implicite. Les enfants prennent la place de

---

[1] Marie Rose-Moro, *Parents en exil. Psychopathologie et migrations*. Presses universitaires de France, 2011.
[2] Source: Elisabeth Bizouard-Reicher

parents. En thérapie, l'objectif s'énonce ainsi: ''chacun doit prendre sa place''.

b) Mythe de non-survie sans les parents dans un monde étranger, hostile, plein de dangers
   Assurer à tout prix la sécurité des enfants en renforçant le cocooning. Leur faire comprendre qu'on n'est en sécurité qu'en famille.

c) Mythe du devoir omnipotent qui engendre le sens de responsabilité parentale exacerbée
   En tant que parent, on est **responsable** *de tout ce qui arrive aux enfants, d'où un contrôle des influences extérieures et une résistance au changement.*

Ce modèle familial a des retombées pour les enfants :

- Non-différenciation FAMILLE-SOCIÉTÉ :
  absence de fidélité affectueuse parents-enfants: *Adam nomme son père devant les intervenants: ''lui-là, celui-là'';*
- Culpabilisation : *Le vrai* moi *de chacun représente le germes d'une révolte qui reste sécrète. Isolement: ''dans la famille, on ne discute pas''; chacun reste avec ses malheurs*
- Rapport public/privé perturbé : *C'est toujours l'école qui a tort: les enfants changent 3 écoles dans deux ans. A. n'arrive pas à adopter la bonne distance face aux autres: recherche d'attention excessive à l'école; en thérapie, il monopolise la parole et prétend être plus âgé que son âge réel. Rôle de passeur du père, perturbé.*

Les mouvements autarciques de la famille sont dus au vécu discriminatoire sur le marché d'emploi (père), dans la rue (contexte de la Charte de la laïcité, 2007), à l'école (regard posé sur la mère voilée; réaction des enfants) et à la solidarité familiale (famille nucléaire) face au rejet de la société d'accueil.

Ainsi, la famille en immigration est profondément bouleversée. Selon Ondong, *immigrer*, c'est

« quitter ou perdre l'enveloppe des lieux, des sons, des goûts, bref: des sensations de toutes sortes qui constituent les premières empreintes sur lesquelles s'est établi le codage du fonctionnement psychique et des enveloppes de sens. » ; tandis que *émigrer*, c'est « tenter de reconstruire, seul ou en famille, en l'espace de quelques années, ce que des générations d'autochtones ont lentement élaboré et transmis. (…) C'est tenter d'acquérir, en peu de temps les codes essentiels balisant et réagissant la culture d'accueil.» [3]

La place du père :

La première cause du repli familial est le changement dans le rôle du père (son rôle de passeur, de contact avec l'extérieur). Or, au Maghreb, le groupe familial est dominé par le mâle. Le devoir du père maghrébin est de protéger, nourrir et entretenir la famille (élargie); en échange, ''il a la main haute et exclusive sur toutes les affaires. Même sans travail ou sans richesse, le père garde son statut et son autorité, porté par le cadre social.''

Actuellement, dans la société d'accueil, on observe la prise en charge de la fonction paternelle par la société laïque et non plus par une autorité religieuse. Ce rôle des institutions laïques est à la fois incompris, craint et contesté dans le discours des parents: l'école est inadéquate, même dangereuse (éducation sexuelle); la crainte de la DPJ étalée sur toute la communauté immigrée. Une des solutions à ce problème est la valorisation des organisations religieuses de la communauté d'origine et exerçant dans la société d'accueil.

Le père est en questionnement identitaire. Il s'est trouvé en décalage par rapport à la société d'origine suite à son immigration en France et déçu suite à son immigration au Canada. Aussi, il est affecté par la symbolique de la *hiérarchie familiale*: de par son origine sociale modeste, (classe populaire), ses méthodes parentales semblent discréditées aux yeux de la mère, qui réclame de lui des solutions miracle. Cartésien

jusqu'à la rigidité, il adopte des méthodes d'encadrement non-traditionnelles (bricolage) , mais les applique sans constance. Nomme l'évolution de son encadrement parental ''en dents de scie.''

Dans ce contexte, un repli familial s'opère. Le pays d'origine est ''intériorisé'' par les parents comme seul repère et seule source des valeurs. Les fêtes familiales sont restreintes à la famille nucléaire: la fête de l'Aid est la seule où la famille fait des rencontres. Adam reproduit à sa manière, par son comportement délinquant, les expériences de rupture en présence dans la famille. Il affirme par rapport à ce mode de vie: ''On n'est pas comme les autres'' et rêve d'amener à la maison des amis d'autres origines (ou les visiter). De plus, les activités parascolaires proposées aux enfants ont lieu uniquement dans la communauté musulmane (scouts, etc.)

Étant donné que l'identité se sent menacée, les stratégies identitaires des enfants se mettent à l'œuvre. Une crise est parfois la seule amorce d'un important changement; la délinquance, la seule issue. L'affirmation identitaire propre à tout adolescent passe ici par le rejet des valeurs de ses parents en s'appuyant quand même sur sa culture d'appartenance. Le défi de l'adolescent immigrant est double il ''manque cet appui fondamental sur la culture d'origine (car cette dernière l'identifie trop aux valeurs de ses parents, desquels il doit se distancier. Et voici les **solutions** trouvées par l'adolescent immigrant, telles que vues **en clinique**: pathologie de l'agir (déprime, ambivalence); dépression (flou identitaire, découragement, paralysie de la pensée); déni du clivage, de l'impasse identificatoire (registre maniaque)[4]

Dans le cas que nous présentons, Adam conteste l'autorité (maison, école) et se met en danger (à 11 ans, rentrait à 23h00 du soir à la maison).

---

[3] Source: Ondongh-Assalt E., Flot, C.

[4] Saskia von Overbeck Ottino, Jerôme Ottino, *Tribulations identitaires chez les adolescents immigrants.*

Cependant, il fonctionne bien lorsque l'encadrement est constant, sécurisant et mène une relation privilégiée avec le thérapeute principal en clinique (qui est un psychiatre, homme, de la même origine ethnique que ses parents). Adam a été diagnostiqué récemment par un psychologue avec le TDAH (médication à stabiliser). Suite au suivi clinique en ethnothérapie, le thérapeute principal soupçonne plutôt de l'angoisse identitaire (clivage du moi) qui amène un mouvement de polarisation entre ses deux cultures.

En ensemble, les relations intrafamiliales s'améliorent de manière substantielle au fil des rencontres cliniques. Les crises d'Adam ont diminué, une meilleure communication parents-enfants s'est installée. Adam a réussi sont intégration au secondaire et poursuit son progrès scolaire. Les rencontres à la clinique se sont espacées (2 par année) et servent à consolider les acquis.

Conclusion

En clinique, il faut: jouer avec les identifications croisées, reconnaître comme essentiels des allers-retours entre la culture d'origine des parents et la culture d'accueil (double appartenance culturelle pour les enfants); valoriser l'effort des parents de réaliser un bricolage culturel des méthodes parentales; travailler trois niveaux: individuel, familial et culturel et tracer, avec la famille, les frontières entre ces trois plans; sensibiliser les institutions de la société d'accueil aux difficultés vécues par les familles immigrantes; favoriser l'accompagnement de ces familles par la médiation interculturelle.

Le modèle d'ethnothérapie, en combinaison avec des services dans le milieu de vie a permis d'éviter un signalement à la DPJ puisque la famille a reçu tout le soutien pour dire, être entendue et comprendre sa situation autrement.

Bibliographie :

Bizouard-Reicher,, Elisabeth « Famille autarcique et situation d'immigration », *Le Divan familial* 2004/2 (No 13), pp. 111-128;

Moro , Marie Rose-, *Parents en exil. Psychopathologie et migrations.* Presses universitaires de France, 2011.

Von Overbeck Ottino, Saskia ; Jerôme Ottino, *Tribulations identitaires chez les adolescents immigrants,* 2001 (http://www.cairn.info/article.php?ID_ARTICLE= LAUTR_004_0095).

Application to the Motion and Equilibrium of the Planar Kinematic Chains with Rotational Links with Clearances

Jan-Cristian Grigore[1], Alexandru Jderu[2], Marius Enachescu[2*]
[1] University of Piteşti, Piteşti, 110040, Romania
[2] University POLITEHNICA of Bucharest, Bucharest, 060042, Romania
* Corresponding author: marius.enachescu@upb.ro

Abstract: Based on the algorithm presented in our previous papers [1], we now present their applications to a few problems.

1. Application

We consider the planar kinematical chain from the Figure 1

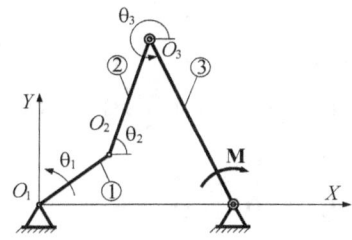

Figure 1. Quadrilateral mechanism with clearances in the kinematical joints O_3, O_4.

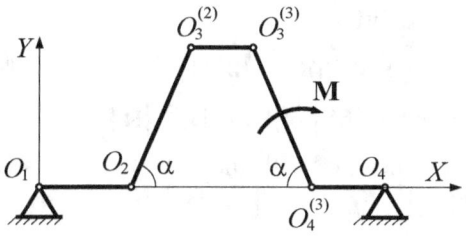

Figure 2.. Initial position of the quadrilateral mechanism with clearances.

Determine the equations of motion and the reactions for the mechanism $O_1O_2O_3O_4$ with clearances in the rotational kinematical joints O_3, O_4, Figure 1, knowing that the element O_1O_2 rotates with constant angular speed ω and the element O_3O_4 is acted by a torque with the moment M. We consider that the elements of the mechanism are homogeneous bar of constant cross section and we consider known the following parameters:

– dimensions $l_1 = O_1O_2$, $l_2 = O_2O_3$, $l_3 = O_3O_4$, $l_1 = l_2$;

– the coordinates X_{O_4} , $Y_{O_4} = 0$ of the articulation O_4;

– the clearances r_3 , r_4 at the joints O_3, O_4;

– the masses m_3 , m_4 and the moments of inertia J_2, J_3;

– the initial conditions, Figure. 2, at $t = 0$:
$X_{O_4} = l_1 + 2l_2 \cos\alpha + r_3 + r_4$, $\theta_2 = \alpha$,

$X_2 = l_1 + \dfrac{l_2}{2}\cos\alpha$, $Y_2 = \dfrac{l_2}{2}\sin\alpha$,

$\theta_3 = -\alpha$, $X_3 = X_{O_4} - r_4 - \dfrac{l_2}{2}\cos\alpha$,

$$\dot{X}_2 = 0 \quad , \quad \dot{Y}_2 = l_1\omega \quad , \quad \dot{\theta}_2 = 0 \quad ,$$
$$\dot{X}_3 = \dot{Y}_3 = 0, \dot{\theta}_3 = 0.$$

Numerical application for $l_1 = 0.2 \text{ m}$, $l_2 = l_3 = 0.6 \text{ m}$, $r_3 = r_4 = 0.005 \text{ m}$, $m_2 = m_3 = 2 \text{ kg}$, $J_2 = J_3 = 0.06 \text{ kgm}^2$, $M = 40 \text{ Nm}$, $\omega = 10 \text{ rad/s}$.

Solution: Based on the previous relations of calculation, we can write in order the equalities $\tilde{x}_2^{(2)} = -\dfrac{l_2}{2}$, $\tilde{y}_2^{(2)} = 0$,

$$\tilde{U}_{2X} = \tilde{x}_2^{(2)} \cos\theta_2 \quad , \quad \tilde{U}_{2Y} = \tilde{x}_2^{(2)} \sin\theta_2 \quad ,$$

$$[\tilde{B}_2^{(2)}] = \begin{bmatrix} 1 & 0 & -\tilde{U}_{2Y} \\ 0 & 1 & \tilde{U}_{2X} \end{bmatrix} \quad , \quad x_3^{(2)} = \frac{l_2}{2} \quad ,$$

$$y_3^{(2)} = 0 \quad , \quad U_{3X}^{(2)} = x_3^{(2)} \cos\theta_2 \quad ,$$

$$U_{3Y}^{(2)} = x_3^{(2)} \sin\theta_2 \quad , \quad x_3^{(3)} = -\frac{l_2}{2} \quad ,$$

$$y_3^{(3)} = 0 \quad , \quad U_{3X}^{(3)} = x_3^{(3)} \cos\theta_3 \quad ,$$

$$U_{3Y}^{(3)} = x_3^{(3)} \sin\theta_3 \quad ,$$

$$D_{3X} = \frac{X_3 + U_{3X}^{(3)} - X_2 - U_{3X}^{(2)}}{r_3} \quad ,$$

$$D_{3Y} = \frac{Y_3 + U_{3Y}^{(3)} - Y_2 - U_{3Y}^{(2)}}{r_3} \quad ,$$

$$\{D_3\} = \begin{bmatrix} D_{3X} \\ D_{3Y} \end{bmatrix}, \quad [B_3^{(2)}] = \begin{bmatrix} 1 & 0 & -U_{3Y}^{(2)} \\ 0 & 1 & U_{3X}^{(2)} \end{bmatrix},$$

$$[B_3^{(3)}] = \begin{bmatrix} 1 & 0 & -U_{3Y}^{(3)} \\ 0 & 1 & U_{3X}^{(3)} \end{bmatrix} \quad ,$$

$$[E_3^{(2)}] = \{D_3\}^T[B_3^{(2)}], \quad [E_3^{(3)}] = \{D_3\}^T[B_3^{(3)}],$$

$$x_4^{(3)} = \frac{l_3}{2} \,, \; y_4^{(3)} = 0 \,, \; U_{4X}^{(3)} = x_4^{(3)} \cos\theta_3 \,,$$

$$U_{4Y}^{(3)} = x_4^{(3)} \sin\theta_3 \,, \; x_4^{(3)} = \frac{l_3}{2} \,, \; y_4^{(3)} = 0 \,,$$

$$U_{4X}^{(3)} = x_4^{(3)} \cos\theta_3 \,, \; U_{4Y}^{(3)} = x_4^{(3)} \sin\theta_3 \,,$$

$$[B_4^{(3)}] = \begin{bmatrix} 1 & 0 & -U_{4Y}^{(3)} \\ 0 & 1 & U_{4X}^{(3)} \end{bmatrix} \quad ,$$

$$D_{4X} = \frac{X_{O_4} - X_3 - U_{4X}^{(3)}}{r_4} \quad ,$$

$$D_{4Y} = \frac{Y_{O_4} - Y_3 - U_{4Y}^{(3)}}{r_4}, \quad \{D_4\} = \begin{bmatrix} D_{4X} \\ D_{4Y} \end{bmatrix},$$

$$[E_4^{(3)}] = \{D_4\}^T[B_4^{(3)}] \quad ,$$

$$[B] = \begin{bmatrix} [B_2^{(2)}] & [0] \\ -[E_3^{(2)}] & [E_3^{(3)}] \\ [0] & -[E_4^{(3)}] \end{bmatrix} \quad ,$$

$$\{C\} = l_1\omega[-\sin\omega t \quad \cos\omega t \quad 0 \quad 0]^T \quad ,$$

$$\{q\} = [X_2 \quad Y_2 \quad \theta_2 \quad X_3 \quad Y_3 \quad \theta_3]^T \quad ,$$

$$\{R\} = [H_2 \quad V_2 \quad N_3 \quad N_4]^T \quad ,$$

$$[M] = \begin{bmatrix} m_2 & 0 & 0 & 0 & 0 & 0 \\ 0 & m_2 & 0 & 0 & 0 & 0 \\ 0 & 0 & J_2 & 0 & 0 & 0 \\ 0 & 0 & 0 & m_3 & 0 & 0 \\ 0 & 0 & 0 & 0 & m_3 & 0 \\ 0 & 0 & 0 & 0 & 0 & J_3 \end{bmatrix}$$

and, since the matrix $[M]$ is invertible, it results that the equation $[M]\{\ddot{q}\} = \{F\} - [B]^T\{R\}$ separates into the equalities

$$\{R\} = [[B][M]^{-1}[B]^T]^{-1}\{[B][M]^{-1}\{F\} - \{\dot{C}\} + [\dot{B}]\{\dot{q}\}\}$$

$$\{\ddot{q}\} = [M]^{-1}\{\{F\} - [B]^T\{R\}\}.$$

Further on, with the aid of the notations $\{f\} = [f_1 \quad f_2 \quad \dots \quad f_6]^T = [M]^{-1}\{\{F\} - [B]^T\{R\}\}$, $p_i = \begin{cases} q_i, \text{if } 1 \le i \le 6 \\ \dot{q}_i, \text{if } 7 \le i \le 12 \end{cases}$, one obtains from (1) the systems of first order differential equations $\dfrac{dp_i}{dt} = \begin{cases} q_i, \text{if } 1 \le i \le 6 \\ \dot{q}_i, \text{if } 7 \le i \le 12 \end{cases}$,

which, by numerical integration using the

fourth order Runge–Kutta, leads us to the results captured in Figures. 3–12

From the Figure 3 we observe that the variations of the kinematical parameters O_2, O_3 in the case with clearance are small relative to the case without clearance. These variations diminish when the clearances r_3, r_4 diminish.

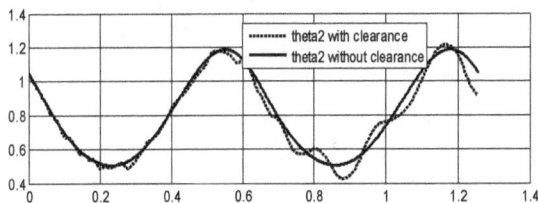

Figure 3. Time history of the parameters q_2 in the cases without and with clearances.

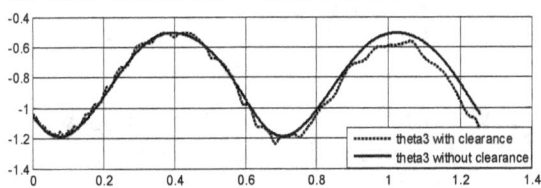

Figure 4. Time history of the parameters q_3 in the cases without and with clearances.

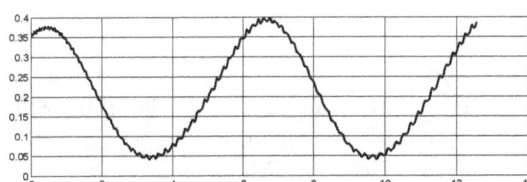

Figure5. Time history of the parameter X_2.

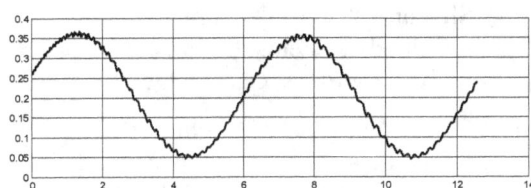

Figure 6. Time history of the parameter Y_2.

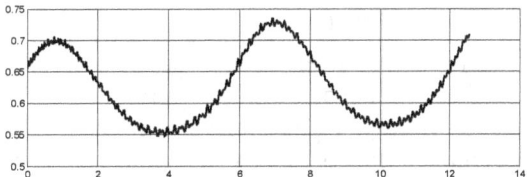

Figure 7. Time history of the parameter X_3.

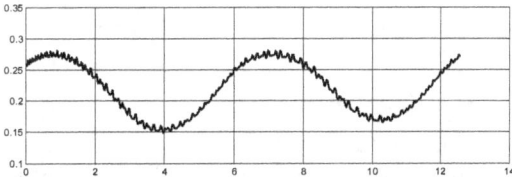

Figure 8. Time history of the parameter Y_3.

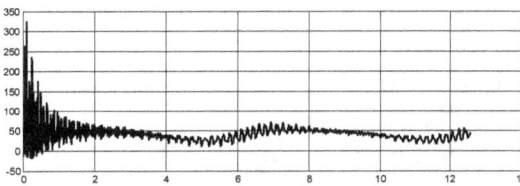

Figure 9. Time history of the reaction H_2.

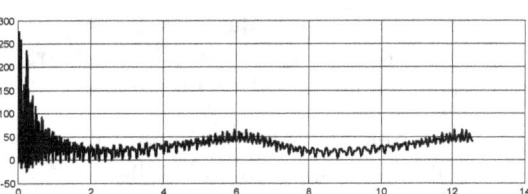

Figure 10. Time history of the reaction V_2.

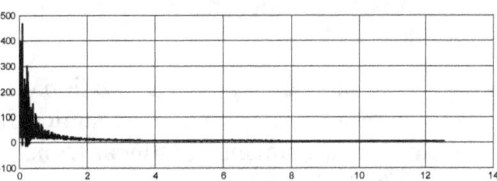

Figure 11. Time history of the reaction N_3.

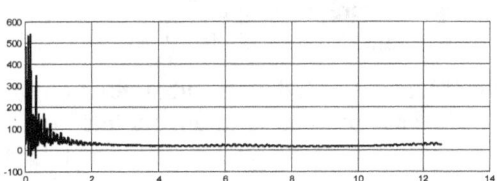

Figure 12. Time variation of the reaction N_4.

2. Conclusions

In this paper we presented one application concerning the motion and the equilibrium of the planar chains with rotational linkages with clearances. The applications are complete and numerically solved.

The numerical applications solved here confirm the statements mentioned on the algorithm presented in our previous papers.

Acknowledgement

This work was supported by Romanian Ministry of Education and by Executive Agency for Higher Education, Research, Development and Innovation Funding, under projects PCCA 2-nr. 166/2012 and ENIAC 04/2014.

References

1) Grigore, J.-C, Jderu A., Enachescu M., Matrix of Constraints for the Motion of the Planar Kinematic Chains with Rotational Links with Clearances, The 39th ARA, 28 – 31 July 2015, Franscati, Rome, Italy, Proceedings American Romanian Academy of Arts and Sciences ISBN: 978-1-935924-18.

2) Amirouche, F., Fundamentals of multibody dynamics. , Birkhänser, Boston, Berlin, (2004).

3) Constantinescu, G., Teoria sonicității, Editura Academiei R.S.R, Bucureşti, (1985).

4) Erkaya, S., Uzmay, I., Investigation on effect of joint clearance on dynamics of four-bar mechanism. Nonlinear Dyn., 58, 179-198, (2009).

5) Flores, P., Ambrósio, J., Revolute joints with clearance in multibody systems. Comput. Struct. 82, 1359-1369, (2004).

6) Flores, P., Modeling and simulation of wear in revolute clearance joints in multibody systems. Mechanism and Machine Theory, 44, 1211-1222, (2009).

7) Grigore, J.-C., Contribution to the dynamical study of the mechanisms with clearances. Doctoral thesis, University of Piteşti, (2008).

8) Pandrea, N., The dynamic calculation of mechanical torque converter „ G. Constantinescu " IFToMM Int. Symp. SYROM`89 pag. 673-679, Bucharest, Romania,(1989).

9) Pandrea, N., Popa, D., Mechanisms. Technical Publishing, Bucharest, (2000).

10) Penestri, E., Valentini, P., P., Vito, L., Multibody dynamics simulation of planar linkages with Dahl friction, Multibody Syst. Dyn. 17, 321-347, (2007).

11) Pfeiffer, F., Glocker, C., Multibody dynamics with unilateral contacts. Wiley, New York (1996).

12) Ravn, P., A continuous analysis method for planar multibody systems with joint clearance. Multibody Syst. Dyn. 2, 1-24, (1998).

13) Samanta, B., Mukherjee, A., Deb, K., Bond graph adaptive modular approach to analysis of planar mechanisms, World Congress Mechanisms and Machine Theory, vol. III, pag. 439-443, Sevilla, Spain, (1987).

14) ShabanA A., A., Dynamics of multibody systems. Cambridge University Press, Cambridge, 2005

15) Stoenescu, E., D., Marghitu, D.B., Dynamic analysis of a planar rigid-link mechanism with rotating slider joint and clearance, J. Sound Vib. 266, 394-404, (2003).

Effect of Nanoparticles Shape on the Efficiency of Hybrid Solar Cells

Mirela-Ionela MIHAI[1], Ruxandra VIDU[2*], Adrian BADEA[1]
[1]University Politehnica Bucharest Romania,
[2]University of California Davis, Dept. Material Science and Engineering, United State
*Corresponding author: rvidu@ucdavis.edu

Abstract: In recent years, the solar cell research interest has been focused on hybrid solar cells that use the advantage of mixed active layer such as semiconductor nanoparticles and polymeric organic materials. The goal of the solar research groups is to increase the efficiency of solar cells, their lifetime and cost of fabrication. For example, the solar cells based on silicon has long lifetime, good efficiency, but high fabrication costs. Solar market shows a great interest in hybrid solar cells due to their increased yield, extended lifetime compared to organic cells, and ease in fabrication with great potential in reducing the manufacturing costs. In this paper, the effect of various nanostructures on the performance of hybrid solar cells was studied. A relationship between the nanoparticle shapes and solar cell efficiency was developed, which showed that the aspect ratio of nanoparticles was an important factor in designing solar cells with improved efficiency.

1 Introduction

The energy demand increases each year. This growth has imposed an increase in both energy production from conventional sources and development of new technologies. Additionally, the global warming caused by the increase in CO_2 concentration [1] has forced the legislative to take action and promote the research and development of the technologies in renewable energy. Solar energy is one of the greatest renewable energy sources because Earth receives from the sun enough energy each hour to meet all the energy needs around the world in a year. Currently, photovoltaic (PV) cells provide only 0.04 % of the total energy produced per year from all the energy sources [1].

The European Renewable Energy Council (EREC) estimates that the renewable energy will increase up to 50% by 2040 [2]. This ambitious target challenges and creates competitive markets for solar energy research and requires a high scale solar energy deployment. Presently, the renewable energy source contributes with 20% at the global level of total energy consumption, while the rest of 80% comes from primary energy sources. For instance, in the 2013 calendar year, the total power output of photovoltaic panel (PV) in the world was around 160 billion kW/h, which means about 0.85% of the total energy demand in the world [3]. The European Photovoltaic Association (EPIA) estimated that the global PV production will increase up to approximately 2,646 TWh/year by year 2030. Table 1 presents the statistics of the renewable energies including the estimated values by the year of 2040 [2].

Data presented in Table 1 predicts that the renewable energy based on PV in the year 2040 will increase up to 13%. Researchers are working to create PV cells with high light absorption, high conversion efficiency, and extended life cycle [4]. Presently, the best performing solar cells are multi-junction solar cells that reached 44.7 %. Among single junction solar cells, the best performing are the solar cells based on Si, followed by thin film cells and hybrid and organic cells [4]. Organic solar cell (OPV) is a type of cell that is based on conductive organic polymers or small organic molecules designed to capture the light absorption to produce electricity. The OPV were discovered in 1987 and their efficiency was just 2.1% [5]. In 2000, the research work on OPV escalated due to their high potential of large mass production at a low cost. OPVs can be produced using inexpensive technologies such as roll-to-roll processes, where the layers were printed on a flexible substrate.

Moreover, materials used in the OPV fabrication are not toxic [5]. The OPV are now in the competition for a market share, reaching an efficiency of 12% in 2013 [6]. The "Heliatek" company alleged that the efficiency of organic solar cells is estimated to reach 15% [6].

Table1. Statistics and predictions of renewable energy technologies based on the "Renewable Energy Scenario to 2040", by the European Renewable Energy Council (EREC) [2]:

Years	1996 - 2001	2001 - 2010	2010 - 2020	2020 - 2030	2030 - 2040
Biomass	2%	2.20 %	3.10 %	3.30 %	2.80 %
Large hydro	2%	2%	1%	1%	0%
Small hydro	6%	8%	10%	8%	6%
Wind	33%	28%	20%	7%	2%
PV	25%	28%	30%	25%	13%
Solar thermal	10%	16%	16%	14%	7%
Solar thermal electricity	2%	16%	22%	18%	15%
Geothermal	6%	8%	8%	6%	4%
Marine (tidal/wave/ ocean)		8%	15%	22%	21%

Then, dye-sensitized solar cell with an efficiency of 11-12% has been developed, followed by the tandem organic solar cells, inorganic solar cells and quantum-dots (QDs). In 2010, the National Renewable Energy Laboratory (NREL) researchers have improved efficiency and performance of the solar cells containing PbS QDs, by adding metal oxides such as: molybden oxide (MoO_x), vanadium

oxide (V_2O_x) in the active layer of the solar cell to improve heterojunction solar cell consisting of ZnO/PbS [7]. The efficiency of the cell has reached 4.4%. In 2011, University of Toronto obtained a solar cell based on PbS colloidal quantum dots (CQDs) mixed with organic ligand with an efficiency rate of 6% (AM 1.5 conditions) [8].

The solar research community continue to search for ways to improve solar cell efficiency while keeping the cost low to fabrication. Hybrid solar cells (HSC) were developed to combine the high efficiency of thin films with the ease of fabrication of organic cells. This paper presents a study of the nanoparticle structure and size used in HSC, based on the most recent developments in the HSC technologies. In the study of HCS performance, both the nanoparticle synthesis methods and the shape and the size of nanoparticles are discussed.

2   Hybrid solar cells

Hybrid solar cells (HSCs) are a potential candidate for commercial PVs due to their fast optical absorption, ease of fabrication, low production costs, and a favorable combination of organic and inorganic semiconductors [9]. HSC has an active layer composed of a blend of organic materials such as polymers and inorganic nanoparticles, which improve the absorption of light and enhance the electrons transport from cell to collector [1]. These cells typically consist of a glass substrate, transparent conductive oxide layer (ITO), poly (3, 4-ethylenedioxythiophene) PDOT:PSS, nano-particles, polymer blend and a metal cathode as shown in Figure 1.

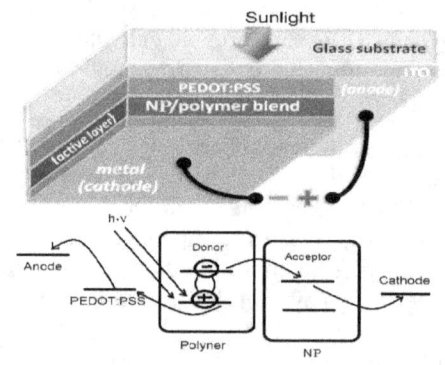

Figure 1. Schematic representation of HSC device (top) and solar cells works to collect the electron (bottom).

The PDOT:PSS polymeric blend forms a layer that helps the transport of electrons to anode.

A critical detail in the composition of HSC is represented by the diversity of shapes of nanoparticle that may be used. A certain shape of nanoparticles may increase the efficiency and the performance of hybrid solar cells but there is no direct relationship demonstrated for them. For example, Lazaro A. et al [10] investigated the role of the PbSe nanostructure shape by comparing elongated and spherical quantum dots (QD) nanoparticles. They found out that the high aspect ratio of elongated particles increases the efficiency of light absorption compared to spherical (dot) nanoparticles [10].

There is a vast diversity of shapes and sizes of nanoparticles that may have a direct and critical impact on the performance of hybrid solar cells. In addition, the efficiency of HSC is influenced by the NP synthesis method as well as device fabrication technique.

3. Nanoparticle for HSC and their Synthesis Methods

3.1. CuInS$_2$ (CIS) nanocrystals

Colloidal synthesis was used to fabricate dispersed particles suspended in a liquid medium. Although it is difficult to control the morphology of semiconductor nanostructures, tailored nano-crystals can be produced by simply changing the synthesis parameters (i.e. reaction time and temperature). In the case of CuInS$_2$ colloidal synthesis, CuInS$_2$ and oleylamine, trioctyl-phosphine oxide (TOPO), was injected in a mixture by 1-dodecanethiol, tert-dodecanethiol [11]. This solution was maintained at 240 °C, resulting in the formation of elongated CIS nanoparticles that had approximately uniform size distribution. Oleylamine acts as a solvent, but also as monomer to stabilize the particles. The nanoparticles have a diameter of 19.1 nm and a length of about 44.8 nm. Transmission Electron Microscope (TEM) analysis was used to study the growth mechanism of CIS nanocrystal during chemical reaction. It was observed that 75 seconds after injection, nanoparticles size increased, resulting in a square-like shape (Figure 2). After 80 seconds, the CIS particles reached an elongated shape, which then

increased along the (002) axis and had an orientations perpendicular to the axis (100) [11].

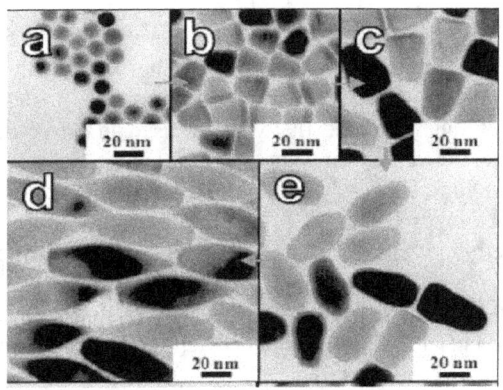

Figure 2. TEM of CIS(CuInS$_2$) NP [11]: a) Cu$_2$S particles before the reaction; b-d) particles of Cu: In, in ratio 1:1 during synthesis after 75 s (c), 80 s (e) and 90 s (d) particles has a perfect elongated shape [11].

CIS elongated shape had an optical absorption between 400 nm and 900 nm. Because of the versatility in shape induced by synthesis process (Figure 2), CIS nanoparticles were used as thin films and hybrid devices in solar cells research.

3.2. CdSe Nanoparticles

CdSe nanocrystal: Brandeburg et al. [12] have synthesysed CdSe nanoparticles by colloidal method. The HSC device was fabricated on ITO substrate, where a PEDOT: PSS layer was applied by spin-coting. Subsequently, an active layer made of CdSe: P3HT was applied and treated at 140 °C for 10 min. An Al layer deposited by evaporation was used as a cathode [12]. The size of CdSe nanoparticles was 2.3 - 10 nm and the wavelength was between 400 nm to 600 nm. The efficiency of the device was 0.45% [12].

CdSe nanorods: Farva et al. [13] obtained CdSe nanorods with controlled size and shape using a slightly modified hot-injection method. After synthesis, CdSe nanorods were annealed at 350 °C for 30 minutes to investigate the influence of air - annealing process on the structural and optical properties of nanorods [13]. During annealing, the nanorod size increased from 4-5 nm to 10-12 nm in diameter, and from about 20 nm to 23-25 nm in length. Additionally, the heat treatment influenced the photoluminescence (PL) spectra due to improved crystallinity after annealing. PL emission peak shifted after annealing from 625 nm (1.98 eV) to 628 nm (red shift) [13].

CdSe tetrapod: Complex nanowire structures that form both hierarchical branches and multi-component heterostructures of engineered material combinations, are likely to be used in harvesting and conversion of solar energy. Nanoparticles with tetrapod shapes were obtained for semiconductors of groups II and VI [14]. CdSe nanoparticles with tetrapod shapes can be obtained by chemical synthesis, because the form of nanocrystals (NCs) can be controlled by varying synthesis parameters. Manna *et al.* [14] obtained tetrapod shapes from CdSe nanocrystals seeds of 10 - 15 nm in size (Figure 3). This 3D shape has an important role in electron transport, because the three arms of the tetrapod are anchored in the cathode, and the fourth arm is directed upwards, reaching the active layer. Additionally, CdSe with a tetrapod shape has a good optical absorption of light, which can be exploited in hybrid solar cells [1]. For instance, the PCE for a cell made of CdSe tetrapod, pyridine-capped and PCTDTBT polymer was about 3.19 %, with an external quantum efficiency (EQE) of 55%, and a wavelength of 630 - 720 nm [1].

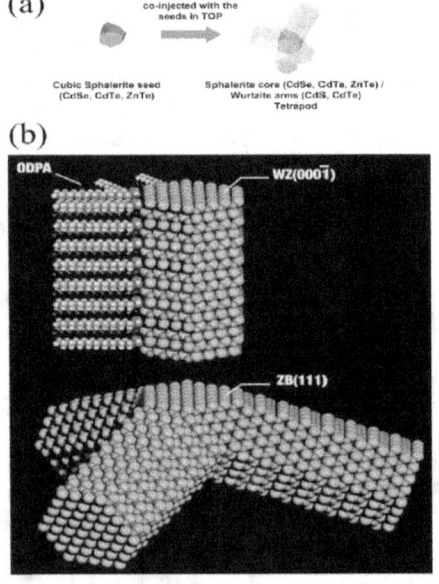

Figure 3. Illustration of tetrapod structure obtained by seeded growth method [1, 22].

Similar to CdSe tetrapod, ZnO tetrapods called "fourlings" were obtained in vapor phase [15]. In this case, polycrystalline ZnO nucleus led to twinning mechanism. This formation mechanism involves stacking fault between the central nucleus and each branch [15].

3.3. TiO$_2$ Nanoparticles

TiO$_2$ crystalline nanoparticles: TiO$_2$ rods are generally used in HSC due to their low cost and good power conversion efficiency (PCE) of about 4%. Guttmann *et al.* [16] obtained crystalline TiO$_2$ nanoparticles for a HSC. This device contained solar cells that are made of FTO/TiO$_2$ rod/ blend/ PEDOT: PSS/Au [16], where the active layer consisted of TiO$_2$ rods embebed in P3HT polymer. Figure 4 shows TiO$_2$ crystalline rods with a diameter to length ratio of 1 to 10, which were obtained from TiCl$_4$ precursor. The solar cell based on the polymeric TiO$_2$ active blend was tested under 1.5 AM conditions. The HSC had an efficiency of 0.13% and filling factor of 31% [16].

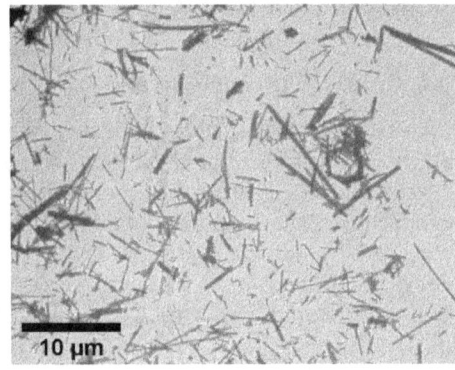

Figure 4. TEM image of TiO$_2$ rods [16].

TiO$_2$ nanorods: Hybrid solar cells based on TiO$_2$ nanorods with a diameter size between (7±1) nm and (19±3) nm [24] were synthesis by non-hydrolytic methods. TiO$_2$ nanorods have quantum efficiencies of up to 15% measured at AM 1.5. The solar PCE was η = 0.49%. These nanoparticles, are not toxic and they have a good physical and chemical stability. TiO$_2$-based HSC was made using ITO/PEDOT: PSS/blend/Al. This blend was formed by TiO$_2$ nanorod and P3HT. After annealing, an improved yield of separation charge was observed. TiO$_2$ nanorod has an important role in blend by forming traps to capture photon. After annealing, the device tested in AM 1.5 spectrum, 50 MW/cm^{-2} had a PCE by $\eta = 0.49\%$ [24].

TiO$_2$ nanoparticles: Typically, the main method for producing metal oxide nanoparticles is the sol-gel process. This process starts with a colloidal solution (sol) which acts as a precursor for the gel [17]. In fact, the sol is a solution that contains colloidal particles (inorganic or organic) suspended in a liquid. For instance, TiO$_2$ colloidal nano-particles can be obtained from a solution containing a mixture of ethanolamine (TEOA), ethoxide (TEO) and distilled water at room temperature, to which NaOH solution is added to adjust for pH. Then, an amine such as trido-decylamine (TDA) is added. The solution is placed in Teflon-lined container for 24 h at 100; then the temperature is increased to 140 °C and kept for 72 h. This process helps the nucleation and growth of TiO$_2$ nanoparticles [18]. TiO$_2$ nanoparticles obtained by sol-gel have elongate shape, with a diameter of 30 - 40 nm and length of 15 - 30 nm. When these TiO$_2$ nanoparticles were used in DSSC, the efficiency was 5.04% with a filling factor (FF) of 0.45% [18].

3.4. Quantum Dot (QD) structures: PbSe-QD, CdSe colloidal QDs, CdSeS-QD, etc.

Quantum Dot (QD) are mainly used in DSSC due to their good absorption power (about 12% [19, 20]) and inexpensive manufacture costs. For example, electrophoretic deposition is a process of depositing layers on a substrate/electrode, from a solution that contains nanoparticles/QDs under an applied electric field [14]. In solvents such as toluene and acetonitrile, QDs are negatively charged, which means that QDs will be attracted to the positive electrode. Electrode materials used in HSC are mesoscopic oxide TiO$_2$, ZnO, SnO$_2$, and Nb$_2$O$_5$ [15], used as thin films of 5 to 10 μm thickness over optically transparent electrodes. There are several methods of making QD films [14-16], as shown in Figure 5.

Several QDs material systems have been studied for their possible adoption in HSC. An example of QD which are often used in HSC is PbSe quantum dot PEDOT: PSS layers [21]. The PbSe QDs have been obtained with a diameter of 4.5 nm. The HSC had a filling factor (FF) of 45.5%, but low efficiency. W. Zou et al. [22] has synthesized colloidal CdSe QDs with oleic acid ligand. The

CdSe QDs had diameters between 3.1 and 3.9 nm with a deviations ± 0.2 nm, and an optical absorption spectral of about 450 nm. Kumari et al. [23] have demonstrated that CdSe QDs (5 ~ 7 nm) with TOPO and oleic acid (OA) have a good surface passivation and optoelectronic properties.

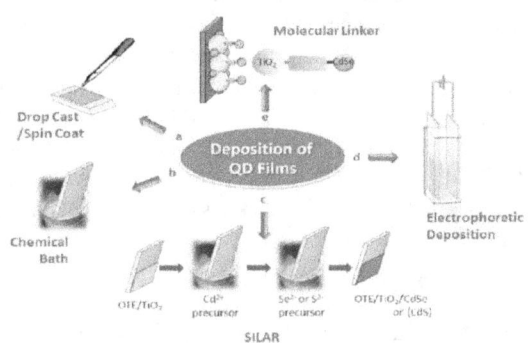

Figure 5. Schematic illustration of deposition of QD films obtained by the method presented in Ref [11]: a) drop cast/spin coat, b) chemical bath, c) SILAR, d) electrophoretic deposition; e) molecular linker.

The CdSe (TOPO) QDs had an absorption of wavelength between 493 nm to 506 nm. However, the efficiency of the HSC device made with CdSe (TOPO) QDs was very low. For example, CdS/CdSe QDs were doped with Mn to improve electrical properties and light absorption through wavelength control [20]. The efficiency conversion for an incident photon in Mn - doped CdS/CdSe increased to 68% and 80%. Using a molecular linker is another method to improve the performance of QDs/ DSSC, through inverted CdS/CdSe QDs, using organometallics at high temperature, and then synthesized with TiO$_2$. The efficiency for this system can reach 5.32% [20].

3.5. CdS nanorods

CdS nanorods can be grown by hydrothermal process, where the nanoparticles used as seeds are electrochemically deposited. An example of this process is provided by Chen et al. [24] who fabricated CdS nanostructures for optical and electrical devices. They obtained CdS nanorod arrays with various shapes and sizes with a wide band gap absorption of light. CdS nanorod structures are used in HSC due to their capability to improve the power conversion efficiency in solar cells. HSC nanostructured device obtained by Chen *et al.* [24] consisted of a layered structure with the following sequence: ITO/CdS nanorod arrays/MEH-PPV/Au. The CdS nanoparticle layer that acts as seed layer was obtained by electrodeposition directly on ITO. CdS nanorods obtained by hydrothermal process had a hexagonal structure and sizes from10 nm to 150 nm [24]. The photoluminescence of CdS nanorod arrays depends on the reaction temperature. The band-edge of CdS nanorod array decreases with increasing the temperature from 10 K to 300 K. The power conversion efficiency in this particular device ITO/CdS nanorod arrays/MEH-PPV/Au was 0.34%±0.06% and the thickness MEH - PPV layer was about 350 nm [24].

3.6. ZnO Nanostructures

ZnO has a large band gap of 3.37 eV and it is used in HSC as an alternative to other oxides to ensure a better open-circuit photovoltage. Among various techniques developed to produce ZnO nano-structures, electrochemistry is one of the techniques that offers innovative approaches to produce complex nanostructures with interesting properties.

ZnO nanorods: Rusen *et al.* [25] obtained a core-shell hybrid nanomaterial by electrodepositing ZnO nanorods over an electroconductive material with a photonic crystal structure (PC). The ZnO nanorods electrodeposited on the hybrid Cu/polymer core-shell structure had diameters of 50-100 nm and lengths of about 300-350 nm [25]. Another example was demonstrated by Chang *et al.* [26]. They fabricated films of ZnO nano-particles array as seeds using self-assembled polymeric hollow particles over ITO using hydrothermal method. The ZnO nanorods had hexagonal shape, with a diameter between 50 nm and 100 nm, a deviation of ± 80 nm, and arranged in an array. ZnO array films were prepared in various growth conditions with wavelengths of about 400 nm to 1000 nm [26].

ZnO nanowires: Vapor-liquid-solid (VLS) is the main techniques to grow semiconductor nanowires. The growth of one-dimensional structures involves the presence of a seed that melts, forming a catalytic liquid alloy. The liquid –solid interface is the place where nucleation occurs, while the adsorption of gas happens at the liquid-vapor interface, supersaturating the liquid alloy. The

nanowires develop by a continuous nucleation and growth at the liquid-solid interface [27]. ZnO nanowire [28] for dye-sensitized solar cells (DSSC) were grown by VLS. The ZnO nanowires with 130 nm in diameter were formed perpendicular to the F:SnO$_2$ (FTO) glass substrate. The efficiency of a ZnO NWs based DSSC device was between 1.2 and 1.5%, while the efficiency of TiO$_2$ NWs-based DSSC was 5-6% [28].

3.7. InP core - shell nanopillar

Other photovoltaic cells (PV) such as arrays of InP nanopillars (NPL) were studied by Fan *et al.* who used metal organic vapor phase epitaxial method (MOVPE) and vapor-liquid-solid (VLS) [29]. This method results in p-n InP core-shell nanopillars by growing first p-type InP on a substrate, followed by epitaxial growth of n-type InP layer. During the growth, SiO$_2$ was introduced over p-type InP. The PV cell obtained using the p-n InP core - shell nanopillar array had a filling factor of about 27%, and an efficiency of about 3.37% [29].

3.8. Ge nanopillar

Another interesting example is a structure with dual-diameter nanopillar array, which was obtained by template synthesis in anodized aluminum oxide (AAO) membrane [29]. Ge nanopillars have a diameter by 60 nm at one end and adsorb the light at wavelengths between 450 nm and 900 nm. The diameter at the base has 130 nm and adsorb the light at wavelengths between 300 nm and 900 nm. The dual - diameter pillar array has a good optical absorption and efficiency close to 99% [28, 29].

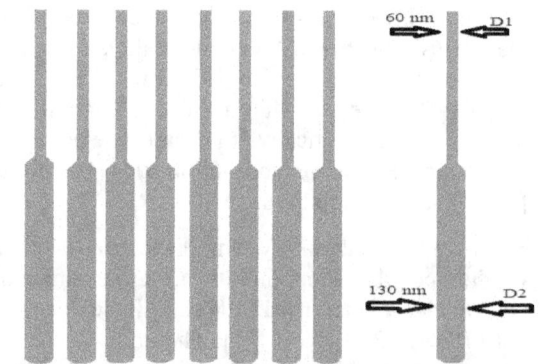

Figure 6 Illustration on Ge dual-diameter nanopillars that were grown in AAO [29].

3.9. Nanotube and Nanocable Structures

Template synthesis for nanotube and nanocable structures is mainly used in combination with electrochemical deposition. In this method, one side of a porous membrane is covered with a metal film that acts as a cathode in an electroplating system. Various membranes can be used, but the membranes with transversal pores are preferred such as polycarbonate track-etched membranes (PCTE) and anodized aluminum oxide (AAO). Martin *et al.* [32] first used Au tubes and an polymer polypyrrole with an template membrane as polycarbonate which has a diameter 10 nm with a pore densities about 10^9 pore/cm^2, using electro-chemical deposition [32].

Other remarkable example is a nanocable (core/shell) structure of CdTe/CdS which was obtained by *Vidu et al.* [31] using template synthesis techniques. This CdTe/CdS nanocable structure may provide higher efficiency than thin film configuration, and a total power output that is 1.5 to 3 times higher than current technologies. This columnar architecture (Figure 7) has high surface area, superior light trapping and minimizes the recombination of electrons due to the high spect ratio metalic core [30].

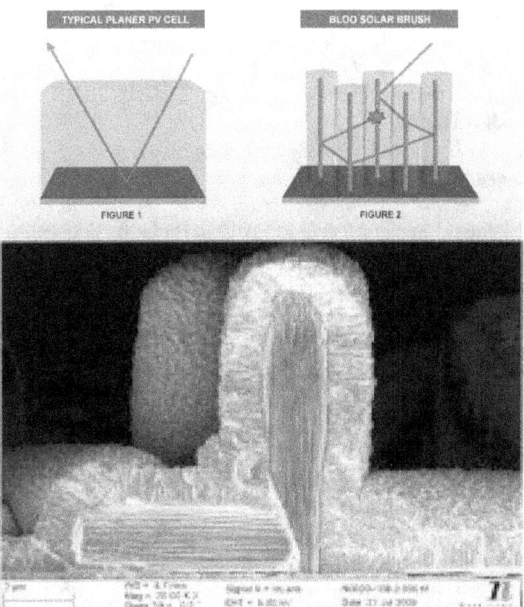

Figure 7. Nanocable (core – shell) structure of CdTe/CdS for the 3D generation solar cells [R. Vidu].

3.10. Carbon nanotube

Carbon nanotube (CNT) [32] may be used in solar cells fabrication, where they replace the ITO glass substrate using a roll-to-roll printing process to decrease the cost of manufacturing. An example is the use of single wall carbon nanotube (SWNTs) in a film for anode, and a bulk - heterojunction (BHT) PEDOT: PSS for HSC device [32]. The SWNTs was used with an PEDOT: PSS polymer and has a PCE of 2.5 % [32]. It was also used as electrode in solar cell dye sensitized or organic solar cells.

4. Nanoparticles for HSC

Based on the efficiency data available in the literature for HSC, Figure 8 was generated to correlate the shape of nanoparticles use in HSC and conversion efficiency. The highest performance was obtained for CIGS nanowires based solar cell produced by template synthesis with an efficiency of 6.18%. HSC dye - sensitized solar cells based on TiO$_2$ quantum dot (QD) shows an efficiency of 5.32% compared to nanotube (4.24%) and nano-wires (6%). The lowest efficiencies were obtained for HSC made with nanoparticles and nanorods of low aspect ratios.

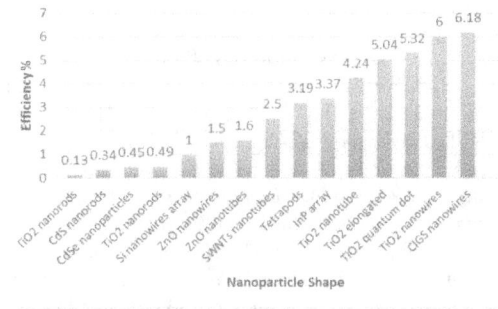

Figure 8. Efficiency of solar cells for various nanoparticle shapes

5. Conclusions and perspectives

Research community has recently devoted a lot of efforts to understand the relationship between synthesis of nanoparticles, their shape and size, and the efficiency of solar cells that uses them. Our study shows that the power conversion efficiency is affected by the form of nanoparticles, which represent the most important part of the hybrid device. Hybrid solar cells have a conversion efficiency of light into electricity rather low compared to Si and thin film cells, but have a higher efficiency than organic cells. In addition, they have the great advantage of being not expensive and to be produced by roll-to-roll printing method. These benefits bring the hybrid solar cells into the investor's attention and improvements in conversion efficiency is expected from HSCs in the near future, when high-aspect ratio nanoparticles are used in the design of cell architecture.

Acknowledgement

The work has been funded by the Sectorial Operational Programme Human Resources Development 2007-2013 of the Ministry of European Funds through the Financial Agreement POSDRU/159/1.5/S/132395.

References

1. A. J. Moule, C. Thambidurai, R Vidu, P.Stroeve, *Hybrid solar cells: basic principles and the role of ligands.* Materials Chemistry, 2012.
2. EREC, *Renewable Energy Scenario to 2040.* http://www.erec.org/fileadmin/erec_docs/Documents/Publications/EREC_Scenario_2040.pdf
3. *www.e-solare.com/produse/schott-220w-poly*
4. *www.cheso.ro/invertoare-solare-power-one-aurora-pvi.php.*
5. Solarmer, *http://www.solarmer.com/about_opv.html.*
6. Badea, A., *Plus Energy Houses -Politehnica Plus Energy Building.* 5th International Conference on Energy and Environment, CIEM 2011, Bucharest, 2011, 2011.
7. NREL, *Improving PbS QD Solar Cell Power Conversion Efficiency to an NREL-Certified 4.4%.* 2012.
8. Tang, J., et al., *Colloidal-QD PVs using atomic-ligand passivation.* Nat Mater, 2011. **10**(10): p. 765-71.
9. Laurentiu Fara, M.R. et al. *"Fizica si tehnologia celulelor solare si sistemelor Fotovoltaice".* Academiei Oamenilor de stiinta din Romania, 2009: p. 80-81.
10. Lazaro A. P. et al, *Carrier Multiplication in Semiconductor Nanocrystals: Influence of Size, Shape, and Composition.* American Chemical Society, 2013.
11. Kruszynska, M., et al., *Synthesis and Shape Control of CuInS2 Nanoparticles.* Journal of the American Chemical Society, 2010. **132**(45): p. 15976-15986.
12. J.E.Brandenburg, et al., *Influence of particle size in hybrid solar cells composed of CdSe nanocrystals and poly (3-hexylthiophene).* PHYSICS, 2011. **110**.
13. Farva, U. et al, *Colloidal synthesis and air-annealing of CdSe nanorods.* Mat Lett, 2010, **64**(13) 1415.
14. Manna, L., et al., *Controlled growth of tetrapod-branched inorganic NC.* Nat Mater, 2003. **2**(6): p. 382-5.
15. Bierman, M.J. and S. Jin, *Potential applications of hierarchical branching nanowires in solar energy conversion.* En & Envir Sci, 2009. **2**(10): p. 1050-1059.
16. Lechmann, M.C., et al., *Comparison of Hybrid Blends for Solar Cell.* Energies, 2010. **3**(3): p. 301
17. CJGWS, B., *"Sol-gel : Science: The Physics and Chemistry of Sol-Gel Processing.".* Acad Press, 1990.
18. Noshin Mir, M.S.-N., *Effect of tertiary amines on the synthesis and photovoltaic properties of TiO2 nanoparticles in dye sensitized solar cells.* Electrochimica Acta, 2013. **102**: p. (274– 281).
19. Kamat, P.V., *Quantum Dot Solar Cells. The Next Big Thing in PVs.* J.Phys Chem Letts, 2013. **4**(6): p. 908-918.
20. Halim, M.A., *Harnessing Sun's Energy with Quantum Dots Based Next Generation Solar Cell.* Nnanomaterials., 2012.
21. Thomas Dittrich n, A., AhmedEnnaoui, *Concepts of inorganic solid-state nanostructured solar cells.* Solar EnergyMaterials&SolarCells, 2011.
22. Wei Zou et al., *Fabrication of surface-modified CdSe quantum dots by self-assembly of a functionalizable comb polymer.* Society of Chemical Industry, 2010.
23. Kusum Kumari et al, *Effect of surface passivating ligand on structural and optoelectronic properties of polymer, J Physics D, Applied Physics, 2008.
24. Chen, F., et al., *Large-scale fabrication of CdS nanorod arrays on transparent conductive substrates,* Solar Energy, 2011. **85**(9): p. 2122.
25. Rusen, E., A. Mocanu, and R. Somoghi, *Core-shell hybrid material for ZnO nanorods generation.* Colloid and Polymer Science, 2012. **290**(18): p. 1937-1942.
26. Chang, C.-J. et al, *Light-trapping effects and dye adsorption of ZnO hemisphere-array surface containing growth-hindered nanorods.* Coll & Surfaces 2010, **363**(1-3): p. 22-29.
27. Joshi, R.K. et al, *Assembly of one dimensional inorganic nanostructures into functional 2D and 3D architectures. Synthesis, arrangement and functionality.* Chemical Society Reviews, 2012. **41**(15): p. 5285-5312.
28. Yu, M., et al., *Recent advances in solar cells based on 1D nanostructure arrays.* Nanoscale, 2012. **4**(9):2783
29. Rehan Kapadiaa, Z.F., KuniharuTakei, AliJaveya., *Nanopillar photovoltaics:Materials,processes, and devices.* www.elsevier.com/locate/nanoenergy, 2011.
30. Feng Gao, S.R.a.J.W., *The renaissance of hybrid solar cells: progresses, challenges, and perspectives.* Energy & Environmental Science, 2020.
31. Ku, J.R., et al., *Fabrication of nanocables by electrochemical deposition inside metal nanotubes.* JACS, 2004. **126**(46): p. 15022-15023.
32. Steve Park, M.V.a.Z.B., *A review of fabrication and applications of carbon nanotube film-based flexible electronics.* Nanoscale, 2012.

Investigations on sintered materials for automotive component manufacturing

Ildiko Peter[1*], **Mario Rosso**[1]

[1] Politecnico di Torino, Department of Applied Science and Technology

Corso Duca degli Abruzzi 24

Torino, Italy

*ildiko.peter@polito.it

Abstract: The present paper presents a study on sintered materials with lower wear rate, constant friction coefficient and high durability and thermal stability connected to a good self-brazing capacity. Copper and Iron based friction materials were sintered and the influence of the composition on the friction and wear properties were studied. Employing a Copper based material with a moderate graphite and ceramic content the desired properties are reached.

1 Introduction

Generally, the friction materials have a complex composition containing both metallic and non-metallic elements. Each constituent has a significant utility during the braking route and the right selection of the elements constitutes a critical phase in the growth of any commercial product. The most important components for brake pads typically contain:

- the frictional additives, which govern the frictional properties of the brake pads including a combination of abrasives and lubricants;
- fillers, which increase the production of the brake pads;
- binders, to keep the components made up the brake pad together;
- reinforcing fibres, guaranteeing the right mechanical strength to the whole piece.

The types of the different elements in a brake friction material are determined considering the friction force, noise affinity, aggressiveness against gray cast iron rotors, wear, brake induced vibration. During the time, many studies have been carried out to investigate different features related to the improvement of the brake performance [1-4].

Ceramic fibres are the most suitable to be used as reinforcing material and glass fibres have also been proposed to be used in reinforcing materials, due to their appropriate thermal resilience, with high melting point and low conductivity, compared to asbestos [5-7]. The concept of use of TiC to reach higher hardness, higher melting point and higher abrasion resistance, as well as the possibility of the self-lubricating effect of graphite in case of lead free copper contact materials reducing the strength of the composites have been extensively studied during the years. Metallic chips or granules are commonly used as reinforcing fibres, like steel, brass and copper. The weakness of using steel is related to their corrosion, when used close to the seaside surroundings. The most important advantage of employing metal fibres is due to their high thermal conductivity. The utility of the binder is to maintain stable the structural integrity of the brake pads under mechanical or thermal stresses. It must show high heat resistance, since if it does not remain structurally continuous at all times during braking, the other constituents will disintegrate. The presence of fillers is necessary in order to increase the manufacturability and to moderate the overall cost of the brake pad and they play an important role in changing some characteristics of the brake friction material. Generally, the choice of the filler material depends on the particularity and type of the components. Generally, the frictional additive modifies the friction coefficient and the wear rates and can be classified in two groups:

1. lubricants, stabilizing the developed friction coefficient during braking, particularly at high temperature; generally graphite and metal sulphides are used for such purposes;

2. abrasives, increasing the friction coefficient and the wear rate; they remove iron oxides and other undesirable surface films developed during braking from the counter friction material; hard particles of metal oxides and silicates are the mostly used abrasives [8-12].

In the present paper, Cu and Fe-based friction materials will be investigated and the optimization of their composition will be performed to achieve higher tribological properties, compared to those produced using polymer matrix materials. Information about the exact composition of the optimized friction material studied will not be make known in the paper.

2 Experimental procedure

Sintering has been performed using different powder mixture made of: iron, copper, bronze, graphite, SiC, Al_2O_3, kindly offered by Höganäs, Sweden and Makin Metal Powders, UK and their size has been selected in line with the industrial application request. The mixed powders have been pressed and then have been positioned on a C72 high carbon steel discs and simultaneously have been bonded to a steel backing during sintering. The samples have been prepared by traditional powder metallurgy method. The cold-compacted samples have been sintered using a tubular furnace (with a heating rate of 10°C/min, T= 900°C-1000°C and a holding time 15 min-45 min and a protective N_2 atmosphere). Cooling to room temperature has been reached naturally. Optical (OM, MeF4 Reichart-Jung) and Scanning Electron Microscopy (SEM, Leo 1450VP) observations, compositional analysis (EDS, Oxford microprobe) and tribological tests have been performed imitating the "repeated single braking" conditions used industrially. The contact pressure varies from 4 MPa to 8 MPa, while the sliding velocity varies between 1÷10 mm/s. After 50 cycles, the reduction of the depth has been measured in 3 points and the wear has been calculated using a semi empirical formula, reported in Equations 1 and 2.

$$\mu = (265/p*r_m*L)/(0.206/p*r_m) \quad (1)$$

where: μ: friction coeffcient, p: mass applied on the samples (kg), r_m: average radius of the wear track (cm), L: length of braking (mm);

$$u = 10^3 *(A*h)/(E*N) \quad (2)$$

where: u: specific wear ($cm^3/10^{-6}$ kg*m), A: contact area (cm^2), E: kinetic energy of the apparatus (J), h: reduction of the depth of the sample tested (mm), N: number of braking.

3 Results and Discuss

Sintered-metal friction materials have been prepared with a variable compositions reported in Table 1, and their properties have been compared to those manufactured with polymer matrix materials.

Table 1. Composition of the friction materials investigated.

Samples	Composition (wt%)
A	55% Fe + 33% Cu + 7% bronze + 5% graphite
B	52% Fe + 31% Cu + 7% bronze + 5% graphite + 5% SiC
C	57% Cu + 35% Fe + 7% bronze + 1% graphite
D	56% Cu + 34% Fe ++ 7% bronze 3% graphite
E	55% Fe + 30% bronze + 5% graphite + 1% Al_2O_3 + 9% $ZrSiO_4$
F	30% Fe + 55% bronze + 5% graphite + 1% Al_2O_3 + 9% $ZrSiO_4$
G	33% Cu + 45% bronze + 2% graphite + 20% $ZrSiO_4$

During sintering, the compacted samples are simultaneously bonded to a steel backing. The first step of the research is dedicated to optimize the consolidation method and the sintering, using the mixtures A-D reported in Table 1. The second step makes use the mixtures E-G reported in Table 1.

High amount of bronze (due to its good joint potentiality and good tribological behavior) and ceramic material has been used. As well, in this condition the possibility to exploit the self-brazing capacity of bronze with no any use of liquid binding agent is possible. Different kinds and quantity

(< 25wt%) of abrasives (SiC, Al2O₃, ZrSiO₄) have been used and the comparison on their effect on the performance of the friction material has been completed. Microstructural characterization and performance tests, including the determination of the friction coefficient and wear rate have been carried out.

Because of the different nature of the parteicles, the pressure acts and deforms in a different way the grains according to their hardness. The use of different mixtures and the experimentally evaluation of their performances have been carried out with the purpose to reproduce as much as possible the industrially employed conditions. Generally, as Fe content increases the density of the green increases as well: this is because of their larger size and for the fact that porous nature has higher tendency for compression. Cu reduces the amount of fade due to its thermal diffusivity by preventing hot spots in the friction interface. Use of SiC involves lower density, but because of its hardness, it can confer higher fragility to the material and generally lower compaction rate. The optimal pressure which produces a high density compacts has been fixed to 800 MPa. The optimal sintering temperature (975°C) has been selected according to the Sn-Cu and Cu-Zn phase diagrams. At this temperature 55 wt.% of the metal is liquefied. In all cases about 10% of small porosity persists, which guarantees a good workability of the piece.

The sintering of the Fe-based materials has been realized and a non-uniform compositional distribution of the different phases (Fig.1.1) has been obtained showing a high content of Fe (grey area Fig.1.2) immersed in an ample content of Cu and bronze, with quite the same appearance and colour on the SEM microstructure (Fig.1.2). Porosities have been closed by the liquefied bronze. Soldering of the sintered material to the steel is not strong enough and a fracture occurs, as reveals the micrograph in Fig.1.3. The evolution of the microstructure is obtained by adding SiC particles to the mixture and reducing the metal content, maintaining the same metals ratio.. An efficient soldering between the two parts (Fig.1.4) and the presence of a high residual porosity (Fig.1.5) has been obtained. A non-homogeneous composition and in the metal matrix high presence of Fe has

been detected. SiC particles have been integrated within the metal matrix (Fig.1.6).

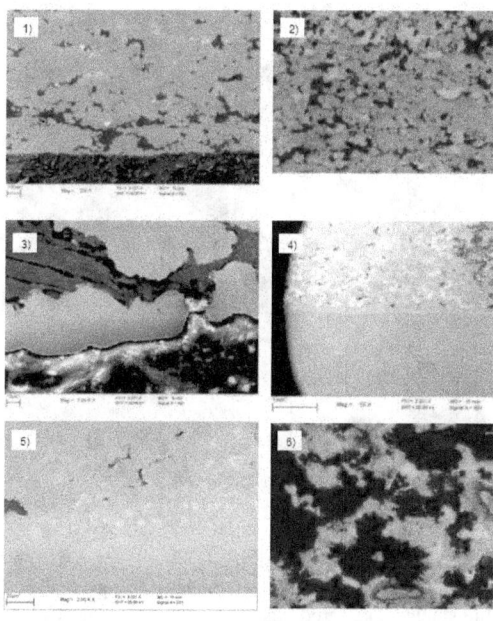

Figure 1. Microstructure of the sintered material: Sample 1 (1, 2) and details of the joining zone between the steel and sintered material (3) and Sample 2 (4, 5) and details of the joining zone between the steel and sintered material (6)

The Cu-based sintered materials are regular with a low porosity, nevertheless an insufficient joining of the parts has been realized (Fig.2.1) with an evident presence of voids and some cracks. In this case, synthetic glue has been used to facilitate the joining during sintering. As expected a high Cu and bronze presence and a reduced Fe and graphite content (Fig.2.2) favor a uniform development of the microstructure (Fig.2.3).

Improving the graphite amount to 3% compared to the previously shown samples no significant variation has been detected as the homogeneity of the samples regards and an improvement of the joining between the sintered part and steel occurs (Figs.2.4 and 2.6). A partial soldering takes place (Fig.2.5) which can be solved by increasing the holding at the maximum temperature. During tribological characterization, failure of samples takes place. The samples with a higher content of bronze and ceramic material (Samples E-G, Table 1) have been prepared.

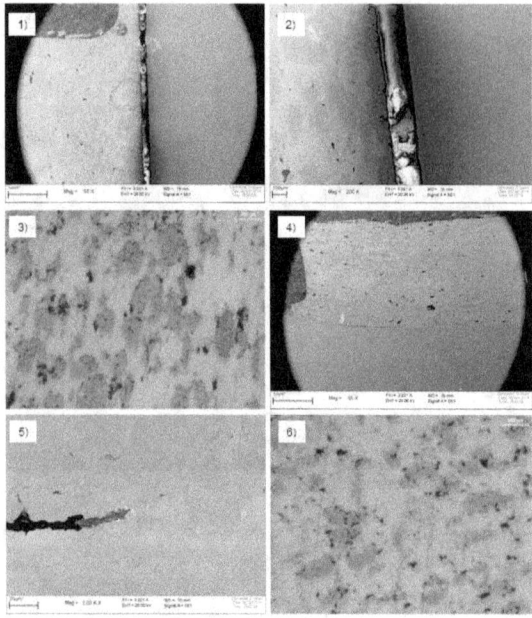

Figure 2. Microstructure of the sintered material: Sample 3 (1, 3) and details of the joining zone between the steel and sintered material (2) and Sample 4 (4, 5) and details of the joining zone between the steel and sintered material (6).

Uniform distribution of the ceramic particles inside the metal matrix is favored in case of low Fe and high bronze content (comparison of Figs.3.1 and 3.2), even if the porosity results superior in this case on the whole surface.

No discontinuity or crack development has been observed at the interface between the sintered material and the steel support demonstrating a good self-brazing capacity of the mixture.

In the case of the sample containing Cu with high bronze and ceramic particles (Sample 7) a uniform diffusion of the ceramic particles within the metallic matrix and a continuous interface with development some large porosity (Fig.3.3) can be detected. The thickness of the joining material is uniform and as expected, it is totally made of bronze.

Gradual heating of the samples has been produced at various loads and speeds during two series of braking: at the end of the braking test, the reduction of the depth has been measured in 3 points and the wear has been calculated using a semi empirical formula, reported in the Equations 1 and 2.

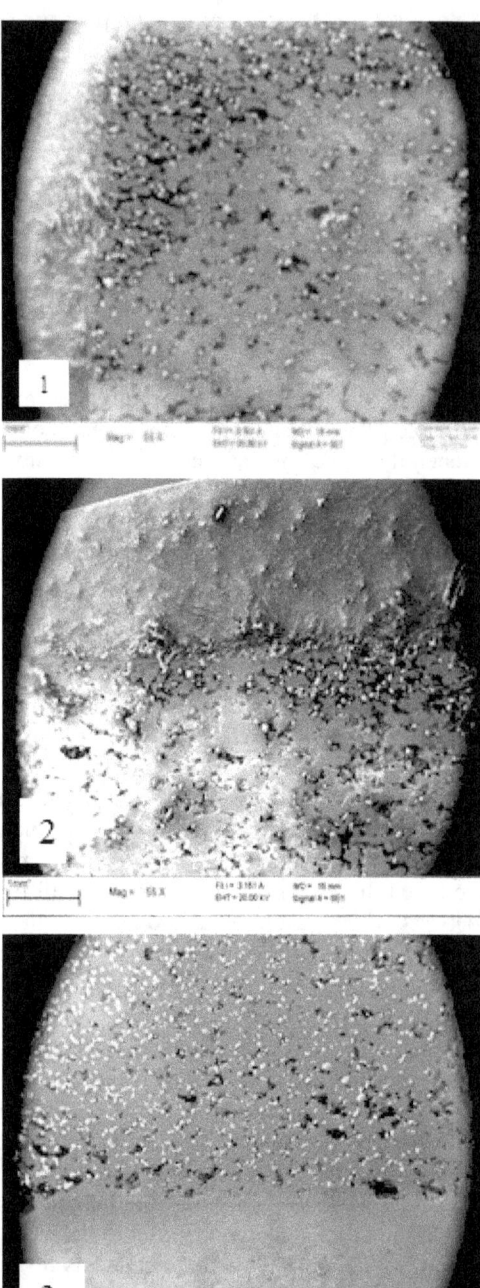

Figure 3. Microstructure of the Samples 5 (1), 6 (2) and 7 (3)

Oscillation on the friction coefficient has been observed. The results are reported in Figure 4. The low graphite content, the metal matrix made of Cu and bronze or Fe and bronze are significant for lowering the friction coefficient in the range of 0.4÷05. As wear rate regards, a comparison between the studied sintered materials and the commercially employed materials has been realized.

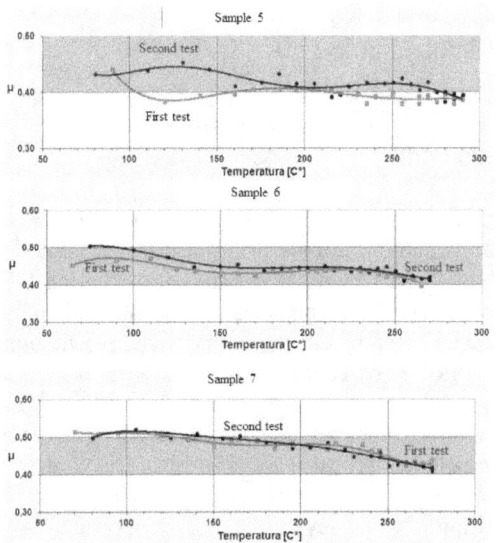

Figure 4. Friction coefficient vs. temperature for the Samples 5÷7

All samples show a lower wear rate compared to the organic material and the average values are below the limit established (1 cm³/10⁶ kg against 2,77 cm³/10⁶ kg for the organic material).

4 Conclusions

Development of sintered friction material was performed, as an alternative solution for the actually used polymer matrix materials. The samples were obtained with a high and constant friction coefficient, a high durability and thermal stability connected to a lower wear rate. Additionally, lack of noise during braking and a good self-brazing capacity were realized. Cu and Fe-based friction materials were sintered and for the optimized compositions, oscillation on the friction coefficient was observed. The friction coefficient of the produced materials fulfills the industrial need: the optimized sintered materials have a lower wear rate compared to those of the actually employed organic material.

References

1) A. Kurt, M. Boz, Materials and Design 26 (2005) 717–721.

2) P. Gopal, L.R. Dharani, Frank D. Blum, Wear 193 (1996) 199-206.

3) I. Gattelli, G. Chiarmetta, M. Boschini, R. Moschini, M. Rosso, I. Peter Solid State Phenomena 217-218 (2014) 471-480.

4) D. Chan, G. W. Stachowiak, Proc. Instn Mech. Engrs Vol. 218 Part D: J. Automobile Engineering (2004) 953-966.

5) N. Gemalmayan, Secret dangerous at air pollution asbestos, Engineering Environmental Science 16 (1983), 32-33.

6) S. Fouquet, M. Rollin, R. Pailler, X. Bourrat, Wear 264, (2008) 850–856.

7) A. Marzocchi, A.E. Jannarelli, D.W. Garrett, Friction for brake linings and the like, U.S Patent 3,967,037. (1976).

8) J.P. Holman, Heat Transfer, 8Tìth edition, McGraw-Hill, Singapore, 641-642.

9) A. Mimaroglu, M. Caliskan, I.Calli, Ind Lubrication Tribol, 53 (2001) 192-197.

10) A. Ravikiran, Influence of apparent pressure on wear behavior of selfmated alumina, J Am Ceram Soc 83 (2000) 1302–1304.

11) G. Strafellini, M. Pellizzari, A. Molinari, Wear 256 (2004) 756-763.

12) A.E. Anderson, ASM Handbook, in: Friction, Lubrication, and Wear Technology, vol.18 ASM International (1992) 569-577.

Distributed Brillouin Sensors for Simultaneous Temperature and Strain Sensing

Ilie Popa[1*], **Alexandru-Alin Jderu**[2,3], **Cristian Livede**[2,3], **Dorel Dorobanţu**[2], **Marius Enachescu**[2]

[1]University of Pitesti, Electronics, Computers and Electrical Engineering Department, Pitesti, Arges, 110040, Romania
[2]Center for Surface Science and Nanotechnology (CSSNT), University Politehnica of Bucharest, Romania
[3]Sc NanaoPro Start MC Srl, Pitesti, Arges, 110310, Romania
*ci.popa@yahoo.com

Abstract: This work makes an overview of Distribute Brillouin Sensors (DBS) with Optical Fiber (OF) aiming specifically at their applications for the temperatures and strain measuring in different structures. Firstly, Brillouin Scattering (BS) effect of light waves through OF, Stimulated Brillouin scattering (SBS) Stokes and anti-Stokes are described. It continues with the DBS networks presentation, the studied methods and techniques, inventorying their main practical applications. It ends with a comparative presentation of the basic parameters for the presented different techniques followed by conclusions.

1 Introduction

Temperature and strain material measurement using DBS is based on BS effect discovered by Léon Brillouin. BS occurs when light transmitted through a transparent material interacts with the propagation time and space generating periodic refractive index (RI) variations due to density variation material. RI of a transparent material changes when the material is deformed by compression-expansion, bending or temperature variations.

The interaction result between the light beam and the propagation material strain lies in the fact that a fraction of the light wave which propagates, changes its frequency and energy along the preferential angles.

If the transport material of light wave is a solid crystal, a chain macromolecular condensed or a viscous liquid, the connections between the atomic frequency and the transport material strain, represented as consisting of quasiparticle, could be: the acoustic oscillation of material mass (called phonons); the electrical charges material displacement (called polarons, in dielectric); oscillating magnetic spin (called magnons, in magnetic materials).

From the solid physics perspective, BS is given by an interaction between an electromagnetic wave and one of three crystalline material quasiparticles mentioned above. Scattering is inelastic: the photon can lose energy (Stokes process) to create one of the three types of quasiparticles, or gain energy (anti-Stokes process) by absorbing one of them. A change of the photon energy corresponds to a Brillouin frequency shift (BFS) proportional to the energy released/absorbed by quasiparticle. Thus, the BS can be used to measure the energy, the wavelength and oscillation frequency of different types of atomic chains.

BS is a result of direct interaction between light waves and the elastic acoustic waves.

Considering that the incident light rays frequency is v, and wavelength is λ ($\lambda = c/v$, where c is the light speed in the crystal) are reflected by a front of plane acoustic waves, their direction changing with angle θ as shown in Figure 1.

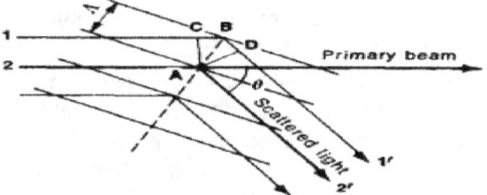

Figure 1. λ_b is Brillouin scattering wavelength.

In order to get a reflected rays maximum intensity in an interference direction, the optical distance CB+DB between rays 1-1' and 2-2', reflected by the adjacent wave fronts, must be equal to λ [1]:

$$2\Lambda \cdot \sin\frac{\theta}{2} = 2n\lambda_a \sin\frac{\theta}{2} = \lambda = \frac{1}{\upsilon} \qquad (1)$$

where: $\Lambda = AB$ is the length of the scattering elastic (hypersonic) wave; n is RI of the OF core; λ_a is acoustic wavelength.

Light waves reflection by acoustic waves is equivalent to the incident light wave modulation with the frequency sound waves.

The frequency shift of light waves BS, υ_b, is equivalent to the acoustic waves frequency, υ_a, as expressed in:

$$\upsilon_b(T, \varepsilon) = \frac{2n}{\lambda} v_a \Big|_{\theta = 180^0} \tag{2}$$

which depends on the acoustic wave velocity in the material which, in its turn, is dependent on the temperature, T, and the relative strait, ε. It is considered, $\theta = 180^0$, because the maximum sensitivity of the sensor network is obtained when the scattered optical light stream is back-propagated. The light scattered frequency relative shift is:

$$\frac{\Delta \upsilon}{\upsilon} = \pm \frac{2 v_a}{c \cdot \sin \frac{\theta}{2}} \tag{3}$$

where, $\Delta \upsilon = \upsilon_b - \upsilon$ is the width of the scattering or amplification Brillouin.

Although the light frequency variation values of BS, Δv are relatively small, they can be acceptably measured by interferometry.

The fundamental parameters used in measurements of T and ε distribution along the OF are: the BFS and the *power exchange* (*the Brillouin signal power*). For this, the networks of the optical DBS were developed.

The development of lasers has improved, not only the possibilities observation BS, but also, led to the discovery of Stimulated Brillouin Scattering, which is distinguished by high intensity and more features quality enhanced.

Spontaneous BS is determined by the density fluctuations caused by the acoustic waves pressure propagated in OF, the acoustic waves being generated by the molecular thermal agitation. The light frequency is Doppler shifted depending on the speed of acoustic waves in the OF (several GHz shift in the glass). The spontaneous BS is very weak (30 dB weaker than Rayleigh scattering), SBS becoming of great importance.

In the case of intense radii (for example laser light) passing through OF, the light beam electric field variations may produce acoustic vibrations in

the material by the electrostriction and the ray undergoes a BS due to these vibrations, usually in the opposite direction of the input beam phase. This process is known as *Stimulated Brillouin Scattering* (SBS). If the input wave power is greater than the minimum threshold, then SBS occurs and because the minimum threshold is approximately 5mW, it can be said that the OF SBS effect occurs at low power levels.

Any optical intensity dynamic change can induce a pressure wave by electrostriction. Such a model of intensity variation may result from interference of two light waves with different counter-propagated wavelengths in FO. In this case, an optical-beat signals within OF is generated. It creates a pressure wave having the same frequency with optical-beat signal, schematically showed in Figure 2. [2]

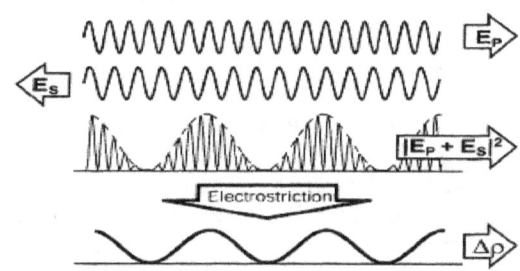

Figure 2. Generation of the pressure wave by electrostriction.

BS can be Stokes type when the wave frequency is shifted lower, and anti-Stokes, when the frequency is shifted higher, as showed in Figure 3 [2].

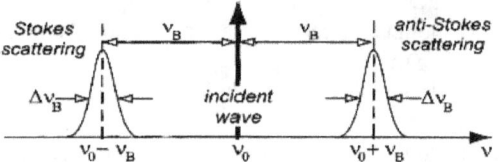

Figure 3. Typical Brillouin back-scattering radiation spectrum in OF. Important parameters are: the BFS (v_B) and the Brillouin linewidth (Δv_B).

2. Distributed Brillouin Sensors for Simultaneous Temperature and Strain Sensing

2.1 General information

The DBS technology with OF is one of the most important areas of sensors OF that provides measurements of temperature, strain, vibrations etc., in industrial, military and civilian installations, with high resolution, on long distances [3]. A unique feature of OF distributed systems is their ability of continuous reading, almost in real time, the measured parameter being as a function of position, in any point along the fiber sensing, covering large infrastructures and distances [4].

The existing technologies use different approaches for distributed sensors, which practically, refer to the phenomenon of light scattering, interferometry, Bragg grid, and the optical loss in quasi-distributed sensors [5, 6].

To analyze and estimate the detection performances are using the following parameters: *spatial resolution; measurement accuracy; sensing range; measurement-acquisition time.* These factors are, generally, linked together and a factor improving can lead to the deterioration of one or more of the others.

An example for the BS dependency of temperature and strain into a standard Single Mode Fiber (SMF) with BFS about 10 to 11 Ghz at 1550nm is showed in Figure 4 [7].

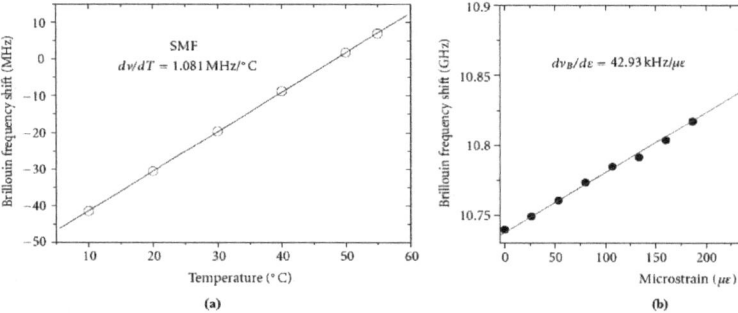

Figure 4. The BS frequency shift dependence on the temperature and strain. Experimental data.

Based on BS effect, are developed following configurations of the distributed fiber optic sensors:
1. Brillouin Optical Time-Domain Analysis (BOTDA);
2. Brillouin Dynamic Grating Distributed Sensing (BDG-DS);
3. Brillouin Optical Time-Domain Reflectometry (BOTDR), OTDR based on spontaneous BS;
4. Brillouin Optical frequency-domain analysis (BOFDA);
5. Brillouin Optical Correlation-Domain Analysis (BOCDA);
6. Brillouin Optical Correlation-Domain Reflectometry (BOCDR);
7. Brillouin Echo Distributed Sensing (BEDS).

2.2 Optical time-domain approach

The time-domain approach for the BFS measuring in distributed OF is based on the use of a pulsed light beam (probe), which interacts with a continuous beam of light counter- propagated (CW-pump). At any moment during the downward pulse propagation in the fiber, SBS effect occurs only in the fiber region where the two beams overlap. CW intensity received at the section where the pulse is launched is monitored as a function of time. This footprint time provides spatial information along

the fiber, because the time-of-flight pulse can be used to convert the spatial coordinate in a temporal one. The Brillouin interaction will take place in a certain section along the fiber if the two counter-propagated beams present a frequency shifting corresponding to the local BFS.

Based on the loss effect of the pump beam intensity (Brillouin loss), Horiguchi et al., developed BOTDA distributed detection technique in 1989 [8]. According to this technique, a laser is placed at each end of the OF sensor. One emits a pulsed light (called probe or Stokes light with v_s frequency) launched at the distance $z = 0$ (start of the fiber), while the other laser emits a continuous light (CW-called pump with frequency $v_p > v_s$) incident from the opposite end, $z = L$, L being the fiber length.

The equations for determining the strain, ε and temperature, T are:

$$\begin{cases} \varepsilon = \dfrac{\upsilon_B(\varepsilon) - \upsilon_B(0)}{C_\varepsilon} \\ T = \dfrac{\upsilon_B(T_0) - \upsilon_B(T_0) + C_T \cdot T_0}{C_T} \end{cases} \quad (4)$$

where: C_ε and C_T are constants that depend on the OF properties; T is measured temperature; T_0 is the reference temperature.

Figure 5 shows schematically the BS spectrum according to the BOTDA technique.

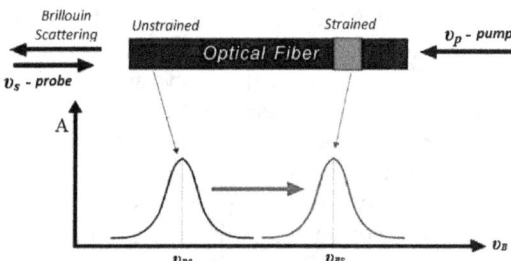

Figure 5. BS spectrum in BOTDA technique.

Measuring time in a traditional BOTDA system is of minutes order and it depends on the length of the OF. This is a serious disadvantage for dynamic monitoring of the structures health.

When the pulse was reduced to 1ns, which is equivalent to 10 cm resolution, the weak Brillouin signal, due to the short interaction, significantly reduced Signal/Noise Ratio (SNR), and therefore the strait and temperature achievable resolutions are reduced. To solve this problem, a differential pulse-width pair Brillouin optical time-domain analysis (DPP-BOTDA) for high spatial resolution sensing [9] was proposed. This method uses two separate different pulses: a long pulse (a few tens of nanoseconds) with a small pulse-width difference (a few nanoseconds) to map the Brillouin Gain Spectrum (BGS) of the sensing fiber. The differential BGS can be obtained by subtracting between the two BGSs, and its spatial resolution is determined by the pulse-width difference of the two separate long pulses. The DPP-BOTDA provides several advantages over conventional BOTDA: (1) Narrowband BGS (a few tens MHz) and high spatial resolution (smaller than 1 m) can be obtained simultaneously. (2) The differential BGS provides stronger signal intensity and thus better SNR than that of directly using the narrow pulse in BOTDA when the pulse-width difference of the two long pulses equals to the narrow pulse width.

2.3 Brillouin Dynamic Grating Distributed Sensing

Brillouin optical fibers sensitive grating (FBG) type are silicon fibers in which sensitive elements intrinsic form of grating (grill) are photo-engraved. Practically, a periodic structure is engraved in the OF core and this periodic structure will reflect a specific optical wavelength that depends on the periodicity [4]. Varying periodicity, the wavelength will vary, too. This period depends on the ambient temperature and strait and therefore it is the basis for a simple sensor (for example, a fiber grid) that can easily be interrogated.

The basic principle concerning the operation of the system FBG sensors is to monitor wavelength shift of the Bragg signal reflected by the measured changes. Bragg wavelength λ_B, or a grating resonance condition is expressed as:

$$\lambda_B = 2n\Lambda \tag{5}$$

where, Λ is the periodicity of the grid and n is the RI of the OF.

When such a device is illuminated by a broadband light source, a narrow band spectral component of the Bragg wavelength is reflected by the grating. This spectral component misses in the transmitted light. The bandwidth of the reflected signal depends on several parameters, including the length of the grid, but typically is about 0.05 - 0.3nm in most sensor applications. Disturbance (for example external deformation and temperature variations) of grid results in a shift of Bragg wavelength grid can be detected in each spectrum reflected or transmitted.

The grids can be: uniform, when the grid periodicity is uniform along the axis of the fiber core and the modulation degree of the RI is constant; non-uniform, when the grid periodicity is non-uniform along the axis of the core and the modulation degree of the RI need not be constant.

In more recent studies, the concept of the Brillouin dynamic gratings (BDG-DS) in single-mode fiber with the polarization maintaining (Differential Pulse-width Pairs - DPP) [10] was implemented. In this concept, acoustic waves generated in the SBS process by optical waves (pump waves) in a single polarization, are used for reflecting an orthogonal polarized wave (probe wave) at a different frequency from the frequency of the pump.

2.4 Brillouin Optical Time-Domain Reflectometry

BOTDR is a coherent detection method that uses a pulsed light. This light is launched into OF for generating spontaneous BS. As illustrated in Figure 6 [7], back-propagated light is measured with a coherent receiver by mixing the spread signal and local oscillator (LO) signal [11]. Because the back scattered signal power is small, the attenuation of the fiber can cause a negative effect on the quality of the measurement. To compensate this disadvantage, the coherent detection mode is used. To extract the local frequency shift, the back-scattered signal is optically mixed with the laser signal CW, and the detected pulsed signal is then electrically mixed with the microwave local oscillator signal. The recorded signal is proportional with optical oscillation amplitude at the microwave oscillator frequency, but its dynamic range decreases with the fiber length.

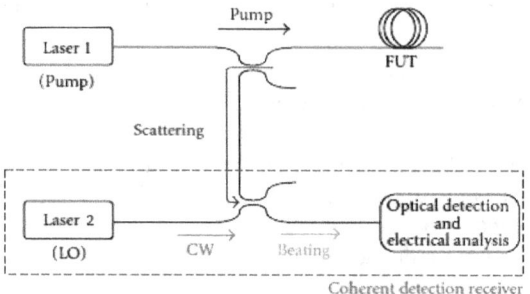

Figure 6. Typical configuration for a BOTDR system. FUT - fiber under test (distributed sensor).

The main disadvantages of this method are: spatial resolution ≥1m; frequency shift is simultaneous dependent by temperature and the longitudinal strait; it is necessary to introduce an electronic filtering to remove the Rayleigh signal.

2.5 Brillouin Optical frequency-domain analysis

An alternative approach for the BS shift measurement is BOFDA. This method is based on measuring a complex baseband transfer function relating to the pump counter-propagated waves and the Stokes waves along the fiber amplitude [12]. Such a transfer function is measured for a pump-probe frequency deviation range, so that, the fiber BFS could be recovered from each location.

An important advantage provided by BOFTA systems in comparison to the BOTDA ones is the operating possibility in narrowband. But a disadvantage of BOFDA is the relatively long measurement time, because it requires multiple measurements at different frequencies to obtain baseband transfer function with a sufficient spectral resolution. During the measurement, OF temperature and strait must be constant not to affect the measurement reliability.

2.6 Brillouin Optical Correlation-Domain Analysis

The based approach on optical correlation gave birth to the BOCDA technique, and this is based on the correlation between the probe and pump CW light waves which generate SBS. The correlation based on CW technique realizes the selective-position generating of SBS by controlling the correlation between the light waves of the pump and counter-propagated probe and simultaneously frequency modulation of them by a sine wave which generates the correlation with the periodic peaks [13].

The disadvantage of BOCDA system is limited measuring range due to the periodic nature of the detection position, the measuring length is shorter than other Brillouin sensors, the transducer is more complex and post-processing is more laborious, increasing the measurement time.

3. Applications and performance

The applications of sensitive distributed OF extend from the health monitoring systems to technical structures, in research and medicine (medical instrumentation). They are part of devices for measurement of: temperature and strain, pressure in enclosures, high voltage transformers, electric power lines etc. They can detect: fire, leakage of gases and liquids in pipes, intrusions in perimeters and enclosure, defects in structures etc. They become indispensable in monitoring and diagnostics applications for: means of conveyance (planes, ships, cars); damage from earthquakes; deterioration of large structures (bridges, dams, buildings, tunnels).

The main performance of systems based on Brillouin distributed OF sensitive achieved over the last three years, are shown in Tables 1 and 2.

Table 1. Performance chart of Brillouin distributed fiber sensors based on light backscattering, out of [14].

Parameter	BOTDR	BOTDA	Brillouin Grating
Spatial resolution	meter	cm to meter	cm
Sensing range	Tens of km	Up to hundreds of km	Tens of meters
Measurement time	1-5min	2-5 min	+10min
Temperature, strain	Yes	Yes	Yes
Temperature accuracy	2-3°C	1-2°C	1°C
Strain accuracy	60με	20με	10με
Dynamic measurement	No	400Hz	No

Table 2. Performance chart for Brillouin distributed OF sensors, extracted out of [15].

Parameter	DPP-BOTDA	Brillouin Grating	BOTDR
Spatial resolution	2 cm (2 km); 2 m (150 km)	1–2 cm	~1 m
Sensing range	150–200 km	20 m	20–50 km*
Measurement time	2–5 min	10+ min	1–5 min
Temperature and strain	Yes	Yes	Yes
Temperature accuracy	1–2 °C	1 °C	2–3 °C
Strain accuracy	20 με	10 με	60 με
Dynamic measurement	yes	No	No
Calibration	**	**	**
Light source requirement	Two DFB lasers (frequency locking of two lasers is required)	Three lasers (frequency locking of two lasers is required)	One narrow linewidth laser
Detectors	Broadband	Broadband	High sensitivity
Detection scheme	Direct	Direct	Coherent

* - Presented by Yokogawa Electric Corporation; ** - Determined by fiber property.

4. Conclusions

Performances of different techniques and methods are markedly different, each of them can be used according to specific application types.

Performances obtained until now have allowed the realization of functional equipment and systems in practical activity, but studies aim at the methods and techniques diversifying, including a combination of Brillouin, Raman and Rayleigh effects in order to improve the performance and diversify the applications.

Acknowledgement

This work was supported by Executive Agency for Higher Education, Research, Development and Innovation Funding and by Romanian Ministry of Structural Funds, under projects ENIAC 04/2014 as well as FOSLAB.

References

1) Still T, *High Frequency Acoustics in Colloid-Based Meso and Nanostructures by Spontaneus Brillouin Light Scattering*, Springer Hardcover, 2010

2) Aldo Minardo, *Fiber-optic distributed strain-temperature sensors based on stimulated Brillouin scattering*, PhD Thesis, Seconda Universita' Degli Studi di Napoli, 2004

3) Anatoli A. Chtcherbakov, Pieter L. Swart, and Stephanus J. Spammer, *Mach-Zehnder and modified Sagnac-distributed fiber-optic impact sensor*, APPLIED OPTICS, Vol. 37, No. 16, (1998), p. 3432

4) Luc Thévenaz, *Brillouin dynamic grating distributed sensing (BDG-DS), Brillouin distributed time-domain sensing in optical fibers: state of the art and perspectives*, Front. Optoelectron. China 2010,3(1): 13–21

5) C. Wang and K. Shida, *A novel multifunctional distributed optical fiber sensor based on attenuation*, Proc. IMTC'06, pp. 2018-2023, April, 2006

6) Anbhawa Nand, *Distributed Fibre Bragg Grating based Sensors, Integrated sensing using chirped optical fiber gratings*, PhD Thesis, Victoria University Australia, 2007

7) C. A. Galindez-Jamioy and J. M. Lopez-Higuera, *Brillouin Distributed Fiber Sensors: An Overview and Applications*, Hindawi Publishing Corporation, Journal of Sensors, Volume 2012, Article ID 204121,17pages, doi:10.1155/2012/204121.

8) A. Zornoza, R.Perez-Herrera, C.Elosua et al., *"Long-range hybrid network with point and distributed Brillouin sensors using Raman amplification,"* Optics Express, vol. 18, no. 9, pp. 9531–9541, 2010.

9) W.Li, X.Bao, Y.Li, and L.Chen, *"Differential pulse-width pair BOTDA for high spatial resolution sensing"* Optics Express, vol.16, no.26, pp.21616–21625, 2008.

10) K. Y. Song, *"Operation of Brillouin dynamic grating in single mode optical fibers,"* Optics Letters, vol. 36, no. 23, pp. 4686–4688, 2011.

11) Y. Lu, H. Liang, X. Zhang, and F. Wang, *Brillouin optical time-domain reflectometry based on Hadamard sequence probe pulse*, Proceedings,the 9th International Conference on Optical Communications and Networks (ICOCN '10), pp. 36–38, October 2010.

12) D. Garus, T. Gogolla, K. Krebber, and F. Schliep, *"Brillouin optical-fiber frequency-domain analysis for distributed temperature and strain measurements"*, J. Lightwave Technol., vol. 15, pp. 654-662, 1997.

13) K.Hotate and M.Tanaka, *"Distributed fiber Brillouin strain sensing with a 1-cm spatial resolution by correlation-based continuous-wave technique"*, IEEE Phton. Technol. Lett., vol.14, no.2, pp.179-181, 2002.

14) Zengguang Qin, *Distributed Optical Fiber Vibration Sensor Based on Rayleigh Backscattering*, PhD Thesis, University of Ottawa, Canada, 2013.

15) X. Bao, L. Chen, *Recent progress in distributed fiber optic sensors*,Sensors,vol.12, no.7,pp.8601-8639,2012.

Potential Controlled Co-Ni Nanowires with Compositional Gradient

George Tepes, Maria Diana Vranceanu, Cosmin Mihai Cotrut, Dionezie Bojin

University POLITEHNICA of Bucharest, Faculty of Materials Science and Engineering, Bucharest, 060042, Romania.

Ruxandra Vidu*

University of California, Davis, Department of Chemical Engineering and Materials Science, Davis CA 95616, United States of America

Corresponding Author: rvidu@ucdavis.edu

Abstract: Co-Ni nanowires with composition gradient were grown by template assisted electrochemical deposition. Using PCTE membrane to grow nanowires makes the deposition of an alloy inside the pores more challenging because the growth is affected by the transport of the electroactive ions into the pores. A novel step-wise deposition-stripping strategy was used to grow NWs with controlled composition along the length of nanowire while maintaining constant the electrolyte composition. A detailed electrochemical analysis was performed to understand and control the Co stripping process, and to demonstrate the applicability of the proposed method. Redox reactions allow us to create custom deposition-stripping programs that enable us to better select and control the potential used to grow the nanowires of controlled composition. Samples were analyzed using scanning electron microscopy (SEM) and EDS. Quantitative analysis performed on samples has revealed that the designed stripping technique and selected potentials resulted in Co-Ni nanowires with up to 3 times more Ni then the samples obtained from similar electrolyte.

Keywords: nano, anomalous deposition, Ni, Co, stripping

1. Introduction

The interest in nanomaterials is motivated by their unique properties that can be applied to electronic devices, nanosensors, imaging devices and data storage. High versatility of nanostructures is due their physical properties, high surface to area ratio, and electronic properties [1-6].

Nanowires (NWs) with good magnetic properties and electrical conductivity are obtained from Co-Ni alloys [2-12]. However, the control of the NW composition is challenging in the iron group alloys due to the anomalous deposition [9, 10, 12-14], which presents an obstacle to the straightforward codeposition of Co-Ni alloys. Anomalous deposition occurs when the less noble of the two codepositing species deposits at an extremely higher concentration than the less noble species.

This paper aims to better control the chemical composition of Co-Ni NWs. In order to obtain the desired physicochemical properties, an advanced and improved control over the final NWs concentration is needed. Stripping technique can be used to restrict the anomalous deposition and to allow for a preferential oxidization of the less noble metal, which is Co in this case.

Generally speaking, to control the composition of Co-Ni nanowires, the deposition may be performed in a single electrolyte containing Co and Ni ions [1, 2, 8, 9, 11, 12, 15-21]. Another way to control the deposition, which is less common, is to perform the deposition in alternating electrolytes containing either Co or Ni [4, 6, 22]. Unlike the deposition in two electrolytes, the deposition in a single electrolyte results in an alloy whose composition can be modified by adjusting the electrolyte concentration [1, 2, 8, 9, 11, 12, 15-21]. This method allows for a better control of the composition by managing the deposition protocol (i.e. high frequency pulses, adjusting the deposition time, potential and rate).

Another advantage of this deposition in a single electrolyte is that segments of various compositions can be obtained. Alloy concentration in each segment is enforced by the concentrations of depositing ions in the electrolyte and the applied potential. The length of the segments is adjusted by changing the time of deposition at a certain potential. The main disadvantage of this method is that it is generally more difficult to obtain segments containing only one depositing element.

In the literature, the most common used ratio of Co:Ni in the solution is 1:2, and the most common deposition potentials for Co and Ni are -0.4 and -1.5V, respectively [2, 9, 12, 23]. However, Co and Ni show anomalous deposition that gives a different ratio of Co and Ni in the deposit [17-19]. Therefore, the compositional control in co-deposition is very important while the deposition in nanostructured template imposes additional challenges.

In this paper we applied pulse deposition because the relaxation times ("off time") allows for a more complex processes that may help us control the composition of the deposit. Relaxation times allow ions to diffuse to the electric double layer, thus preventing the depletion of depositing ions. Due to the difference in ion velocities, the relaxation time allows for Ni atoms to "relax" while Co atoms are oxidizing.

2. Sample Preparation

All samples and electrolytes were obtained using high purity materials and chemicals, and using a working protocol that helped to obtain high reproducible results. Specific protocols were developed for template preparation, sample preparation, electrochemical treatment and NWs growth.

To prepare the template for NWs growth, we first sputter Au on one side of the PCTE membrane. Next, the membrane was immersed in n-Butanol solution and sonicated for 5 minutes. After sonication, the template is placed with the gold side on a copper tape and sandwiched in between two isolating tapes. A small hole in the tape exposes the sample and allows the ions to reach the gold electrode through the template pores. These samples are used as Working Electrodes (WE) in a 3-electrode electrochemical cell that are connected to a potentiostat/galvanostat PARSTAT 4000 (Princeton Applied Research AMETEK, USA). A calomel electrode was use as a Reference Electrode and a platinum rod as a Counter Electrode.

To assure the reproducibility of the NWs results, an electrochemical treatment in 50 mM H_2SO_4 is conducted. This treatment consists of 2 steps: a Cyclic Voltammetry (CV) step that helps with the Au film cleaning, and a potentiostatic (PT) step. The main aim of this procedure is to clean the Au film surface and to favor Au atoms surface diffusion to result a flat surface [24-31].

3. Experiments

A series of CVs were conducted in order to understand the electrochemical behavior of Au in various electrolytes such as: blank solution (40g/l H_3BO_3), Ni (44 g/l $NiSO·6H_2O$ + 40 g/l H_3BO_3) and Co (20 g/l $CoSO_4·7H_2O$ + 40 g/l H_3BO_3) electrolytes. Electrodeposition in pulses was performed using a PCTE membrane to grow Co-Ni NWs. The electrochemical behavior of Au in each of the 3 solutions was investigated to observe if there are any surface reactions in that potential range. For each solution, there were recorded 2 cycles.

Characterization of the Co-Ni NWs was performed using a scanning electron microscope equipped with an EDS. First the PCTE membrane is dissolved in a concentrated Cl_2CH_2 solution. Then, samples are washed with Milli-Q water and dried with N_2.

4. Results and Discussions

4.1 Electrochemical characterization

CVs were performed in a blank solution (BS), Co and Ni electrolytes.

Figure 1a shows the CV curves of Au in blank solution (BS), where there are no depositions. Analysis of CV in BS show no faradaic reaction in the scanned potential range. Fig. 1 b,c shows the potentials that redox reactions start and end for each electrolyte, i.e. Co and Ni ion containing electrolytes. These potentials are then used to design an analytical program for obtaining a certain chemical composition gradient in nanowires.

The electrochemical parameters for nanowire growth were chosen based on the analysis of cyclic voltammetry. The CVs are obtained by scanning the potential between two selected values to study the redox reactions that may occur on the electrode in the selected electrolyte. We selected several potential (between -1.1 V and 0, 0.1 and 0.2 V) and scanning speed of 2, 5 and 50 mV/s, (for instance, Figure 1 presents only at 50 mV).

In Ni and Co electrolytes, as the applied potential moves to more negative values, the current increases due to the electrochemical reaction and deposition of atoms on of the working electrode surface. Figure 1 shows that Co reducing begins around -0.58 V while Ni reduction begins around the -0.37 V.

After repeating each experiment twice to check the reproducibility the CV curves were used to determinate the potential steps for our designed experiments.

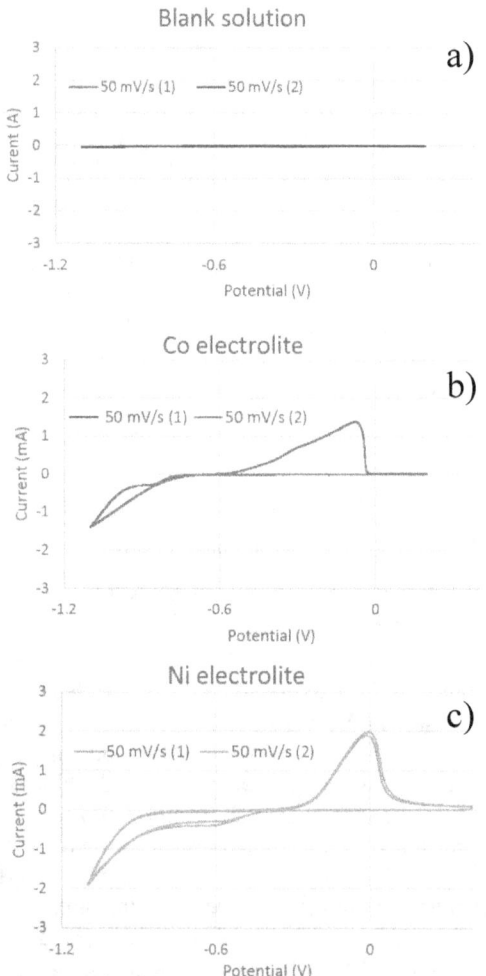

Figure 1. Cyclic voltammetry of Au in blank solution (1), Ni (b) and Co (c) ion containing solution at a sweep rate of 50mV/s.

4.2 NWs growth

We applied two electrochemical experiments to grow NWs with custom compositional gradient on the longitudinal axis. The electrochemical experiments were set one with stripping off (OFF) (Figure 2) and one with stripping on (ON) (Figure 3). After the NWs growth, the samples were prepared for SEM and EDS analysis.

For both OFF and ON, the final pulse is set to allow for Co deposition, so that at the top of the NWs, Co concentration is almost the same. Figures 2.a and 3.a present the charts of the applied potentials and of the recorded current during the NWs growth.

4.3. Scanning electron microscopy

SEM images that were taken on the Co-Ni NWs samples (Figure 2, b) - d) and Figure 3 b) - d)) offer valuable information on NWs dimension and surface topography. At a first glance, SEM images show that the NWs obtained with ON are shorter than the ones obtained with OFF, at a same growth time. SEM measurements showed that the NWs obtained in stripping ON conditions are 2.3 ± 0.3 µm long compared with OFF that have grown in to mushrooms (>6 µm). Mushroom morphology is a common morphology of overgrown nanowires that continue to grow after they reach the surface.

NWs diameter is 150 ± 20 nm which is 40% larger than the measured membrane pore diameter, i.e. 113.7 ± 25nm. This difference between the NW diameter and the diameter of the pore has been previously reported [23] as being caused as a result of swelling of pores during the hydrophilic treatment in n-Butanol.

To better understand the chemical composition profile obtained in NWs, EDS analysis was conducted on samples obtained in both sets of experiments. Results are shown in Table 1 and 2. Each NW was scanned and measured in 3 points, at the ends and in the middle.

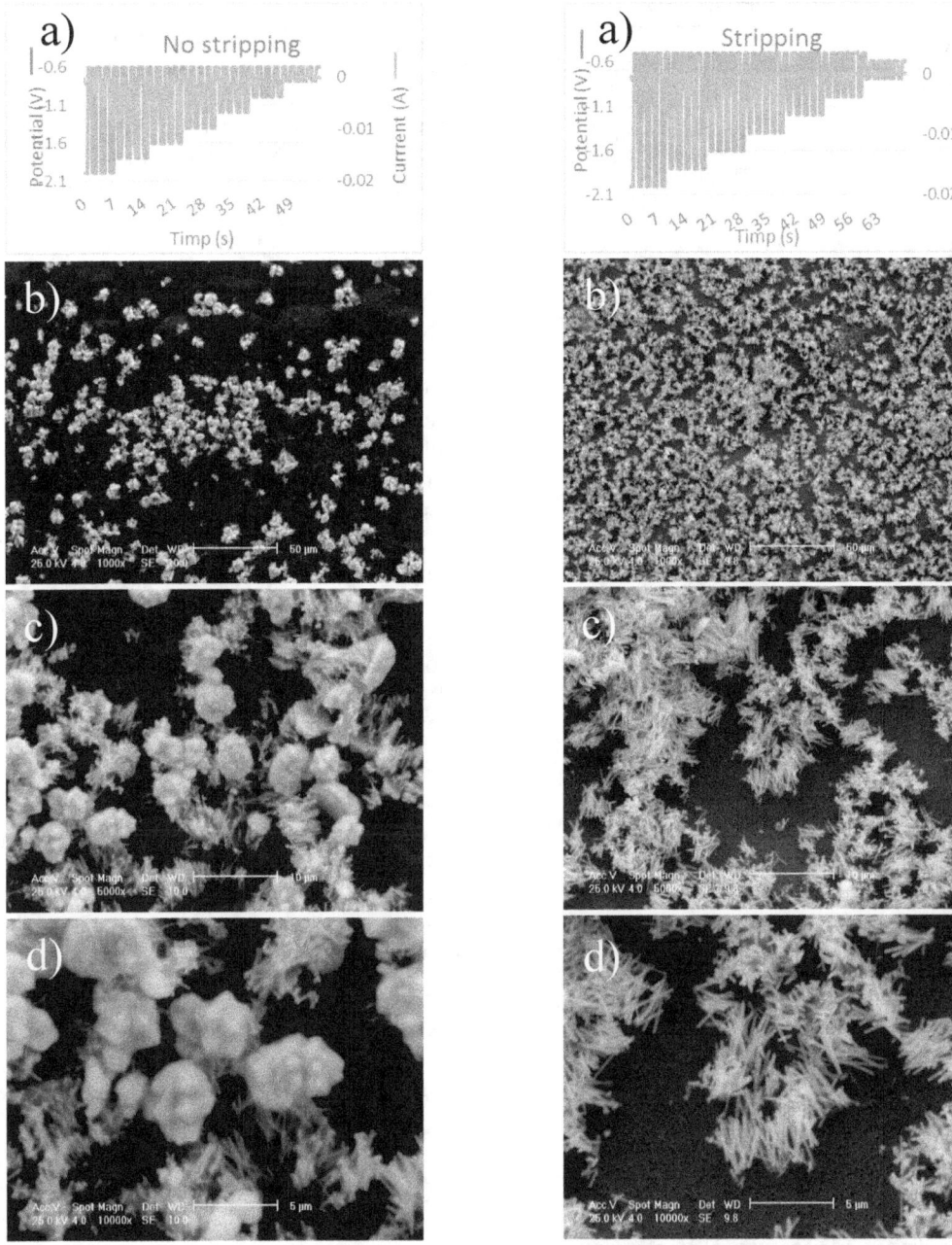

Figure 2. a – Chart of the applied potentials (blue) along with the recorded current (orange) vs time;

b, c, d – SEM images of NWs obtained with stripping OFF.

Figure 3. a – graphical representation of the applied potentials (blue) along with the recorded current (orange) vs time;

b, c, d – SEM images of NWs obtained with stripping ON.

Table 1: Stripping ON

Experiment No.	Co	Ni
1	20.45	79.55
2	33.67	66.33
3	69.69	30.31

Table 2: Stripping OFF

Experiment No.	Co	Ni
5	28.71	71.29
6	36.64	63.36
7	67.93	32.07

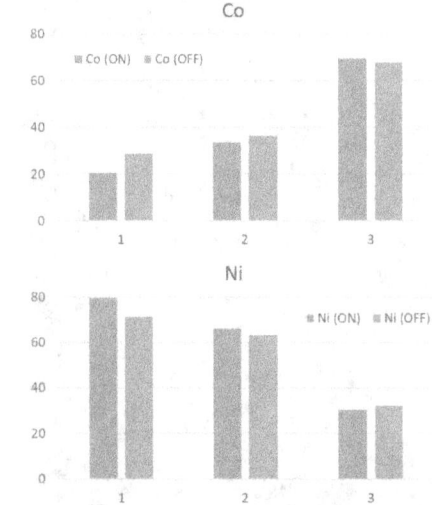

Figure 4. Comparison of Co and Ni concentrations measured in NWs

According to EDS results, NWs sample that are obtained using ON presents an increased Ni concentration at one end, i.e. 79%, compared to the NWs obtained with OFF, i.e. 71%. In the literature, using electrolytes with much higher Ni:Co ratio, such as 21:1/Ni:Co (i.e. 1,250 M $NiCl_2 \cdot 6H_2O$ and 0,06 M $CoCl_2 \cdot 6H_2O$) [23] or 30:1/Ni:Co (i.e. 300 g/l $NiSO_4 \cdot 7H_2O$ and 10 g/l $CoSO_4 \cdot 7H_2O$) [4],

similar Ni concentration difference in NWs was observed.

Additionally, the EDS measurements show that Ni concentration decreases along the length of the NWs in both ON and OFF growth situations. These results may be due to the difference in diffusion rates of Co and Ni at more positive potentials, where Co has higher diffusion rate than Ni.

5. Conclusions

Ni-rich NWs have been obtained using same Co:Ni ratio as reported in the literature. Using the same electrolyte that Prida *et all* [2, 9] reported to obtain Co reach (Ni poor NWs (46%)), we have obtained Co poor NW (Ni reach NWs (79%)) using a pulse electrochemical schedule.

The composition of the NWs as measured by SEM/EDS is variable along the longitudinal axes. The pulse electrochemical techniques used in these experiments resulted in Ni reach NWs obtained in a low concentration Ni electrolyte. This pulse technique can be used to overcome the anomalous deposition that may appear in Ni-Co electrolytes and opens new ways to control the deposition and growth rate.

Acknowledgment

The work has been funded by the Sectorial Operational Program Human Resources Development 2007-2013 of the Ministry of European Funds through the Financial Agreement POSDRU/159/1.5/S/134397.

References

1. Chen, M., C.-L. Chien, and P.C. Searson, *Potential Modulated Multilayer Deposition of Multisegment Cu/Ni Nanowires with Tunable Magnetic Properties.* Chemistry of Materials, 2006. **18**(6): p. 1595-1601.
2. García, J., et al., *Template-assisted Co–Ni alloys and multisegmented nanowires with tuned magnetic anisotropy.* physica status solidi (a), 2014. **211**(5): p. 1041-1047.
3. Hussain, M., et al., *Fabrication and temperature dependent magnetic properties of Ni–Cu–Co composite nanowires.* Physica B: Condensed Matter, 2015. **475**: p. 99-104.
4. Pereira, A., et al., *A soft/hard magnetic nanostructure based on multisegmented CoNi nanowires.* Physical Chemistry Chemical Physics, 2015. **17**(7): p. 5033-5038.
5. Vivas, L., et al., *Magnetic anisotropy in CoNi nanowire arrays: analytical calculations and experiments.* Phys Rev B, 2012. **85**: p. 035439.
6. Wang, H., et al., *Effect of sub-layer thickness on magnetic and giant magnetoresistance properties of Ni–Fe/Cu/Co/Cu*

multilayered nanowire arrays. Chinese Journal of Chemical Engineering, 2015. **23**(7): p. 1231-1235.

7. Shahzad Khan, B., et al., *Effect of workfunction on the growth of electrodeposited Cu, Ni and Co nanowires*. Materials Letters, 2014. **137**: p. 13-16.

8. Vilana, J., E. Gómez, and E. Vallés, *Electrochemical control of composition and crystalline structure of CoNi nanowires and films prepared potentiostatically from a single bath*. Journal of Electroanalytical Chemistry, 2013. **703**: p. 88-96.

9. Prida, V., et al., *Electroplating and magnetostructural characterization of multisegmented Co54Ni46/Co85Ni15 nanowires from single electrochemical bath in anodic alumina templates*. Nanoscale Research Letters, 2013. **8**(1): p. 263.

10. Vazquez-Arenas, J. and M. Pritzker, *Steady-state model for anomalous Co–Ni electrodeposition in sulfate solutions*. Electrochimica Acta, 2012. **66**: p. 139-150.

11. Hamrakulov, B., et al., *Electrodeposited Ni, Fe, Co and Cu single and multilayer nanowire arrays on anodic aluminum oxide template*. Transactions of Nonferrous Metals Society of China, 2009. **19, Supplement 1**: p. s83-s87.

12. Ghahremaninezhad, A. and A. Dolati, *A study on electrochemical growth behavior of the Co–Ni alloy nanowires in anodic aluminum oxide template*. Journal of Alloys and Compounds, 2009. **480**(2): p. 275-278.

13. Chung, C.K. and W.T. Chang, *Effect of pulse frequency and current density on anomalous composition and nanomechanical property of electrodeposited Ni–Co films*. Thin Solid Films, 2009. **517**(17): p. 4800-4804.

14. Oriňáková, R., et al., *Influence of pH on the electrolytic deposition of Ni–Co films*. Thin Solid Films, 2008. **516**(10): p. 3045-3050.

15. Dolati, A. and S.S. Mahshid, *A study on the kinetics of Co–Ni/Cu multilayer electrodeposition in sulfate solution*. Materials Chemistry and Physics, 2008. **108**(2–3): p. 391-396.

16. Nasirpouri, F., et al., *GMR in multilayered nanowires electrodeposited in track-etched polyester and polycarbonate membranes*. Journal of Magnetism and Magnetic Materials, 2007. **308**(1): p. 35-39.

17. Rohan, J.F., et al., *Coaxial metal and magnetic alloy nanotubes in polycarbonate templates by electroless deposition*. Electrochemistry Communications, 2008. **10**(9): p. 1419-1422.

18. Azarian, A., et al., *Field emission of Co nanowires in polycarbonate template*. Thin Solid Films, 2009. **517**(5): p. 1736-1739.

19. Mathe, V.L. and A.D. Sheikh, *Magnetostrictive properties of nanocrystalline Co–Ni ferrites*. Physica B: Condensed Matter, 2010. **405**(17): p. 3594-3598.

20. Ohgai, T., *Magnetoresistance of Nanowires Electrodeposited into Anodized Aluminum Oxide Nanochannels*, X. Peng, Editor. 2012. InTech. p. 101-125.

21. Monzon, L.M.A., et al., *Fabrication of multisegmented magnetic wires with micron-length copper spacers*. Electrochemistry Communications, 2013. **36**: p. 96-98.

22. Maleak, N., et al., *Fabrication and magnetic properties of electrodeposited Ni/Cu nanowires using the double bath method*. Journal of Magnetism and Magnetic Materials, 2014. **354**: p. 262-266.

23. Hansal, W.E.G., et al., *Pulse reverse plating of Ni–Co alloys: Deposition kinetics of Watts, sulfamate and chloride electrolytes*. Electrochimica Acta, 2006. **52**(3): p. 1145-1151.

24. Vidu, R. and S. Hara, *Surface alloying at the Cd/Au(100) interface in the upd region. Electrochemical studies and in situ EC-AFM observation*. Journal of Electroanalytical Chemistry, 1999. **475**(2): p. 171-180.

25. George Tepes*, A.A.M., Maria Diana Vranceanu, Cosmin Mihai Cotrut, Dionezie Bojin, Victor Kuncser, Ruxandra Vidu. *Influence of the electrochemical treatment on the magnetic properties of nanowires*. in *American Romanian Academy*. 2015. National Institute of Nuclear Physics, Frascati, Roma.

26. Ruxandra Vidu, S.H., *In situ electrochemical atomic force microscopy study on Au,,100.../Cd interface in sulfuric acid solution*. American Vacuum Society.

27. K. Kubo*, N.H., S. Hara, *Decay of nano-islands on Au(1 0 0) electrode in sulfuric acid solution with Cl⁻ anions*. Applied Surface Science, 2004. **237**: p. 301–305.

28. Nobumitsu HIRAI, H.O.a.S.H., *In Situ Electrochemical Atomic Force Microscopy with Atomic Resolution of Fe(110) in Sodium Sulfate Aqueous Solution*. ISIJ International, 2000. **40**(7): p. 702 - 705.

29. Nobumitsu Hirai, H.T., Shigeta Hara, *Enhanced diffusion of surface atoms at metalrelectrolyte interface under potential control*. Applied Surface Science, 1998: p. 506–511.

30. Nobumitsu Hirai, K.-i.W., Akiko Shiraki, and Shigeta Hara, *In situ atomic force microscopy observation on the decay of small islands on Au single crystal in acid solution*. Journal of Vacuum Science & Technology B, 2000. **18**(7).

31. Nobumitsu Hirai, M.Y., Toshihiro Tanaka, Shigeta Hara, *Decay of nano-islands on the surface of a Au(111) electrode in contact with sulfuric acid solution*. Science and Technology of Advanced Materials 5, 2003: p. 115–118.

Traian Vuia's contribution to the development of propulsive power

Ioana Ionel

Politehnica University Timisoara, Faculty for Mechanical Engineering, Bv. M. Viteazu 1, 300222, Timisoara,

ioana.ionel@upt.ro

Abstract: The paper describes, based on recherché of documents, the invention of Traian Vuia consisting of the steam generator, meant to offer in very short time, a sufficient propulsive power. The construction was considered and still is a remarkable contribution to the steam boilers development, basically applying enhanced combustion, heat transfer principles, as well as forced circulation.

*Keyword*s: steam generator, efficiency, enhanced combustion, heat transfer

1. Introduction: The boiler was a necessity

Among other inventions, Vuia designed also a steam generator, patented in many countries of the world in the interwar period. Even since 1902, when he arrived in France, Traian Vuia was preoccupied with building a suitable motor for driving his heavier than air flying machine and capable of taking off using its own propulsive power.

The engine with carbon dioxide, designed and built by Vuia for driving the airplane with which on March 18, 1906, on the field of Montesson, he achieved the epochal flight, did not satisfy him because it was not strong enough, was not very reliable while functioning, had poor efficiency and the working fluid used was consumable. Not even the internal combustion engines, with either spark ignition or compression, met Vuia's requirements since, being at the beginning of their development, they realized low unitary power, had poor efficiency and a large weight compared to the power output. Due to this, Vuia considered the best thermodynamic agent was high pressure and temperature steam, aimed to relax in a turbine, which would drive the propeller of the plane. The exhaust steam was to be condensed and the condensation was to be returned to the steam generator.

This original idea of using a closed thermodynamic cycle involved the realization of an energy engine group composed mainly of a steam boiler, a turbine, a condenser, pumps, blowers, etc. All this together were to weigh so little and have small dimensions in order to enable installation on a plane, which was supposed to lift off by itself and fly safely. It was hard to believe that this could be done at that time and therefore this idea was regarded with restraint and countered by quite many.

2. Concept

How was it possible to rapidly make a steam generator, which to be put into operation in 1-3 minutes, to produce steam at p = 100-120 bar and t_s = 400-500 0C, to have an hourly flow rate of about 10 times its own weight, considering that at that time, the vast majority of boilers operated at a pressure of p = 10 bar, producing saturated or slightly overheated steam, contained a very large volume of water and were put into operation in 2 - 4 hours? But, still, it was possible!

At the core concept of this steam generator, Vuia sensed the need to introduce three fundamental ideas that had just revolutionized the technique of constructing steam generators: (i) *enhanced combustion*, (ii) *increase heat transfer* and (iii) *forced circulation boiler*.

Obviously, each of these ideas would require an extensive description, but, the following section aims to point out some characteristic features in order to make clear the leap in the technical thinking back then, realized by introducing Traian Vuia's innovative ideas in the construction of steam generators.

Accelerated combustion is characterized by a high combustion speed. To achieve this goal, Traian Vuia equipped his steam generator with a burner of a special construction very similar to a carburetor (Figure 1).

When it starts, the fuel mixture is ignited by a spark plug. Combustion is going on inside a refractory steel furnace tube, highly alloyed with Cr and Ni, which during operation is heated to incandescence. In these circumstances, the rate of the chemical reaction of oxidation of the combustible elements is determined only by the constant of the reaction rate K, whose temperature dependence is expressed by Arrhenius law:

$$K = K_0 \cdot e^{-\frac{E}{RT}}$$

(1)

where: K_0 is the pre-exponential factor, considered by the kinetic theory as a size proportional to the total number of collisions between the molecules of the substances that take part in the reaction;

E - Activation energy in W/mol;

R - Universal constant for gas, in W/(mol/K);

T - Absolute temperature, in K.

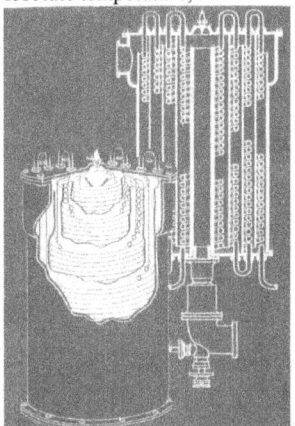

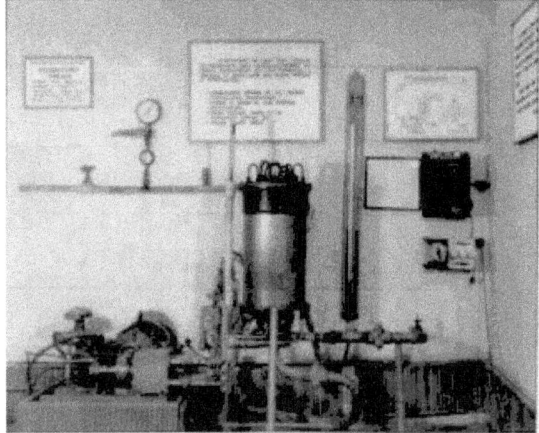

Figure 1. View and section through Vuia's steam generator, [1], [4].

Equation (1) shows that the reaction rate increases along with the temperature of the reaction system. If for small values of T, the growth of the reaction rate is modest, after exceeding a certain threshold temperature, the reaction rate increases very rapidly after an exponential law. This is precisely the situation in which combustion occurs in the furnace tube of Vuia's boiler, where the average temperature stabilizes around T_m = 1800 - 2000 K, and the thermal load reaches about q_f = 460 kW/m^3.

The intensification of the heat transfer from the flue gases to the water and steam flowing in the pipes was sensed and carried out by Vuia using forced convection at high speed and counter current circulation of working fluids. Indeed, the mathematical expression of the coefficient of heat transfer (k_1) written for a tubular heating area (Figure 2)indicates that in order to increase the value of k_1, the value of the coefficient of thermal

convection α_1 from the flue gases to the pipe walls must be increased. This is because the heat resistance, conductive through the wall $\left(\dfrac{d_1}{2\lambda}\ln\dfrac{d_1}{d_2}\right)$ and convective from the wall to the fluid flowing through the pipes $\left(\dfrac{1}{\alpha_2}\right)$ is lower than the thermal resistance of the flue gases at the wall $\left(\dfrac{1}{\alpha_1}\dfrac{d_1}{d_2}\right)$

$$k_1 = \cfrac{1}{\cfrac{1}{\alpha_1}\cfrac{d_1}{d_2} + \cfrac{d_1}{2\lambda}\ln\cfrac{d_1}{d_2} + \cfrac{1}{\alpha_2}}\left[\cfrac{W}{m^2 K}\right]$$

(2)

For a given report of the pipe diameter, the increase of α_1 leads directly to increasingits value.

It is known that, in the case of forced convection,invariant Nusseltis dependent on invariant Reynolds and Prandtl:

$$N_u = f\left(R_r^n P_r^m\right)$$

(3)

or explained:

$$\frac{\alpha d}{\lambda} = C\left(\frac{wd}{v}\right)^{0,8}\left(\frac{v}{a}\right)^{0,4}$$

(4)

Whereas the average values of cinematic viscosity v, thermal diffusivity a, and thermal conductivity λ remain practically constant, equation (4) indicates that the value of the convection coefficient increases along with gas velocity w at power 0.8. If in the case of conventional boilers the flue gases circulation rate varies within limits 6- 14 m/s, in the case of Vuia's boiler, this size has a value of 120-150 m/s. This leads to a growth of the convection coefficient value of about 8-10 times, compared to the values found for conventional boilers.

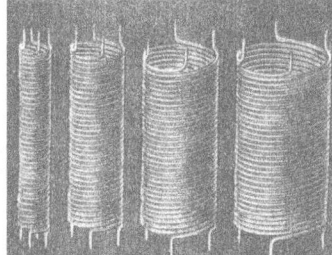

Figure 2. Concentric heating surface made of spirally wound pipes [1], [4].

Among the possible variants of boilers with natural circulation of the water in the vaporizer, with multiple forced circulation and with forced circulation, Vuia chose the latter one for his boiler. Forced circulation boilers have the heating surface formed by one or more very long serpentine pipe with a small diameter of 10-20 mm. At one end of the coil, it enters feed water, and, at the other end, it comes out superheated steam. The processes of steam heating, vaporization and superheating occur one after another, along the coil; the border between the economizer, vaporizer and super heater is sliding, depending on how it is achieved the equality between the heat flow received from the flue gases and the heat flow necessary for the transformation offered water into superheated steam.

Traian Vuia designed and realized five steam generators, having 2 to 32 coils, made of pipe with d = 10 mm, spirally wound in four concentric cylinders. In Table 1, the main dimensions of Vuia's steam generators are presented, and in Figure 3 a section is indicated, which shows the arrangement of the surfaces, as well as the direction of counter current circulation of the flue gases, water and steam, respectively.

The heating surface is composed of coils connected in series and coaxial with a central furnace tube. The metal body of the boiler, which is made of heat-resistant tin, has a cylindrical shape and is closed with two lids at the ends. The gas channels have annular section and are shaped by cylindrical tin obstacles (coaxial with the central tube). They are fixed in an alternating way by removable joints or by welding on the top and bottom lids of the boiler. In the centre of the top lid, a spark plug is set up to ignite the fuel when started.

The advantages of Vuia's boiler are: simple construction, very easy to perform repairs (replacement of coils), quick start-up (minutes), low thermal inertia, high efficiency (95 %), very high productivity (30 kg steam/(m²h)), reduced size. Its disadvantages are: relatively low rate, it requires a good automation and good quality fuel.

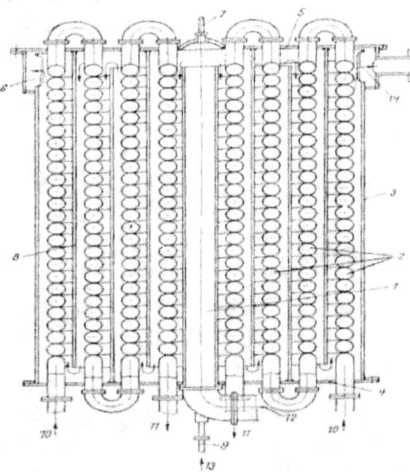

Figure 3. Vuia's four coil boiler. Functional outline: 1 – furnace tube; 2 – coil; 3 – jacket; 4 – bottom lid; 5 – top lid; 6 –flue gases collector; 7 – spark plug; 8 – obstacle; 9 – injector; 10 – admission feed water; 11 – steam exit; 12

– air exit; 12 – fuel entry; 14 – gases exit [1], [4]

The boiler is generally used to produce saturated steam of 1.2-100 bars, and sometimes to produce hot water to heat urban centers or for naval needs.

Considering that the amount of water found at a certain moment in time in a coil is very small (about 300 g), the boiler is actually free of thermal inertia, so it can be put into operation in very rapidly (2-4 minutes). Furthermore, there is no danger for a coil to break, because the small amount of vapors, formed suddenly due to pressure drop, does not have the destructive effect met for example in the case of boilers with high water content.

Table 1. Dimensions of Vuia's steam generators

Type	Body diameter [mm]	Height [mm]	Volume [m³]	Furnace tube (height x diameter) [mm]	Steam flow [kg/h]	No. of coils
A	275	550	0.030	50 x 500	210	2
B	340	700	0.063	65 x 650	420	4
C	420	900	0.125	80 x 800	840	8
D	480	1115	0.200	100 x 1000	1680	16
E	560	1365	0.360	125 x 1250	3360	32

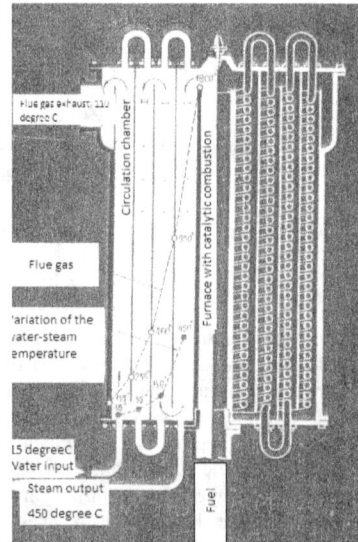

Figure 4. Section through Vuia's steam generator [1], [4]

Forced circulation boilers can operate only at pressures higher than 100 bar, in which the circulation of water as vapor takes place without variation too much the specific volume. As a result of high operating pressures, the thickness of the pipes' metallic walls is increased, but this increase is insignificant when the pipe diameter is small. This is the reason why Vuia chose the diameter of the coils to be only of 10 mm, achieving a boiler weight approximately 10 times lower than for conventional boilers (Figure 4).

The validity of Traian Vuia's innovative ideas applied in the construction of steam generators was verified and confirmed by the experimental research undertaken by PhD. Eng. G. Brola in the *Laboratoire des hautes températures de Paris* of the Faculty of Sciences, University of Sorbonne, which was led by professor Gustave Ribaud, member of the French Academy of Sciences. The measurements done showed that the combustion was going perfectly, with very low air excess, the obtained heat was transferred almost entirely to water and steam. The flue gases were going out of the boiler at temperatures $t_{gc} < 60°$ C, if the supply water temperature was of $t_{w1} = 20$ ° C. Under these conditions, the gross thermal efficiency of the boiler was $t_b > 95$ %, value not found in the case of other boilers.

In some of his research, Vuia was helped by Gavrilă Brola from Banat, with whom he began collaborating in 1934. G. Brola, together with professor G. Ribaud, continued the research started by Vuia, and, in 1952, they made a forced circulation generator, based on Vuia's principle. The generator, exhibited at the International Exhibition on Combustion in Paris, 1957, was manufactured in series in France, Belgium, England and Germany.

Traian Vuia, together with another collaborator, Emmanuel Yvonneau, patented several types of steam generators. The first patent was issued on January 21, 1928, by the French Ministry of Commerce and Industry (no. 661254), the second patent on December 22, 1928 (no. 680567) and the third on November 12, 1932 (no. 740226).

After Traian Vuia donated to the Romanian state the right to use his patents in 1950 in the Vulcan plant in Bucharest and starting with 1953 in the Energy Institute of the Academy, extensive research was carried out to determine the conditions of operation, reliability and endurance. Due to the difficulties encountered regarding both automation and the lack of consumers of a low flow of superheated steam at extremely high pressures, the principle of forced circulation was abandoned and boilers with multiple forced circulation were made instead. They had a circulation of 1 or 3 t/h saturated steam at p = 17 bar and were used in the food industry, the light one and not only. Hot water boilers of 2325 kW (2Gcal/h) were also realized which were used to equip numerous district heating micro plants (Figure 5).

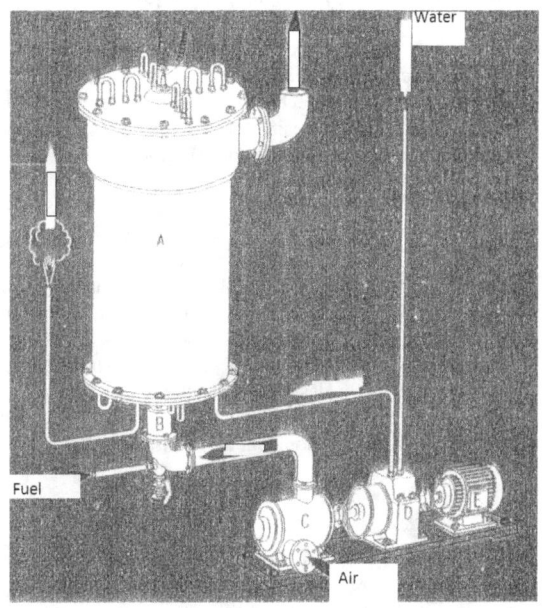

Figure 5. Vuia's boiler plant: A – steam generator, B – carburetor, C – blower, D – circulation pump, E – electric engine. [1], [3]

3. Conclusion: Vuia worked for the glory of humanity

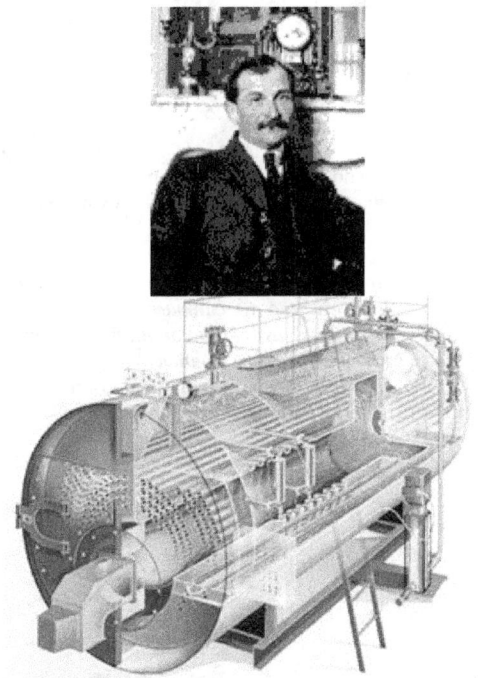

Figure 6. Vuia with the model of his invented plant,

"What does it matter who did this thing, it is important that it exists." Quote from Vuia [2], [3], [5]

Today, looking retrospectively, Vuia's steam generator seems a remarkable creation in the field, which opened new roads in techniques related to combustion, heat transfer and steam production. His humble personality remains in our memory (Figure 7).

The FLIGHT magazine, in the issue from March 30, London, 1956, on page 366 presented: "Vuia has been described by those who knew him as a very modest man. Indeed, he never made any other claims for *his own efforts that they had contributed to the firm establishment in 1906 - 1907 of powered flight as a practical proposition.* His inventiveness has been shown, and another legacy is the design of a steam generator. He was a very worthy pioneer, as much for his vision, as for his part in the earliest development of the airplane" [2].

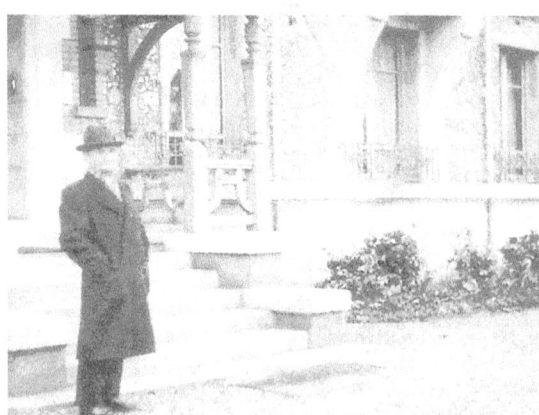

Figure 7. Vuia in front of his home, by 1939, Garches: "I do not work for my personal glory, but for the glory of human genius". Traian Vuia – own translation, [6], [7]

"How lucky would mankind be, if there were many nations to have given to it – compared to the number of inhabitants - as much as the Romanian nation gave to it in the last 120 years." Henri Coandă
"Your value lies in what you are and not in what you have" – (Thomas Edison).

May his memory be eternal (Figure 8)!

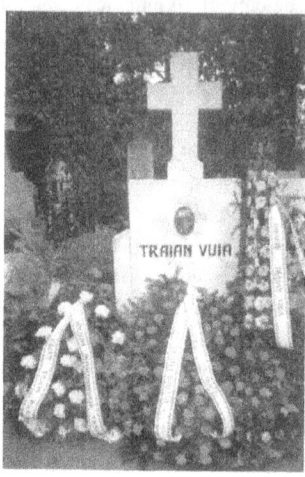

Figure 8. The grave og Traian Vuia in Bellu Cemetery, Bucharest

References

[1] M.-Ctin, Dianu, Aparate termice: cazane, Bucureşti, Editura Ministerului Administraţiei şi Internelor, 2009
[2] *** Dictionar de personalitati, www.euroavocatura.ro/dictionar/319652/Traian_Vuia, accessed March 2016
[3] *** Documentary films, "Conceived in Romania (Seria de filme documentare intitulată „Gândit în România")
[4] A. Metianu, Traian Vuia Steam generator with catalytic combustion (Generatorul de abur Traian Vuia cu ardere catalitica), Editura Tehnica, 1957
[5] J. Simmons, "100 cei mai mari savanţi ai lumii", Editura Lider, 1996
[6] D. Antoniu, I. Buiu, D. Haditca, R. Homescu, G. Cicos, Traian Vuia, Viaţa şi opera (monografie bilingvă, română şi engleză), Editura Anima Bucuresti, 2013
[7] N. Iorga, Oameni care au fost, editura pt literatura, 1967

Comparison on depollution capabilities of two iron based magnetic nanopowders

Cristian Pantilimon[1], **Andra Predescu**[1*], **Ruxandra Vidu**[2], **Ecaterina Matei**[1]
and **Cristian Predescu**[1]

[1] University Politehnica, Centre for Research and Eco-Metallurgical Expertise, 060042, Bucharest, Romania

[2] University of California Davis, Department of Chemical Engineering and Materials Science, Davis, CA 95621, USA

*e-mail: andra.predescu@ecomet.pub.ro

Abstract:

Removal of toxic metals from waste waters is one of the main concerns for environmental protection due to the high toxicity combined with their very long life span and slow biodegradability. Heavy metals such as Cr (VI) are very hazardous to the health of humans, flora and fauna because once they are absorbed into a living organism they are eliminated very slowly and become part of the food chain. The removal of such toxic materials has been tested through the use of magnetic nanoparticles in order to obtain a clean depollution of waste water. The comparison of two such materials, magnetite and maghemite was presented. The synthesis procedure for both materials was compared, their morphologies analyzed by XRD, TEM and SAED and their adsorption capabilities were evaluated at various pH values. The purpose of this study was to determine the differences in performance of the two magnetic nanopowders in terms of Cr (VI) removal from waste waters.

1 Introduction

The removal of toxic metals from waste waters has been an important subject for environmental protection due to their high toxicity to both natural and anthropic environments. Most of these metals are not biodegradable and because of their long lifespan and the possibility of accumulation in living organisms they may cause various diseases and disorders, becoming a risk to both the environment and human health [1]. For these reasons, the removal of metals such as Cd, Cr, Ni, and Cu has been researched in order to improve the water purification techniques.

Several removal methods have been documented such as adsorption, chemical precipitation, ion exchange, filtration, membrane separation and reverse osmosis [2, 3]. Due to the

developments in nanotechnology, nanomaterials have been tested for the removal of heavy metals from waters because of their high specific surface area and adsorption capabilities. Magnetic Fe oxide nanoparticles show promise as an efficient material for removing metal pollutants and a high possibility of re-use through magnetic separation with ion exchange [4]. The synthesis of such magnetic nanoparticles can be done through various methods including: co-precipitation, sol-gel synthesis, sonochemical reactions, hydrothermal reactions, flow injection synthesis and electrospray synthesis [4].

The adsorption properties of the materials are directly connected to the particle size and the specific surface area of the powders. A high specific surface area leads to an increase in adsorption.

In this study, a comparison is made between the adsorption properties of 2 magnetic nanomaterials, Magnetite (Fe_3O_4) and Maghemite (γ-Fe_2O_3), and their specific morphologies for a better understanding as to how the adsorption properties of the materials vary in various conditions. The heavy metal studied for removal is Cr (VI).

The adsorbents were obtained through coprecipitation and they were characterized by X-Ray Diffraction (XRD), Transmission Electron Microscopy (TEM) and Selected Area Electron Diffraction (SAED). After being submersed in waters polluted with Cd and Cr, the nanomaterials were washed and the resulting solutions were tested using a molecular absorption spectrometer and an atomic absorption spectrometer. The separation of the magnetic particles from the solution was performed using an external magnetic field.

2 Experimental Procedure

2.1 Synthesis of iron based nanoparticles

The synthesis method used to obtain the magnetite is the conventional coprecipitation method. Commercial reagents with pure analytical grade were selected as precursors. 0.4 mol/L $Fe(NO_3)_3 \times 9H_2O$ and 0.4 mol/L $FeCl_2 \times 9H_2O$ were mixed at a molar ratio of 1:2 in the presence of 0.5 mol/L NaOH in distilled water at room temperature. The pH of the solution was maintained at 10 for 3 h, after which the precipitate was separated by centrifugation and washed several times with distilled water until the pH becomes 7 [6]. In order to prevent agglomeration of the particles, D-sorbitol was added to the solution. [7]

After washing, the magnetite particles were dried in an oven at 60°C. Part of the magnetite was separated and used to obtain maghemite. The procedure involves heating the magnetite at 200°C for 3 hours. After this the precipitate presents a red-brown color, as opposed to completely black, which represents the maghemite formation [8].

2.2 Adsorbent characterization

The crystalline structure was analyzed by X-ray diffraction (XRD) at room temperature using a Panalytical X'PERT MPD equipped with a copper anode which generates Cu Kα radiation (λ = 1.54065 Å) with a 2θ scanning range of 10° to 90°. The particle size and distribution were determined by high resolution transmission electron microscopy (HRTEM). The analysis procedure involved mixing the samples in ethanol, ultrasonication and placement on carbon grids. The samples were analyzed with a TECNAI F30 G^2 transmission electron microscope. The iron based particles were also analyzed by selected area electron diffraction (SAED) [8].

2.3 Adsorption experiments

The Cr(VI) solution was prepared by dissolving K_2CrO_4 into ultrapure water. The pH of the solution was adjusted by using solutions of HCl (0.1 N) and NaOH (0.1 N). The adsorption studies were performed by measuring the initial and final concentrations of the metal with a GBC 932 AB Plus spectrometer (flame atomic absorption spectrometry) with spectral domain between 185 and 900nm. For the Cr (VI) solution analysis, a Cintra 202 GBC spectrophotometer, with spectral domain between 190 and 1000 nm was used. [6]

The removal efficiency was calculated based on the formula [9]:

$$\eta = 100 \times (C_0 - C_e)/C_0 \qquad (1)$$

where C_0 is the initial concentration (mg/L) and C_e is the equilibrium concentration (mg/L).

The adsorbed metal amount at equilibrium is expressed as q_c [4]:

$$q_c = (C_0 - C_e)V/m \qquad (2)$$

where V is the solution volume (L), and m is the adsorbent quantity (g).

The quantity of adsorbents used was 0.1g and the concentrations of metal ions in the synthetic solutions that were tested were selected as 40 and 50 mg/L [6].

3 Results and Discussion

3.1 Characterization of the iron based nanoparticles

The XRD patterns of the powders can be observed in Figure 1. The XRD analysis of the two magnetic nanopowders reveals that the patterns are quite similar. At a closer look, the main differences that can be observed are the variation in peak intensities and the positions of the peaks. In this case the maghemite shows lower intensity than the magnetite.

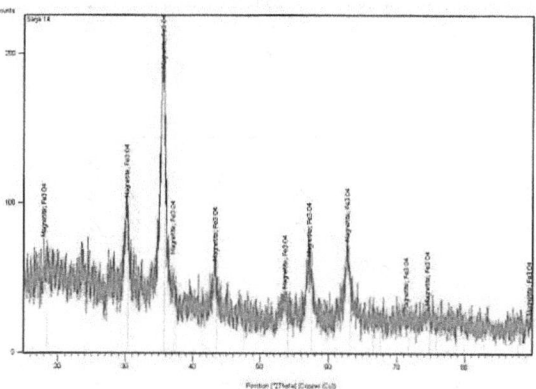

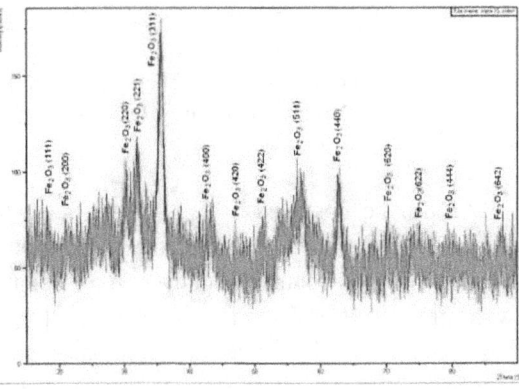

Figure 1. XRD patterns of synthesized magnetite (top) and maghemite (bottom)

The powder morphology was analyzed by transmission electron microscopy, combined with selected area electron diffraction, in order to evaluate the size and the shape of the particles, as well as the structure. Figure 2 shows TEM images of Fe_3O_4 and γ-Fe_2O_3. Both particles show high sphericity and appear to be nanosized.

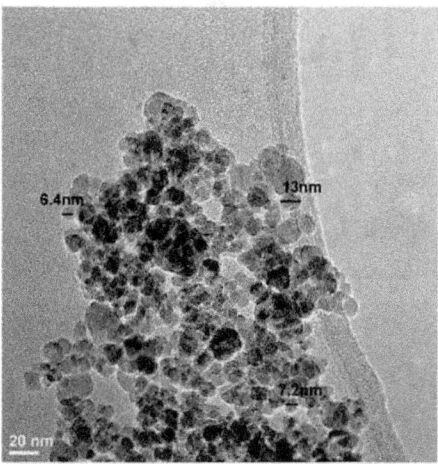

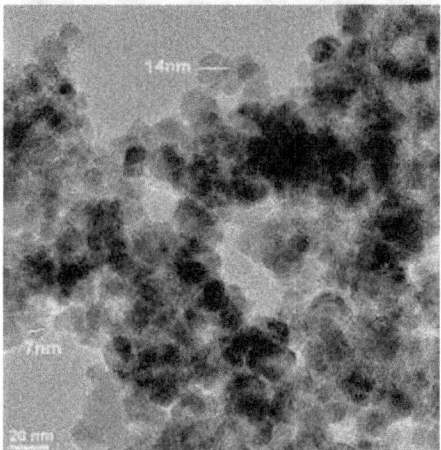

Fig. 2. TEM images of Fe_3O_4 (top) and γ-Fe_2O_3 (bottom)

The nano-scale characteristic of the particles is also demonstrated through the SAED analysis, which displays a ring pattern specific to nanoparticles. Due to the heat treatment applied to magnetite to obtain maghemite, the particles show a small increase in size compared to magnetite, but they maintained the nano-size feature. The SAED analysis associated with the TEM images are presented in Figure 3.

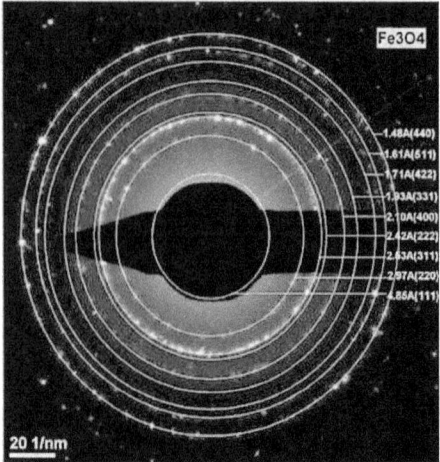

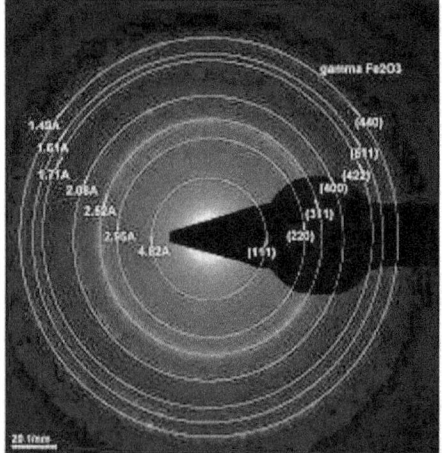

Fig. 3. SAED images of Fe₃O₄ (top) and γ-Fe₂O₃ (bottom)

combined with 50 mg/L and 40 mg/L solution of Cr (VI) respectively [7, 10].

The removal efficiency of the magnetic powders was calculated every 10 minutes for 100 minutes at various pH values and it was tested for various pH values [7, 10]. The results of this analysis is presented in Figure 4.

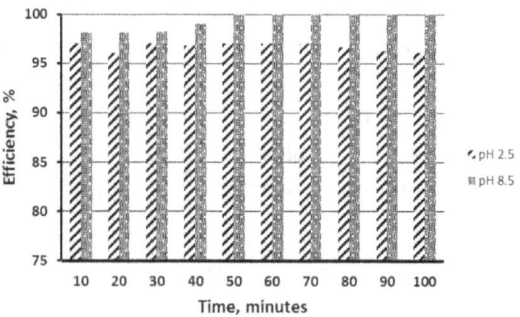

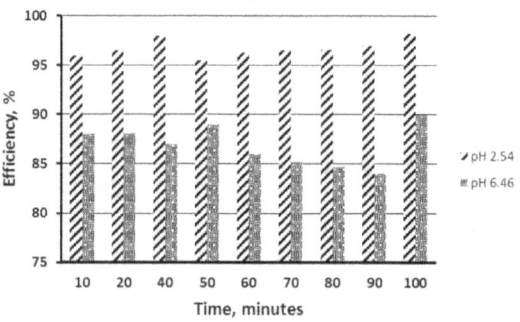

Fig. 4. Removal efficiency for Fe_3O_4 (top) and γ-Fe_2O_3 (bottom) at various pH values

3.2 Adsorption capabilities

The main purpose of this study is to compare the adsorption capacities of magnetite and maghemite particles, previously characterized, in order to obtain a better understanding of the influence of various factors that may have over the depollution properties of the magnetic powders.

The adsorption tests involve the use of 0.1g of each magnetic powder, Fe_3O_4 and γ-Fe_2O_3

As can be observed in Fig. 4, the magnetite displays a higher removal efficiency when used in a more basic environment, but even in the solution with pH 2.5 it shows a high capacity for Cr adsorption. In the case of the maghemite, increasing the pH of the solution to a neutral standpoint leads to the decrease of efficiency to under 90%. Comparing the results of the two solutions for adsorption in acid environment it can be seen that the maghemite displays a slightly higher removal efficiency than the magnetite, but the magnetite shows a high stability in removal.

The magnetite has also been tested for removal of Cd, Ni and Cu, under the same conditions as the Cr (VI), the results showing that the removal efficiency after 10 minutes of the magnetite in solutions with pH 2.5 follows the order Cr>Ni>Cd>Cu, while in the case of using a pH of 8.5 the order changes to Ni>Cr>Cu>Cd. [10].

In case of the amount adsorbed at equilibrium, q_e, the magnetite shows a value of 48.5 mg/g at pH 2.5 and 0.59 mg/g at pH 8.5 after NaOH precipitation [10]. The maghemite displays a q_e value of 19.16 mg/g at pH 2.54 and 17.85 mg/g at pH 6.46 after 10 minutes [7].

4 Conclusions

Both magnetite and maghemite display high removal efficiencies for the depollution of Cr from waste waters at various pH values. The usefulness of the iron based magnetic nanoparticles also comes from their small sizes combined with high specific surface areas and due to the possibility of recovery of the materials and reuse. In the case of magnetite, the pH variation did not have a high impact as in the case of the maghemite. Both magnetic nanopowders may be used for the removal of heavy metals from synthetic solutions.

References

1) C. Balan, C. Cojocaru, P. Bulai, D. Bilba, M. Macoveanu, *Environmental Engineering and Management Journal*, 8 (2009), 225.

2) C. Modrogan, D.G. Apostol, O.D. Butucea, A.R. Miron, C. Costache, R. Kouachi, *Environmental Engineering and Management Journal*, 12 (2013), 929.

3) A. Valipour, S.M. Taghvaei, V.K. Raman, G.B. Gholikandi, S. Jamshidi, N. Hamnabard, *Environmental Engineering and Management Journal*, 13 (2014), 145.

4) X. Wang, C. Zhao, P. Zhao, P. Dou, Y. Ding, P. Xu, *Bioresources Technology*, 100 (2009), 2301.

5) Y.F. Shen, J. Tang, Z.H. Nie, Y.D. Wang, Y. Ren, L. Zuo, *Separation and Purification Technology*, 68 (2009), 312.

6) E. Matei, A.M. Predescu, C. Predescu, M.G. Sohaciu, A. Berbecaru, C.I. Covaliu, *Journal of Environmental Quality*, 42 (2013), 129.

7) E. Matei, A.M. Predescu, A. Predescu, E. Vasile, *Environmental Engineering and Management Journal*, 10 (2011), 1711.

8) E. Matei, A. Predescu, E. Vasile, A. Predescu, *Journal of Physics: Conference Series*, 304 (2011), 012022.

9) M. Ozmen,, K. Can, G. Arslan, A. Tor, Y. Cengeloglu, and M. Ersoz, *Desalination*, 254 (2010), 162.

10) E. Matei, C. Predescu, A. Badanoiu, A. Predescu, D. Ficai, *Environmental Engineering and Management Journal*, 14 (2015), 1001.

Green Methods Used to Enhance Enzymes

Mihaela D. Leonida

Fairleigh Dickinson University, Teaneck, NJ, 07666, USA

mleonida@fdu.edu

Abstract: This project focuses on modifying properties of redox enzymes by using green chemistry. Working hypothesis: upon partially unfolding and subsequent refolding enzymes, new moieties can be embedded in their structure. These may confer additional/beneficial properties to the enzyme. Two methods were used to partially unfold the proteins: a) exposure to ionic liquids and b) application of high hydraulic pressure. The redox enzymes modified by exposure to an ionic liquid were lactate dehydrogenase (LDH), cholesterol oxidase (ChOx), and amine oxidase (AO). AO was also modified by transient exposure to high pressure. These are important oxidoreductases used in clinical laboratory, defense, sports medicine, the food industry. The kinetics of electron transfer is important when a redox enzyme is evaluated for an application because its redox centers are buried in insulating protein resulting in slow electron transfer. Since the rate of electron transfer in proteins decays exponentially with the distance donor-acceptor, we decreased the distances between redox centers within the enzyme by molecular alteration of its 3-D structure in the presence of modifiers. The choice of modifiers was tailored to the characteristics of each enzyme. Following the procedure, additional redox centers were entrapped within the protein structure thereby enhancing it. The enzymes were assayed before and after modification to assess the benefits of the procedures. All modified enzymes (ME) retained activity. The ME were tested as biosensing elements for analytical applications and performed well. The proposed methods are inexpensive, environmentally friendly, and enzyme friendly due to the species used as modifiers.

1. Introduction

Oxidoreductases are needed for analytical applications (in biosensors) and for chiral synthesis of compounds with high enantiomeric purity for the pharmaceutical industry. However the rate of electron transfer in the reactions catalyzed by them is very slow when enzyme electrodes are used. A solution to the problem is the use of mediators to enhance the rate of electron transfer. One successful strategy to achieve this is covalently-binding redox-active centers (the mediators) to sites on the enzyme. This approach is known in the chemical literature as enzyme "wiring" and it typically results in an important loss in enzyme activity.

Ionic liquids (IL) have gained interest lately due to their unique range of physical and chemical properties, notably their solvent capabilities, negligible vapor pressure and thermal stability [1]. Many commonly-used IL have excellent solvent properties for a wide range of organic, inorganic and organo-metallic compounds [2]. Their ability to dissolve proteins is of particular interest to bioorganic catalysis, IL replacing the traditional organic solvents in some applications [3]. Besides being environmentally unfriendly, organic solvents also affect adversely some suspended proteins [4]. In the present study we reversibly denatured redox enzymes by exposure to an ionic liquid 1-ethyl-3-methylimidazolium tetrafluoroborate (emim-BF$_4$). While the enzymes are partially unfolded with the redox centers exposed, they came in contact with species having electron transfer mediating capacity. Upon reversing the denaturation, small mediator molecules were trapped in the tertiary structure of the refolded enzymes which, consequently, become "wired" enzymes.

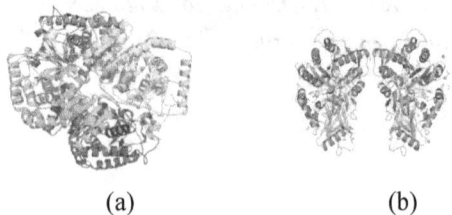

(a)   (b)

Figure 1 (a) LDH from rabbit muscle; (b) ChOx from *Brevibacterium sp.*

In the present study lactate dehydrogenase (LDH, Fig. 1a) and cholesterol oxidase (ChOx, Fig. 1b) were modified by transient exposure to emim-BF₄ with and without flavin adenine dinucleotide (FAD) - the prosthetic group of LDH and ChOx - present. After removal of the IL by dialysis, the activity of the renatured enzymes was assayed spectrophotometrically. Previous studies have shown the properties of IL to dramatically change in the presence of small quantities of water [5]. Thus, a parallel procedure was conducted in the presence of water and the differences in recovered activities were evaluated in terms of structural changes in the modified enzymes (ME).

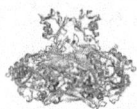

Figure 2. Amine oxidase from *E. coli*

Amine oxidase (AO, Fig.2) is an important oxidoreductase used in analytical applications for clinical laboratory and for the food industry (especially for fish and seafood products). Depending on the natural source, some AO have FAD as a prosthetic group. Some others belong to the copper-enzymes family and are disulfide-linked homodimers. They base their redox properties on a copper (II) ion located near the active site (close to the center of the homodimer) which is coordinated by three histidine residues. Also involved in the redox action of AO is a tyrosine residue that has been modified into topa-quinone (TPQ), which is also a cofactor in copper amine oxidases. As modifiers used during high pressure exposure we investigated several directions due to the versatility of this enzyme. In one experiment we used Cu^{2+} ions and in a parallel one pyridoxal phosphate (PLP, closely related to TPQ, but less expensive), targeting enhanced stability and improved kinetics of electron transfer. Since there are AO species having FAD as cofactor, another experiment used FAD as modifier. Based on new literature mentioning copper enzymes as potential cardioprotective agents for heart patients, a modification of AO was conducted in the presence of α–lipoic acid (LA), agent with strong antioxidant properties and reliable electrochemical activity. This experiment targeted both enhanced kinetics and potential for therapeutic use.

AO was "wired" using another green method as well, reagentless (mechanical) reversible modification of enzyme tertiary structure using high hydraulic pressure. Like the first method (using IL) this environmentally-friendly procedure targeted several AO properties: a) increased stability; b) enhanced kinetics of electron transfer; c) enhancement of antioxidant action. After the modification procedures all ME were lyophilized and stored at -20 °C.

2. Results and discussion

LDH and ChOx were modified in parallel procedures: using different enzyme:IL ratios, with/without FAD present, with/without additional water present. The ME were assayed after overnight dialysis. All modified LDH and ChOx retained activity following the denaturation-renaturation procedures. The results are summarized in Table 1 for LDH and in Table 2 for ChOx.

Table 1. Enzymatic activityof LDH (kU) before and after exposure to emim-BF₄ a) no water added; b) with 5% water added to modifying mixture

a)"Wiring" LDH without water present

Ratio	no water added			
LDH/IL	with FAD		no FAD	
kU/mg	initial	final	initial	final
5.00	0.0692	0.0996	0.0763	0.1465
6.67	0.0377	0.2236	0.012	0.2722
10.00	0.1131	0.2625	0.0151	0.0008

b)"Wiring" LDH in the presence of 5% water

Ratio	with 5% w/v water added			
LDH/IL	with FAD		no FAD	
kU/mg	initial	final	initial	final
5.00	0.1247	0.3163	0.1093	0.3401
6.67	0.1089	1.0298	0.1363	0.6576
10.00	0.1052	1.4534	0.1391	0.2977

c)Enhancement (factors) due to LDH "wiring"

Ratio	no water added		with 5% w/v water added	
LDH/IL	with FAD	no FAD	with FAD	no FAD
kU/mg	Final/ initial	Final/ initial	Final/ initial	Final/ initial
5.00	1.44	1.92	2.54	3.11
6.67	5.93	22.68	9.46	4.82
10.00	2.32	(0.05)	13.82	2.14

The enzyme changes structure upon exposure to IL and then again, upon removal of IL. The transient exposure to IL triggers enzyme denaturation and the removal of the IL by dialysis induces a second alteration, resulting in a refolded enzyme. Lozano *et al.* [6] proposed a molecular mechanism of enzyme stabilization in ionic liquids and attributed stability and increased activity to the preservation of essential water in the enzyme's microenvironment. This study lends support to the concept of an altered enzyme structure formed upon suspension in the IL and a second change upon removal of IL in which the tertiary structure is partially restored. The increase in catalytic activity can be attributed to the combination of two mechanisms: a) an altered 3-D structure with a more favorable spatial arrangement, formed upon removal of IL, with increased exposure and interaction of an increased number of prosthetic groups with the substrate. The increase in activity in all samples supports the concept of residual alteration of LDH into a non-native structure, the increased activity of which is independent of the effects of molecular "wiring". Such alteration can result from interaction of the enzyme with the anion (BF_4^-) [7], which is a chaotrope that affects the structural water of the enzyme [8].

Also at work is the chaotropic cation $emim^+$, and this would explain the appearance of an optimum LDH/IL ratio for enzyme activity, as result of the competition between the two groups for structural water. The second possible mechanism is: b) the entrapment of FAD molecules within the structure of LDH/ChOx affords a molecular "wire" to the enzyme. Entrapping the FAD within the LDH/ChOx requires noncovalent modification of the enzyme structure. Possibly, a more open configuration is achieved during suspension in IL, by virtue of the hydrogen-bonding properties of the anion under investigation (BF_4^-) [6]. Stronger hydrogen bonding anions might have resulted in irreversible loss of catalytic activity due to a permanently bound anion to the peptide backbone of certain enzymes. Upon removal of the IL, a more compact, active, native-like configuration is achieved in which the FAD remaining in the ME is entrapped within the enzyme and enhances the kinetics of the electron transfer which translates into a higher activity for LDH. The more modest enhancement afforded by FAD as a molecular "wire" within ChOx was correlated with the status of FAD present in the native enzymes: noncovalently bound in LDH and both non-covalently and covalently (less stable) bound in ChOx.

Table 2. Characterization of the modified ChOx

Ratio mg ChOx to ml IL	FAD (mg)	Activity ME (% initial)	Increase in activity due to FAD (%)	Increase in peak current (%)
3.8:0.25 No water	0	25.3	-	-
3.8:0.25 No water	0.89	45.8	81.53	74.70
3.8:0.25 No water	0	22.5	-	-
3.8:0.25 No water	1.65	28.9	28.04	33.35
3.8:0.25 5% water	0	35.8	-	-
3.8:0.25 5% water	0.89	46.2	29.11	34.55
3.8:0.25 5% water	0	43.5	-	-
3.8:0.25 5% water	1.65	78.0	79.35	144.86

The lactate biosensors built with ME-LDH showed higher catalytic effects than the non-wired one (using FAD as mediator added to the solution) and linearity in lactate concentration (Figure 3 a and b). The biosensor built with the ME-ChOx showed catalytic effect (Table 2) and linearity in cholesterol concentration as well.

(a)

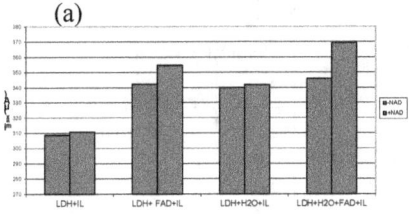

(b)

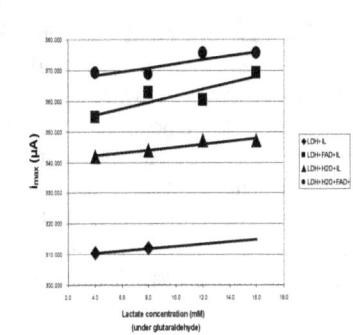

Figure 3. ME-LDH as biosensing element: a) catalytic effect; b) linearity in lactate concentration.

Table 3. Enhancement and stabilization of ME-AO

ME-AO modifier	Activity Dec. 2014 (U)	Activity Sept. 2015 (U)	Residual activity (%)
Native	-	-	7.56
FAD	$5.68*10^{-5}$	$3.90*10^{-5}$	68.7
PLP	$1.78*10^{-4}$	$1,87*10^{-5}$	10.5
CuSO$_4$	$2.08*10^{-4}$	$8.89*10^{-6}$	4.27
LA	$4.44*10^{-5}$	$7.55*10^{-5}$	170.3
FAD+ CuSO$_4$	$2.13*10^{-4}$	$3.27*10^{-4}$	153.3
FAD+LA	$8.88*10^{-5}$	$1.19*10^{-4}$	133.8
PLP+ CuSO$_4$	$1.47*10^{-4}$	$1.78*10^{-5}$	12.1
PLP+LA	$1.51*10^{-4}$	$2.25*10^{-6}$	1.5

Following modification in the presence of one modifier or a binary combination thereof, all ME-AO retained activity. The results are presented

in Table 3 together with the remarkable values of the activity and residual activity after 9 months. The most enhancement was afforded by the modification done in the presence of PLP+CuSO$_4$. Taking into account that a biosensor based on ME-AO modified with PLP + CuSO$_4$ showed good catalytic effect and linearity as a function of benzylamine concentration, this enzyme-friendly procedure for AO "wiring" shows promise for analytical applications.

Table 4. Activity of AO modified at 325 MPa

Modifier/ Modification time	Activities (U)	Adjusted Activity* (U)	Residual Activity (vs. Native) %	Residual Activity (Modifier Effect) %
Native	$5.033*10^{-5}$	$1.678*10^{-4}$	-	-
Native /1 h	$2.862*10^{-5}$	$9.540*10^{-5}$	56.87	-
FAD / 1h	$2.466*10^{-5}$	$8.220*10^{-5}$	48.99	86.16
PLP / 1 h	$2.345*10^{-4}$	$7.817*10^{-4}$	465.9	819.4
CuSO$_4$ / 1 h	$1.890*10^{-4}$	$6.300*10^{-4}$	375.6	660.5
LA / 1 h	$1.292*10^{-4}$	$4.307*10^{-4}$	256.8	451.6
Native / 30 min	$5.404*10^{-5}$	$1.801*10^{-4}$	107.4	-
FAD / 30 min	$1.087*10^{-4}$	$3.623*10^{-4}$	216.0	201.2
PLP / 30 min	$3.944*10^{-4}$	$1.315*10^{-3}$	783.6	729.8
CuSO$_4$/ 30 min	$7.952*10^{-4}$	$2.651*10^{-3}$	1580.0	1471.4
LA / 30 min	$2.537*10^{-4}$	$8.457*10^{-4}$	504.1	469.4

*Compared to the native AO after compression followed by lyophilization.

Using another environmentally- and enzyme-friendly approach, AO was modified through a reagentless (mechanical) procedure where high hydraulic pressure (325 MPa) was applied transiently with/without modifier present. Two parallel modifications were conducted using 2 different compression times, 1 h and 30 min, respectively. The same single modifiers were used (FAD/PLP/ CuSO$_4$ /LA). The results are presented in Table 4. The relative activities were calculated in relation with the native enzyme and also in relation to the native enzyme subjected to high pressure and then lyophilized (adjusted activity, since 70% of activity is lost following lyophilization). All ME (with one exception) showed remarkable residual activities, higher than that of the native enzyme processed at high pressure followed by lyophilization. Similarly to the "wiring" done in the presence of IL, the highest enhancement was afforded by PLP and $CuSO_4$, species intrinsic to AO. Shorter times used to apply pressure resulted in more significant enhancement.

ME-AO were tested (preliminary) for antioxidant action as well. All showed significant effect. The modification done in the presence of LA did not afford the highest effect, as expected Table 5 displays the antioxidant activities of the ME-AO expressed as antioxidant effect equivalent to that of a certain concentration of a Trolox solution (water soluble analog of vitamin E.

Table 5. Antioxidant activity of ME-AO

Modifier / Modification time	Antioxidant activity (mM Trolox)
No modifier / 1 h	1.416
FAD / 1h	1.935
PLP / 1 h	1.221
$CuSO_4$ / 1h	1.188
LA / 1 h	0.117
No modifier / 30 min	2.455
FAD / 30 min	1.513
PLP / 30 min	1.675
$CuSO_4$ / 30 min	0.831
LA / 30 min	1.123

Table 6 shows a comparison between the two green procedures for oxidoreductase enhancement proposed herein. While both afforded enhancement to AO, the modification conducted by applying high hydraulic pressure for shorter time (30 min) seems better if expense vs. resulting enhancement is considered. The use of binary combinations of modifiers compared to single modifiers results in less enhancement in ME-AO compared to benefit of using 30 min vs. 1 h exposure to high hydraulic pressure in the presence of one modifier.

Table 6. Comparison between modification methods for AO: using exposure to IL (single/binary modifier) vs. application of high pressure for 1 h/30min.

AO Modifier	Residual Activity % (IL)	Residual Activity % (1 hour)	Residual Activity % (30 min)	Residual Activity % (+ Cu^{2+})*	Residual Activity % (+ LA)*
FAD	54.96	48.99	216.0	206.1	85.88
PLP	171.8	465.9	783.6	142.2	145.7

*Modification using IL exposure

The methods presented for oxidoreductase enhancement resulted in molecularly modified enzymes with strong self-mediating capability. In none of the biosensors studied was ascorbate interference observed. The fact that the procedures are environment-friendly and the level of enhancement they afforded (activity and stability) recommends them for use in analytical applications. The use of high hydraulic pressure for enzyme "wiring" is a very inexpensive, green procedure highly recommended by these preliminary data.

Acknowledgment

Thanks are due to Prof. Catherine Royer from RPI for allowing access to the hydraulic press.

References

1) Welton, T., *Chem. Rev.* (1999) 99, 2071.
2) Poole, C.F., *J. Chromatography* (2004) 1037, 49.
3) DiCarlo et al, *Bioelectrochem* (2006) 68,134.
4) Gorman, L.S., Dordick J.S., *Biotechnol. Bioeng.* (1992) 39, 392.
5) Fujita, K. et al, *Biotechnol. Bioeng.* (2006) 94, 1209.
6) Lozano, P. et al.*Biocat Biotransf* (2005) 23, 169.
7) Khmelnitsky, L et al., *Eur. J. Biochem.,*(1991) 198, 31.
8) Zhao, H., *J. Chem. Technol. Biotechnol.,* (2006) 81, 877.

BIOMIMETIC CALCIUM PHOSPHATE COATING OF CoCr ALLOYS

Mariana Prodana[1], Florentina Golgovici[1], Andrada Negru[2], Marius Enachescu[1]*

[1]Center for Surface Science and Nanotechnology,
University "Politehnica" of Bucharest, 060042, Bucharest, Romania
[2]University of Pitesti, 110040, Pitesti, Romania
*e-mail: marius.enachescu@upb.ro

Abstract. CoCr aloys are most common material used as implantable material because of it reduced cost and good mechanical properties. This work presents our results of obtaining biomimetic calcium phosphate coating of CoCr and to characterize the properties of such coatings. The coating was prepared by immersing the CoCr substrates into the simulated body fluid (SBF) containing Ca^{2+} ions in sealed plastic bottles, kept at room temperature for one, fourteen and twenty one days. Detailed characterization including chemical, structural and morphological characterization (SEM, EDS, X-ray diffraction) were perfomed. ICP/MS (inductively coupled plasma mass spectrometer) determinations sustain chemical results put in evidence by Fourier Transformed Infrared Spectroscopy (FTIR), the ions release being much smaller for phosphate coatings formed after longer immersion time in SBF. The hydrophilic/hydrophobic character of the coatings was put in evidence by contact angle measurements (CA).

1. Introduction

Recently, there is a great interest in developing good coatings for implantable alloys that possess low cost and good mechanical properties. Osseointegration of metallic implants in bone is crucial for successful long-term treatment with ceramic prostheses in hip and knee arthroplasty. Cobalt chromium based alloys have remarkable mechanical and tribological properties being considered as the material of choice by some authors [1-3].

Different routes that simulate the bone healing of alloys implants have been evaluated in the last periods of times. Coatings such as hydroxyapatite (HA/ $Ca_{10}(PO_4)_6(OH)_2$) [4,5] has been used as bioactive coating due to the presence of Ca and P elements existing in the inorganic part of the bone [6-9].

Coating metallic implant as hydroxyapatite (HA) or other calcium phosphates with mimetic features of natural bone may reduce metallic ion release acting as a barrier against corrosion in biological media and accelerate the bone formation on the initial stage of osseointegration [10-14]. It is known that hydrophilic/hydrophobic balance may affect the bio performance results as well.

2. Material and methods
2.1. Materials

In the present study, CoCr alloy samples with the chemical composition given in Table 1 were used. Rectangular specimens $10 \times 10 \times 1$ mm in size were cut from a CoCr plate. The pretreatment procedure was that specimens were mechanically polished, cleaned by HF/HNO_3 solution, rinsed with ethanol and deionised water, and subsequently air-dried. The group numbers of the specimens were assigned to the different processing conditions as flows: S1 –CoCr alloy immersed for one day in SBF as a control; S2- CoCr alloy immersed for fourteen days in SBF and S3- CoCr alloy immersed for twenty one days in SBF.

The chemical composition of SBF is as following: NaCl 8,367255g; $NaHCO_3$ 5,534g; $Na_2HPO_4.2H_2O$ 22,165g; $MgCl_2.6H_2O$ 0,51359g; $CaCl_2.2H_2O$ 0,51891g; Na_2SO_4 0,071 for 1 liter of deionized water;

Table 1. Chemical composition of cobalt–chromium alloy

Composition	Percentage amount
Chromium	29.00–31.00
Molybdenum	4.50–6.00
Silicon	0.70–1.30
Manganese	0.50–1.00
Carbon	0.40–0.50
Cobalt	60.0–64.5
Iron, nickel, nitrogen	Traces

2.2. Methods

SEM and EDAX analysis: The microstructure of the hybrid ceramic material was studied by scanning electron microscopy (SEM) using a Hitachi SU8230 Microscope equipped with EDX (Energy Dispersive X ray).

Spectroscopic measurements: the chemical bonds of new hybrid composite material were studied by FTIR which were performed with ATR Perkin-Elmer equipment.

CA investigations: The hydrophilic/hydrophobic balance was evaluated with the contact angle meter CAM 100.

ICP-MS determination: For Co^{2+} and Cr^{2+} ions release determination, an ELAN DRC-e inductively plasma mass spectrometer was used (ICP-MS). All samples (typically: 0.1–1.2 mg) were digested in 100 mL concentrated nitric acid (ULTRAPURE, Fa. Merck). Acid digestion was performed in a well determined volume of HNO_3 65%; after digestion, the samples were diluted 100 times and liquid fractions were analyzed.

XRD analysis: The composition and structure of the films formed on Co-Cr alloys were studied by X-ray diffraction (XRD) technique utilizing a SmartLab X-ray diffractometer (Rigaku, Japan) at 45 kV voltage, 200 mA, with a Cu target; the scanning rate was 5°/min with a step size of 0.01°.

3. Results and discussion
3.1. Spectroscopic analysis and morphology characterization of CoCr alloys

Surface morphology, structural and elemental analysis of coating, composition of calcium phosphate formed on the CoCr alloys were studied using the scanning electron microscope (SEM) described above, i.e., Hitachi SU8230 equipped with EDX at 10kV.

From figure 1 we can see the surface of CoCr alloy with some linear scratches that are specific for metallic sample because of the direction of polishing. On the metallic sample we can see some deposit of NaCl, fact that is confirmed by EDX analysis.

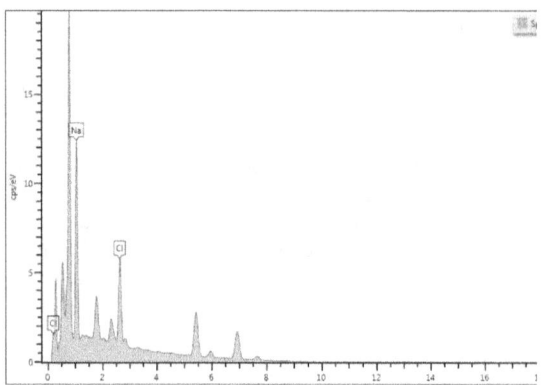

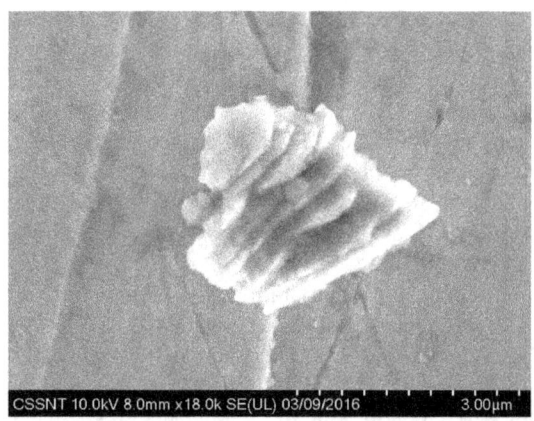

Figure 1. EDXS (top) and SEM (bottom) morphologies of Sample S1

In figure 2 we obtain on the surface of sample S2 morphology like a dandelion flower. From elemental analysis the Ca/P ratio is almost 1, that is specific for brushite ($CaHPO_4·2H_2O$) which is a precursor of hydroxyapatite in bones and teeth, with an important role in bones mineralization. Peaks specific for metallic support appears in the spectra because the layer of brushite is not very dense and homogenous.

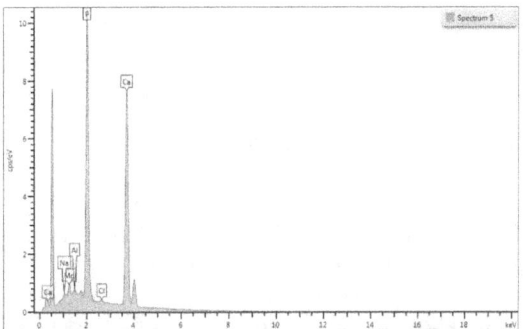

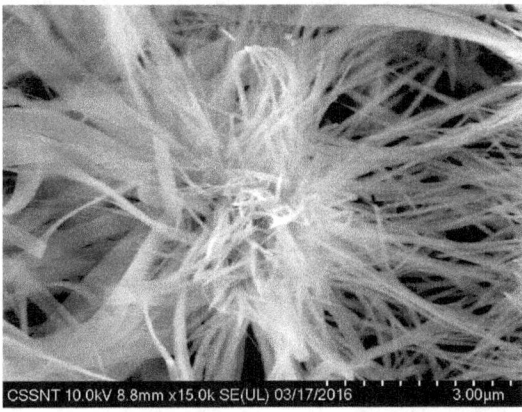

Figure 2. EDX (top) and SEM (bottom) morphologies of Sample S2

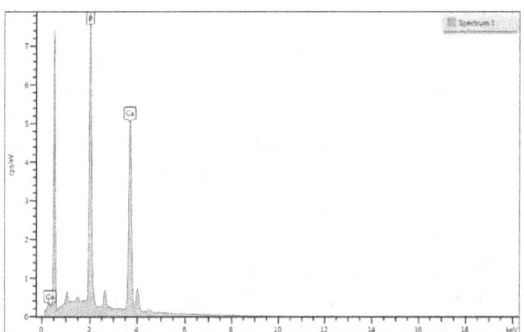

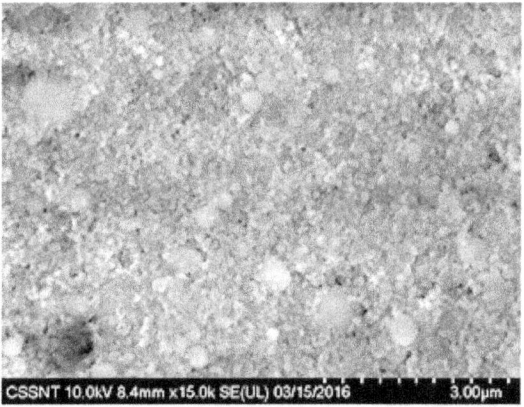

Figure 3. EDX (top) and SEM (bottom) morphologies of hydroxyapatite-sample S3

In figure 3 we observe the morphology of hydroxyapatite. We obtained a quantitative results with a Ca/P ratio of 1.67, fact that is confirmed from X-ray diffraction analysis.

The composition and crystalline structure of the films were studied by X-ray diffraction (XRD) technique. The XRD pattern for films deposited on CoCr alloys immersed in SBF solution at room temperature for 21 days are shown in figure 4. The patterns of the film formed on metallic surface indicate the presence of hydroxyapatite film.

The major peaks in figure 4 were observed at 2θ values of 31.77°, 32.19° and 32.91° corresponding to (-161), (-222) and (-360) planes of hydroxyapatite phase (according to ICDD File 00-076-0694). The XRD patterns of our samples is in good agreement with the literature data.

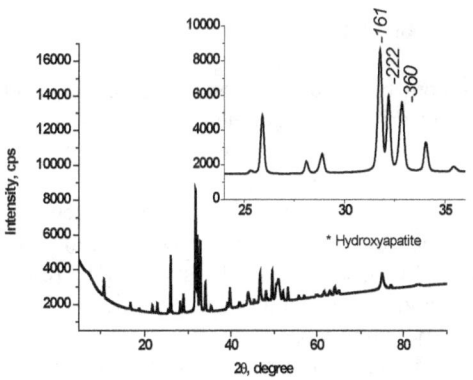

Figure 4. X ray diffraction spectra for sample S3

3.2 FTIR measurements
Sample S3 was analyzed using FTIR and infrared Spectral data.

In the literature, FTIR spectrum of HA coating present peaks at 3453 cm^{-1} and 1640 cm^{-1}. These peaks were identified that are due to the stretching and bending modes of absorbed water [15].

The stretching and bending vibration modes of hydroxyl group were identified at 3270 cm^{-1} and 634.23 cm^{-1}, respectively. PO_4^{3-} stretching and bending modes were seen at 560.13 cm^{-1}, 960 cm^{-1}, 042.71 cm^{-1}, 1099.33 cm^{-1} [16].

This presence confirms the formation of a well crystallized apatite structure. The carbonate band was observed at 1640 cm^{-1}. The peak observed at 2354 cm^{-1} could be assigned to adsorbed carbon dioxide [17].

In figure 5 is present our obtained spectra for CoCr/HA-sample S3.

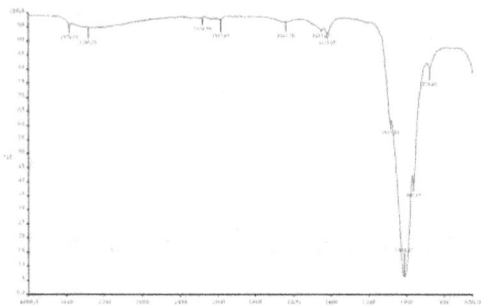

Figure 5. FTIR spectra of sample S3

3.3 CA investigations

The contact angle is a good indicator of the state of wettability of sample surfaces. An equal volume of distilled water was placed on every sample by means of a micropipette, forming a drop or spreading on the surface. After depositing HA, the corresponding surfaces had remarkably reduced contact angles. The sample S1 present a hydrophobic behaviour while sample S2 and S3 has a hydrophilic one.

The hidrophilicity is in a direct relation with the biocompatibility; the smallest contact angle the better biocompatibility.

The values obtained for every sample are summarized in Table 2.

Table 2. Contact angle for samples S1-S3

Sample	Contact angle (degree)
S1	93.24
S2	70.61
S3	56.05

3.4 ICP-MS determination

ICP-MS were performed in NaCl 9%. Samples S1-S3 were immersed in this solution for 24 hours. From ICP-MS data (figure 6) we can observe that after 14 and 21 days of immersion of CoCr alloy in SBF, the quantity of Co^{2+} and Cr^{2+} ions measured after 24 hours of immersion in NaCl 9% decrease compared with the same sample after one day of immersion in NaCl 9%, fact that put in evidence that the hydroxyapatite is stable and is a protective layer against corrosion in human body.

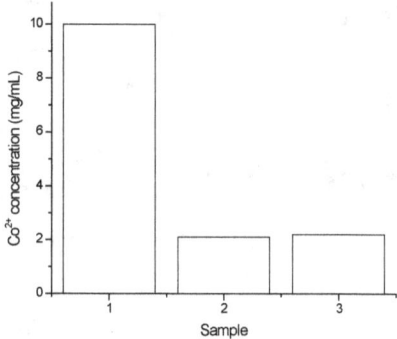

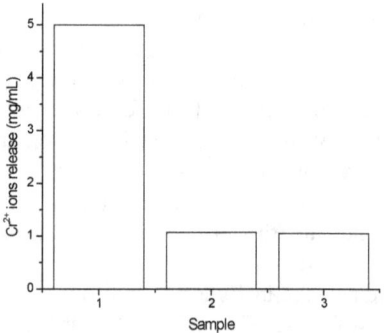

Figure 6. Co^{2+} (top) and Cr^{2+} (bottom) concentrations for samples S1-S3

4. Conclusions

The increase of immersion time of CoCr alloys in SBF lead to a more stable coating. After immersion in SBF after 14 days we obtained a precursor of hydroxyapatite and after 21 days of immersion in SBF we obtained hydroxyapatite with very good results. More investigations are needed to better control the composition of hydroxyapatite precursors deposited on CoCr alloys in order to use them for a particular application in the human body.

Acknowledgements

This work was supported by Romanian Ministry of Education and Scientific Research, as well as by Executive Agency for Higher Education, Research, Development and Innovation Funding, under projects PCCA 2- nr. 66/2014 and PCCA 2-nr. 166/2012.

References

1) A. Chiba, K. Kumagai, N. Nomura, S. Miyakawa, *Acta Mater.* 55 (2007) 1309.

2) N.G. Sotereanos, C.A. Engh, A.H. Glassman, G.E. Macalino, C.A. Engh Jr., *Clin. Orthop. Relat. Res.* 313 (1995) 146.

3) Carl Lindahl, Wei Xiaa, Håkan Engqvist, Anders Snis, Jukka Lausmaa, Anders Palmquist, *Applied Surface Science* 353 (2015) 40.

4) F.H. Lin, Y.S. Hsu, S.H. Lin, J.S. Sun, *Biomaterials* 23 (2002) 4029.

5) X. Liu, P. Chu, C.X. Ding, *Mater. Sci. Eng. Rep.* 47 (2004) 49.

6) L.L. Hench, *J. Am. Ceram. Soc.* 74 (1991) 1487.

7).R.H. Doremus, *J. Mater. Sci.* 27 (1992) 285.

8) T.J. Webster, E.A. Massa-Schlueter, J.L. Smith and E.B. Slamovich, *Biomaterials* 25 (2004) 2111.

9) T.J. Webster, C. Ergun, R.H. Doremus, R.W. Siegel and R. Bizios, *Biomaterials* 21 (2000) 1803.

10) F.Barrčre, CM van der Valk, R.A.J Dalmeijer, G. Meijer, CA van Blitterswijk, K de Groot, P.Layrolle, *J. Biomed. Mater. Res. A* 66 (2003) , 779.

11) F.Barrčre, CM van der Valk, R.A.J Dalmeijer, K de Groot, P.Layrolle, *J.Biomed. Mater. Res. B* 67 (2003) 655.

12) E Aldea, N. Badea, I Demetrescu, *Rev. Chimie* 58, (9) (2007) 918.

13) E. Boanini, P. Torricelli, M. Gazzano, *Biomaterials* 27 (2006) 4428.

14) J. Forsgren, F. Svahn, *Acta Biomaterialia* 3 (2007) 980.

15) H. Najafi, Z.A. Nemati, Z. Sadeghian, *Ceram. Int.* 35 (2009) 2987.

16) C. Kaya, I. Singh, A.R. Boccaccini, *Adv. Eng. Mater.* 10 (2008) 131.

17) Q. Chena, H. Liua, Y. Xinb, X. Chenga, *Electrochim. Acta* 111 (2013) 284.

Adhesive Properties Study of f-SWCNTs:P3OT Nanocomposite Thin Film

J. Alzanganawee[1,2], **O. Brincoveanu**[1,3], **R. Mesterca**[1], **D. Balan**[1,3],
A. Apaz[1,3], **M. Enachescu**[1,4]

[1] *Center for Surface Science and Nanotechnology (CSSNT), University Politehnica of Bucharest, Romania*
[2] *Physics Department, College of Science, University of Diyala, Iraq*
[3] *University of Bucharest, Faculty of Physics, Bucharest, Romania*
[4] *Academy of Romanian Scientists, Bucharest, Romania*
marius.enachescu@upb.ro

Abstract: Topography and 'pull off' curves are two of the most important among the many available AFM modes, which are enabling the morphological and mechanical properties investigations of the surfaces. It is known that nanocomposites synthesizing has attracted the worldwide scientists' interest, due to the enlarged properties of these materials. There are few studies on P3OT:SWCNTs nanocomposites used in OPVs when compared to fullerene derivatives and even fewer focused on mechanical properties, as the main research direction was to increase the efficiency. In this paper, preparation and characterization of the P3OT pure polymer and 12% f-SWCNTs:P3OT nanocomposite, respectively, were reported. The experimental results showed the noticeable effect of the carbon nanotubes addition into the polymer matrix. The surface roughness of the nanocomposite increased due to the presence of the f-SWCNTs, while the adhesion force values decreased. The adhesion force variation distribution becomes more uniform, with the carbon nanotubes addition into the host.

Keywords: topography, adhesion, P3OT, f-SWCNTs, nanocomposite

1 Introduction

With the development of the Atomic Force Microscopy (AFM) besides surface morphological analysis the investigations of various properties over different material surfaces at the nanoscale level, such as elastic and adhesion properties, became easily achievable [1].

Adhesion forces can describe the interconnection at interfaces between single or multi-layer systems, and it may be either due to chemical bonding or mechanical interlocking interactions [2]. It is a known fact that the mechanism behind adhesion differs from macroscopic to microscopic level [3]. At molecular level there is no need for various adhesives or joints to bond two components, rather the objects spontaneously attach to each other and they are taking apart with difficulty. This type of adhesion is called "pre-mature adhesion" and the bond between two objects cannot be broken without the use of force or through some kind of surface contamination [3].

In this paper we report the study of topography, morphology and the adhesive properties on P3OT and 12 wt. % f-SWCNTs – P3OT thin film samples, respectively. In the organic optoelectronic industry, there is an underlying focus on poly(3-octylthiophene-2,5-diyl) (P3OT) and poly(3-hexylthiophene-2,5-diyl) (P3HT) π-conjugated polymers [4]. A special interest has been given to conjugated polymer:fullerene derivatives as the active layer in OPVs in contrast to polymer:SWCNTs hybrid structures. The main focus within the research directed towards OPVs has been on the enhancing the solar cell's efficiency to enable their large scale use, while studies on mechanical properties, such as adhesion and cohesion are scarce, even though they play a vital role in device reliability and stability [5,6,7]. The nanocomposite thin film consisting of the polymer reinforced with carbon nanotubes exhibits new mechanical and morphological properties [8,9]. In the analysed samples, SWCNTs in the polymer matrix could be observed in the acquired topography image and were found to have an influence on both the uniformity of distribution and adhesion force values.

2 Experimental

2.1 Sample preparation

The materials used for sample preparation were commercially available P3OT and SWCNTs. P3OT polymer was used as-purchased from Sigma Aldrich and did not require further processing. SWCNTs were acquired from NanoIntegris Inc, with a purity of approximately 70%, a diameter ranging from 0.9 to 1.7 nm and lengths in the 0.3 - 4 um range. The nanotubes were further processed using a functionalization procedure with 6M nitric acid solution to yield f-SWCNTs.

A three-step procedure was used to obtain the nanocomposites mixture, using P3OT as the polymer matrix and f-SWCNTs (12 wt.%) as the reinforcing phase. Firstly, the nanotubes were dispersed into chloroform solvent at room temperature for 30 minutes with the aid of a high ultrasonic tip, with an energy about 15000 Joule to ensure a better incorporation of the filler into the matrix in the next step. Afterwards, the resulted dispersion was blended with the polymer and finally the nanocomposite solution was sonicated for 1h at 45-50°C to increase the homogeneity of the mixture.

A pre-deposition cleaning of the glass substrates was performed with acetone, isopropyl alcohol and deionized water for 15 minutes using an ultrasound bath. Two thin films were spin-coated for 30 sec, at 1000 rpm, in ambient conditions on the cleaned substrate using the as-purchased P3OT solution and the nanocomposites mixture.

2.2 Sample characterization

Surface topography and adhesion properties of the prepared thin films were investigated using a multi-mode commercial atomic force microscope (NTEGRA – NTMDT), in ambient conditions. For these measurements, carried out in AFM contact mode, commercially available standard cantilever chips with a 0.26 N/m stiffness and monocrystalline silicon tips with radius of approximately 10 nm were used.

The surface roughness and Skewness parameters were determined from the acquired topographic images using the equations (1) and (2), with the help of an image processing software.

$$\mathbf{Ra} = \sum_{i=0}^{N} \frac{h_i}{N}. \tag{1}$$

$$\mathbf{R_{Skew}} = \frac{1}{NR_q^3}\sum_{i=1}^{N}\left(h_i - \bar{h}\right)^3 \tag{2}$$

The adhesion properties measurements, performed in AFM spectroscopy for all the samples, were determined from the 'pull off' region of the force-distance curves, applying a constant force between the tip and the sample. The adhesion force F (nN) between the tip and the sample was calculated using the equation 'F = k x Δd', where k is the spring constant of the cantilever and Δd is the deflection distance. For this purpose, adhesion maps consisting of 400 curves on a 10x10 um size were acquired.

The internal structure of the Nano composite was analyzed using a Hitachi HD 2700 High Resolution Scanning Transmission Electron Microscope (HR-STEM).

3 Results and discussion

The as-prepared nanocomposite thin layers are characterized by their 3D imaged topographies, displayed in figure 1, 2. The noticeable uniformity of the pure P3OT polymer observed in its acquired topography image is confirmed also by its revealed roughness (Ra) of 0.99 nm, while the high surface roughness parameter value (31.89 nm) determined for the 12% f-SWCNTs:P3OT sample is clearly influenced by the presence of the carbon nanotubes into the polymer matrix, which leads to the formation of broad valleys and hills seen in the topography image of the nanocomposite.

Due to the high attraction between them caused by the Van der Walls forces [10], the carbon nanotubes are matched into bundles, as they appear in the 12% f-SWCNTs:P3OT sample's topography image.

On the other hand, the calculated Skewness parameter values for both investigated samples, pure polymer and nanocomposite, of about 0.7 and 0.9, respectively, reveal the comparable positive asymmetries of the heights and depths distributions over the surfaces, which can be explained if we assume a homogenous distribution of the carbon nanotubes into the host. A positive value for the Skewness parameter translates into the prevalence of the peaks present in the topography images [11].

Table 1. Morphological parameters

Sample	Ra (nm)	Rskew	$\overline{F_{ad}}$ (nN)
P3OT	0.99	0.7	16.1
12% f-SWCNTs:P3OT	31.89	0.9	14.4

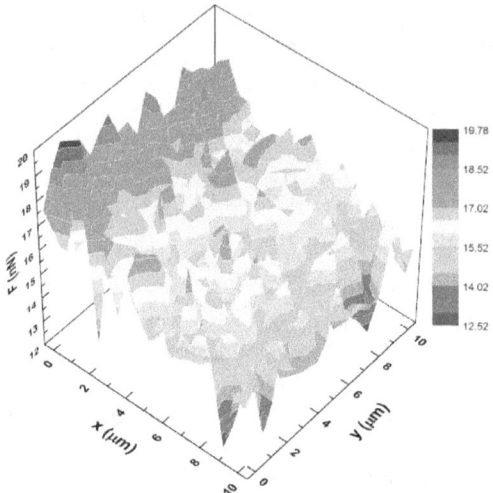

Fig. 3 3D Adhesion force variation map for the P3OT pure polymer

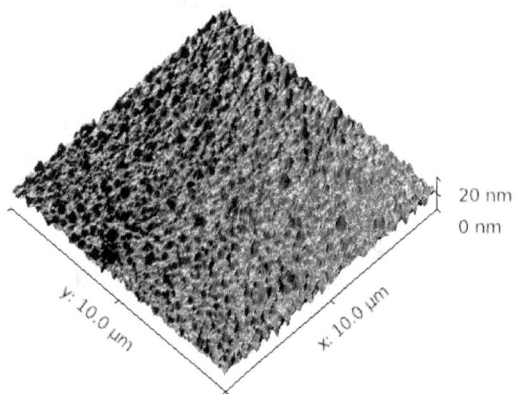

Fig. 1 3D AFM topography image of the P3OT pure polymer

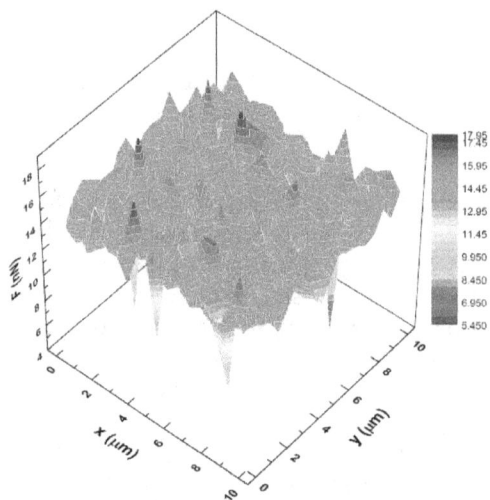

Fig. 4 3D Adhesion force variation map for the f-SWCNTs:P3OT nanocomposite

The changes of the surface properties [12] with the carbon nanotubes addition into the polymer matrix are revealed by the imaged force distance measurements, illustrated in figure 3, 4. As previously was mentioned each adhesion map was plotted as an adhesion force values dependence of the (x, y) position of each from those 400 force-distance curves acquired.

In the adhesion map obtained for the pure polymer adhesion force values, ranged from 12.5 to 19.8 nN, are grouped in large regions with approximately the same value, while the presence

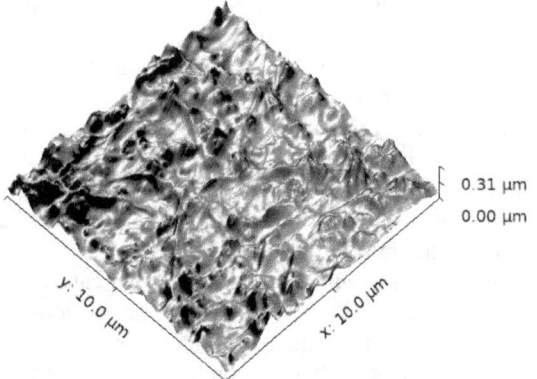

Fig. 2 3D AFM topography image of the f-SWCNTs:P3OT nanocomposite

of the carbon nanotubes bundles into the host leads to a more uniform distribution of the adhesion force values over the nanocomposite surface.

An overview of the obtained adhesion force values for the analysed samples reveals the decisive role of the carbon nanotubes addition. The calculated mean adhesion values (tab. 1) for each map of about 16.1 nN and 14.4 nN for pure polymer and nanocomposite, respectively, reveal the changes of adhesive properties of the surfaces, as a result of carbon nanotubes addition.

The variation of adhesion force is dependent on the contact area (tip-sample interaction), so the increase of the surface roughness can entail the decrease of adhesion force value.

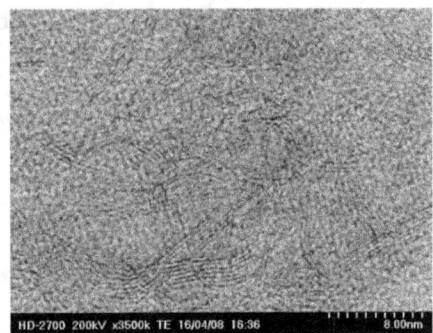

Figure 5. HR-STEM image of f-SWCNTs:P3OT nanocomposite

Figure 5 illustrates the HR-STEM image in which we can observe the presence of both individual carbon nanotubes and bundles. We can notice a high density of the f-SWCNTs bundles into the polymer matrix.

4 Conclusions

Two thin films were obtained by spin-coating technique using a pure P3OT polymer and a mixture of P3OT as the host and 12% mass concentration functionalized single wall carbon nanotubes as the reinforcing phase. In order to examine the effect of the carbon nanotubes addition into the polymer matrix, AFM studies such as topography, roughness and force distance curves were performed.

The experimental results proved the noteworthy influence of the carbon nanotubes on both, morphological and mechanical properties of the analysed samples. The presence of the f-SWCNTs into the polymer matrix leads to an increase of the surface roughness, while the adhesion force values are decreasing.

Also, with the carbon nanotubes addition into the host, the manner in which the adhesion values are distributed becomes more uniform.

Acknowledgment

This work was supported by Romanian Ministry of Education and Scientific Research, as well as by Executive Agency for Higher Education, Research, Development and Innovation Funding, under projects PCCA 2- nr. 66/2014 and PCCA 2-nr. 166/2012.

References

[1] C. Gerber, H.P. Lang, Nat. Nanotechnol. 1, 3-5 (2006)

[2] Costantino Creton, Gwendal Josse, Rebecca Webber, and Dominique Hourdet, April, 10–13, (2006)

[3] Kevin Kendall, Kluwer Academic / Plenum Publisher, New York (2001)

[4] J. AL-Zanganawee, S. Iftimie, T. Mubarak, A. Radu, O. Brincoveanu, S. Antohe, M. Enachescu, Journal of Ovonic Research 2, 95 (2016)

[5] Vitali Brand, Christopher Bruner, Reinhold H. Dauskardt, Solar Energy Materials & Solar Cells 99, 182–189, (2012)

[6] R. Phatak, T. Y. Tsui, and H. Aziz, Journal of Applied Physics, 111(5), 054512 (2012)

[7] Stephanie R. Dupont, MarkOliver, Frederik C. Krebs, Reinhold H. Dauskardt, Solar Energy Materials & Solar Cells 97, 171–175, (2012)

[8] D. Qian, E. C. Dickey, R. Andrews, T. Rantell, Appl. Phys. Lett, 76, 2868-2870 (2000)

[9] E. Kymakis, I. Alexandou, G.A.J. Amaratunga, Synthetic Metals, 127, 59-62 (2002)

[10] J-H, Du, J. Bai, H-M. Cheng, Polymer Letters 1 (5), 253–273 (2007)

[11] Marko Sedlac, BojanPodgornik, Joze Vizintin, Tribology International 48, 102–112, (2012)

[12] H. A. Mizes, K.G. Loh, R. J. D. Miller, S. K. Ahuja, and E. F. Grabowski, Applied Physics Letters 59 (22), 2901-2903 (1991)

The study of p-Si/Al$_2$O$_3$/n-Si (100) sandwiches structures deposited by KrF excimer laser ablation

Calin Moise[1], Oana Brincoveanu[1,2], Adrian Katona[1], Dorel Dorobantu[1], Dionizie Bojin[1], Marius Enachescu[1*]

[1]Center for Surface Science and Nanotechnology, Politehnica University of Bucharest, 060042, Romania
[2]University of Bucharest Faculty of Physics P.O. Box MG-11, Magurele, Ilfov, 077125 Romania
*marius.enachescu@upb.ro

Abstract: Laser ablation is a versatile technique for deposition of metals, semiconductors as well as dielectrics. Our pulsed laser deposition (PLD) system is set up into an ultra-high vacuum (UHV) machine with working pressure of $1.5*10^{-10}$ Torr. In this work we report the successfully deposition of sandwiches structures p-Si/Al2O3/n-Si (100) substrate. The obtained thin layers were characterized by atomic force microscopy (AFM), scanning electron microscopy (SEM) and composition was investigated by energy dispersive X-ray (EDX). Also micro Raman spectroscopy was involved for measurement of the stress in n-Si (100) substrates as well as top deposited p-Si. Depositions were performed at three different values for: distance between target and substrate (3, 4, 5 [cm]), temperature of substrate (400, 500, 600 [°C]), laser pulse energy (400, 500, 580 [mJ]) and laser pulse repetition rate (20, 30, 40 [Hz]). We conclude that the optimally conditions for Al$_2$O$_3$ layer are: 5 cm, 500°C, 580 mJ, 20 Hz and for top p-Si same values except the pulse repetition rate, which is 30 Hz.

1 Introduction

The market demand for thinner field effect transistor (FET), such is now the available 14 nm technology, increased the interest of scientists for buried oxides layers. Following this trend we test our ability to deposed successive layers Al$_2$O$_3$ and p-Si type over a Si n (100) substrate by laser ablation and to find the proper parameters for uniform surfaces.

Laser ablation is a versatile technique for deposition of: metals, semiconductors as well as dielectrics.[1,2]

The alumina (Al$_2$O$_3$) is well known as one of the best insulator with resistivity $1x10^{14}$ Ωcm and is chemical stabile even at high temperature, being suitable for thin buried oxide layer.

2 Experimental and results

Pulsed laser Deposition (PLD) system is a unique versatile research tool. The system offers a broad range of materials and applications. The ability to extend the vacuum capabilities to ultra high vacuum base pressures allows the control of unwanted film impurities. Up to now, our best vacuum level is $1.5*10^{-10}$ Torr. The laser target manipulator accommodates up to four 2" diameter in vacuum which are selectable through the controlling computer. Each of the individual targets can be rotated about its axis, which together with the laser scanning provides a uniform ablation of the target. Using this flexibility, a multitude of thin film structures deposition are possible.

The system consist of three chambers, load lock, growth and the RHEED gun, which serve for: load/unload the targets and samples, deposition and *in-situ* analysis respectively.

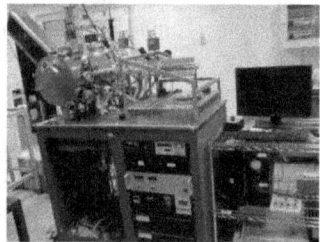

Figure 1. Photo of PLD machine

The ablations were performed with Coherent Compex pro 205 F KrF excimer laser: $\lambda = 248$ nm and pulse duration of 20ns.

First step is the substrate preparation prior to deposition. The commercial (Sigma Aldrich) Si wafer n type (100) oriented surface was treated by HF 10% for 10 minutes to eliminate the native oxide on surface, after was cleaned by acetone and dried in nitrogen flow.

The (100) surface orientation was proved by RHEED investigation (not shown here).

For studying p-Si/Al$_2$O$_3$/n-Si (100) sandwiches structures we must have the possibility to investigate both deposed surfaces, therefore after the alumina deposition the samples were taked out and a Ti foil masck was used for partial covering the surface before the deposition of p-Si layer (Fig. 2).

Depositions were performed at three different values for: distance between target and substrate (3, 4, 5 [cm]), temperature of substrate (400, 500, 600 [°C]), laser pulse energy (400, 500, 580 [mJ]) and laser pulse repetition rate (20, 30, 40 [Hz]).

The rotation speed of the substrate and the time for ablation were kept constant at 10 Rpm and 10 minutes respectively.

In figure 2 we can observe the macroscopic diferences of surfaces obtained at 4, 5 and 6 cm. The left and right parts of samples are uncovered n-Si (100) substrates. Bottom is alumina layer and top is p-Si over alumina.

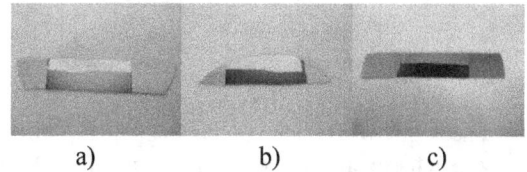

a)   b)   c)

Figure 2. Macroscopic differences of deposition performed at 4 (a); 5 (b); 6 (c) cm respectively

Alumina and p-Si layers show different colours and shining as function of distance between target and substrate.

Nano scale investigations were carried out by AFM (Solver Next) and reveal that optimal value of distance is 5 cm for both deposed layers (Fig. 3). [3]

The surfaces were further analysed by SEM (Fig. 4). Composition and stoichiometry of alumina and Si p layers were proved by EDX (Table 1).

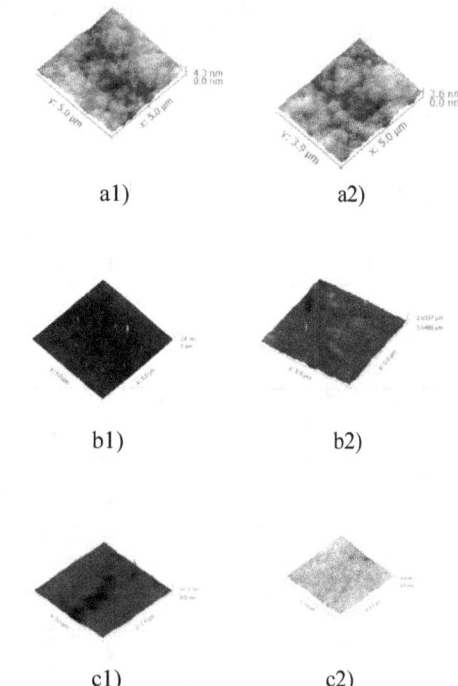

a1)   a2)

b1)   b2)

c1)   c2)

Figure 3. 3D topography images of alumina (1) and top p-Si (2) for: 4 cm a); 5 cm b) and 6 cm c)

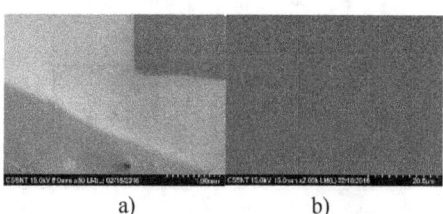

a)   b)

Figure 4. SEM images of all layers a); alumina layer b)

Table 1. The atomic percentages as resulting from EDX analysis proving stoichiometry of alumina layer

	Atomic (%)		
Statistics	O	Al	p-Si
Max	61.96	38.04	98.36
Min	1.57	0.07	81.97
Average	26.43	13.46	

Table 2 show the roughness indicators (RA and RMS) for all the parameters investigated.

Table 2. The first line is corresponding to alumina deposition and second one to top p-Si. The values in the table correspond to: index of sample, distance between target and substrate, pressure before ablation, pressure during ablation, substrates temperature, laser pulses energy, repetition rate of laser pulses and the surface roughness indicators RMS and RA. With yellow are highlighted the optimal values.

No	D (T-S) cm	P_i Torr	P_a Torr	T °C	E mJ	RR Hz	RMS nm	RA nm
4	4	$7.7*10^{-7}$	$4*10^{-6}$	500	580	30	1.125	0.782
		$7*10^{-7}$	$4.8*10^{-6}$	500	580	30	0.524	0.413
8	5	$2*10^{-7}$	$2.8*10^{-6}$	500	580	30	0.7000	0.4239
		$4.7*10^{-7}$	$1.2*10^{-5}$	500	580	30	0.4484	0.3475
5	6	$6.5*10^{-7}$	$2.1*10^{-6}$	500	580	30	0.6159	0.4837
		$5.7*10^{-7}$	$5*10^{-6}$	500	580	30	0.8015	0.6208
7	5	$1.3*10^{-6}$	$9*10-6$	600	580	30	0.799	0.57
		$1.1*10^{-6}$	$1*10-5$	600	580	30	0.449	0.3437
8	5	$2*10^{-7}$	$2.8*10^{-6}$	500	580	30	0.7000	0.4239
		$4.7*10^{-7}$	$1.2*10^{-5}$	500	580	30	0.4484	0.3475
6	5	$8*10^{-7}$	$3*10^{-6}$	400	580	30	1.125	0.782
		$7*10^{-7}$	$1.8*10^{-6}$	400	580	30	0.498	0.397
9	5	$6.4*10^{-7}$	$3.6*10^{-6}$	500	580	20	0.3639	0.2889
		$9*10^{-7}$	$1*10^{-6}$	500	580	20	0.8544	0.6729
8	5	$2*10^{-7}$	$2.8*10^{-6}$	500	580	30	0.7000	0.4239
		$4.7*10^{-7}$	$1.2*10^{-5}$	500	580	30	0.4484	0.3475
10	5	$2.6*10^{-6}$	$2.8*10^{-7}$	500	580	40	0.63	1.06
		$4.1*10^{-7}$	$4.6*10^{-6}$	500	580	40	1.09	0.55
12	5	$1*10^{-6}$	$5.8*10^{-6}$	500	400	30	2.011	0.961
		$2*10^{-6}$	$2.4*10^{-6}$	500	400	30	0.8544	0.6729
11	5	$2.7*10^{-7}$	$6.5*10^{-7}$	500	500	30	0.83	0.65
		$1*10^{-6}$	$1.6*10^{-6}$	500	500	30	0.99	0.72
8	5	$2*10^{-7}$	$2.8*10^{-6}$	500	580	30	0.7000	0.4239
		$4.7*10^{-7}$	$1.2*10^{-5}$	500	580	30	0.4484	0.3475

To test the quality of ours depositions we performed the friction mapping on surfaces by AFM [4]. As can be seen in figure 5 quite uniform values were found for both layers.

For investigate the stress of p-Si deposed layers we used micro Raman spectroscopy. Figure 6 shows such spectrum and the position of peak at normal value 520.8 cm^{-1} unstressed.

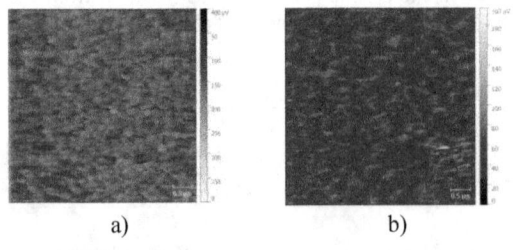

Figure 5. Friction mapping: alumina a) and b) top p-Si

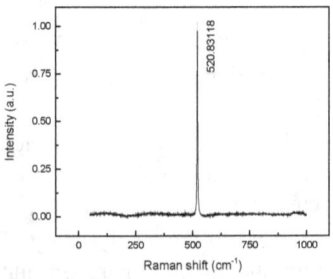

Figure 6. Raman spectrum of top p-Si layer

3 Conclusion

By this work we proved our ability to grow uniform layer buried oxide (alumina) on a n-Si (100) commercial substrate.

Deposited layers were investigated by AFM in contact mode as well as friction mapping and by SEM.

The composition and stoichiometry of Al_2O_3 and top p-Si was proved by EDX analysis.

We conclude that the optimally conditions for Al_2O_3 layer deposition are: 5 cm, 500°C, 580 mJ, 20 Hz and for top p-Si same values except the pulse repetition rate, which is 30 Hz.

The top p-Si layer shows a Raman peak at 520.8 cm^{-1} indicating unstressed deposition.

Further directions for continuing research is to measure the layers thickness and to control it from experimental parameters.

Acknowledgment

This work was supported by Romanian Ministry of Education and by Executive Agency for Higher Education, Research, Development and Innovation Funding, under projects PCCA 2-nr. 166/2012 and ENIAC 04/2014.

References

1) Editor: Phipps Clude "Laser Ablation and its Applications" Springer Series in Optical Science 2007.

2) Stafe Mihai, Marcu Aurelian, Puscas Nicolae "Pulsed Laser Ablation of Solids" Springer Series in Surface Science 2014.

3) A. Moldovan, P.M. Bota, D. Dorobantu, I. Boerasu, D. Bojin, D. Buzatu, M. Enachescu "Wetting properties of glycerol on silicon, native SiO_2, and bulk SiO_2 by scanning polarization force microscopy" Journal of Adhesion Science and Technology, 28, 13, 1277-1287, (2014).

4) A. Moldovan, P.M. Bota, T.D. Poteca, I. Boerasu, D. Bojin, D. Buzatu, M. Enachescu "Scanning polarization force microscopy investigation of contact angle and disjoining pressure of glycerol and sulfuric acid on highly oriented pyrolytic graphite and aluminum" The European Physical Journal Applied Physics, 64, 31302-31308, (2013).

Real-Time Implementation of Nonlinear State-of-Charge Estimation Techniques for Hybrid Electric Vehicles

Roxana-Elena Tudoroiu[1], **Wilhelm Kecs**[2], **Maria Dobritoiu**[3], **Nicolae Ilias**[4], **Valentin-Stelian Casavela**[5], **Dumitru Burdescu**[6] and **Nicolae Tudoroiu**[*7]

[1, 2, 3, 4, 5] University of Petrosani, Petrosani, 332006, Romania.
[6] University of Craiova, Craiova, 200585, Romania.
[7] John Abbott College, Saint-Anne-de-Bellevue, QC H9X 3L9, Canada.
e-mail: ntudoroiu@gmail.com

Abstract: The problem of State-of-Charge (SOC) estimation is crucial in automotive industry for successful marketing of both electric vehicles (EVs) and hybrid electric vehicles (HEVs). Nowadays the wealth of rechargeable Ni-MH and Li-ion batteries is a great prospective to be used extensively to plug-in hybrid electric vehicles (PHEVs), hybrid electric vehicles (HEVs), and battery electric vehicles (BEVs). In addition the Li-ion batteries are much lighter in weight and of reduced size, therefore much easier to be integrated in the vehicle structure in order to provide more power to boost up the acceleration, and to increase the energy efficiency through on-board battery energy storage by means of regenerative braking and mechanical-electric energy conversion. Gradual capacity reduction and performance decay can be evaluated strictly based on the current knowledge of rechargeable battery technology, and consequently is required a rigorous monitoring and a tight control of the SOC level, that is essential for increasing the operating lifetime of the rechargeable batteries. In this paper we present some preliminary results obtained in our research to implement some of SOC estimation techniques for the rechargeable batteries of different chemistry. We investigate the effectiveness of two nonlinear Kalman Filters estimators (Extended and Unscented versions) in order to find the most suitable estimator, more accurate and to perform better. The contribution of this paper is the implementation of these estimators in real time based on a generic model of the battery by means of extensive simulations in a MATLAB/SIMULINK programming environment.

Key words: Hybrid Electric Vehicles, State of Charge, Extended Kalman Filter, Unscented Kalman Filter.

1 Introduction

The State-of-Charge (SOC) of a battery is its available capacity expressed as a percentage of its rated capacity. The problem of state-of-charge (SOC) estimation for both electric vehicles (EVs) and hybrid electric vehicles (HEVs) is of vital importance for accurate estimation of remaining battery capacity and prediction of the instantaneous battery power that can be delivered to the electric engine. Determination of an optimum battery pack management system (BMS) is perhaps the one of the most significant technical issue for the automotive industry to have a successful marketing of HEVs. More particularly, an optimal operating vehicle energy and battery management system (BMS) should have the following features based on

an accurate SOC estimate and a precise available battery power forecast:

a) avoid overcharging and undercharging damages to the battery

b) aggressively use the entire battery pack capacity

c) enhance overall power system performance and its reliability

d) allow the use of smaller and lighter battery packs, and

e) reduced production and service related costs.

These issues are among the main motivating factors for the research proposed in this paper. The necessity of the rigorous consistency requirements envisaged for HEVs do indeed justify the development and implementation of more advanced estimation and prediction techniques for the optimal design of the BMSs. However due to the high complexity, nonlinearity, time-variation, and uncertain behavior of the batteries in HEVs, it is quite difficult to achieve reliable, accurate and satisfactory estimation/prediction performance specifications using model-based approaches that rely on *linear* models or *linearized* models of the batteries of different chemistry, such as techniques that are based on Kalman filtering (KF) Extended Kalman filtering (EKF) [1]. Thus, there is a need to develop nonlinear model-based estimation techniques [1, 2, 3, 4, 5, 6] that rely and are based on the true *nonlinear* model of the process.

Towards this objective, nonlinear model-based techniques have been extensively studied in the literature. In this paper, in addition to the EKF technique, which has already been developed in [1] for SOC estimation in Li-PB batteries, we have developed and implemented a new extension for nonlinear estimation techniques including Unscented Kalman Filter (UKF) for SOC estimation of nickel metal hydride (Ni-MH) batteries. The remainder of the paper is organized as follows. In section 2, the generic Ni-MH battery modeling issues are introduced together with the other batteries modeling aspects. In section 3 are developed the both EKF and UKF filtering nonlinear estimators based on the generic Ni-MH battery model chosen in section 2 in order to estimate the battery SOC. The simulation results for both nonlinear estimators in MATLAB/SIMULINK programming environment are given in section 4. The robustness capability of the both estimators is also evaluated in the same section. Finally, performance comparisons of the both algorithms and conclusions are given in section 5.

2 Ni-MH Battery Model Description

Battery modeling forms the basis of and stands as an useful tool for battery design, manufacturing, and control. It is particularly important for battery characterization (i.e., State-of-Charge (SOC) and State-of-Health (SOH) estimation) and battery management since the model development is logically the first step in developing any system identification and adaptive state estimation algorithms [4]. The most common models can be generally classified into two groups: electrochemical models and equivalent circuit

models. Equivalent circuit models consisting of electrical circuit components such as capacitors, resistors, diodes, and voltage sources, can be readily developed using electric circuit simulation software such as PSpice [7]. Other types of models, given in algebraic or differential equations, may be more suitable for a generic simulation environment such as MATLAB/SIMULINK. Recently, MATLAB also released a generic battery model in its SimPower systems toolbox [7]. Nevertheless, an equivalent circuit model can be easily converted into other model formats. The choice of model representation will be determined by the matter of convenience and the simulation tools available.

National Renewable Energy Laboratory (NREL) also developed an electric circuit model for batteries that is basically a RC network, as shown in figure 1, and is a part of its ADVISOR tool package [7, 8].

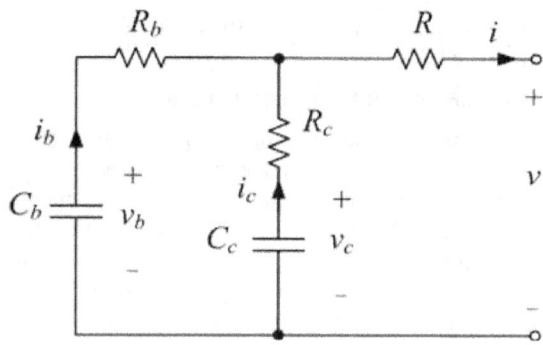

Figure 1. NREL battery model [7, 8]

The model contains two capacitors (C_b and C_c) and three resistors (R_b, R_c, and R). The capacitor C_b models the main storage capacity of the battery. The capacitor C_c captures the fast charge and discharge aspect of the battery and is much smaller than C_b [7]. The model has a big advantage that can be converted into other model formats for the convenience of simulation. A full representation of the battery model in state space is described in detail in [7], and can be considered as a generic representation for all the batteries of different chemistry. In particular, the nickel-metal hydride (Ni-MH) batteries in a tight competition with Li-Ion batteries can be represented successfully by the same generic model with the battery thermal model represented by a lumped first-order equation with linear dynamics [7]. The parameters of the components are functions of the SOC and battery temperature (T). In addition, the resistance also depends on whether the battery is in "charge" or "discharge" mode. The overall SOC is a weighted combination of the states of charge on C_b and C_c [7]. Mathematical modeling is indispensable in this process since a cell model, once validated experimentally, can be used to identify cell-limiting mechanisms and forecast the cell performance for design, scale-up, and optimization. The both algorithms developed in this paper are implemented based on a similar generic Ni-MH battery model. The various modeling aspects and phenomena inside the battery such as thermodynamics, polarizations, hysteresis, etc. have been incorporated into the battery model. Also, to improve the estimation performance, this Ni-MH battery model was combined with the coulomb-accumulation model developed in [7]. A very useful tool for performance evaluation of both nonlinear estimators for different standard driving cycles (current profiles) is provided by Advisor 3.2 software package.

3 Prediction and Estimation Techniques

The ultimate goal of state and parameter system identification is to obtain a mathematical model whose output matches the output of a dynamic process that is subjected to a given input. The solution to the exact matching problem, in general, is extremely difficult. Consequently, for practical reasons the original problem is commonly relaxed to developing a model whose output can be made "as close as possible" (in some metric sense) to the output of the dynamic system. Different methods have been developed in the literature for both linear and nonlinear system identifications and [1, 2, 3, 4, 5, 6]. A common characteristic of most of these methods is the use of a parameterized model where the parameters are recursively updated in real-time to minimize a performance index such as the output identification error. These methods can be broadly classified into two main categories, namely conventional model-based schemes and intelligent and neural network-based schemes. Conventional methods are well established for linear systems and have recently been under investigation for generalization to nonlinear systems. The main disadvantage of these methods is that they are generally applicable and extendable to only a special and quite limited class of nonlinear systems. A good survey of conventional

nonlinear identification methods may be found in [4].

3.1   Extended Kalman Filter Estimator

The Kalman filter is a powerful and popular tool for the stochastic estimation problem that is proposed by R.E. Kalman in 1960 [1]. Consider a *linear* stochastic difference equation:

$$x_{k+1} = Ax_k + Bu_k + w_k \qquad (1)$$

$$y_k = Cx_k + Du_k + v_k \qquad (2)$$

where w_k and v_k are the process and measurement noise, respectively, and are assumed to be independent white Gaussian random process with zero mean value,

$$E[w_n w_n^T] = \begin{cases} \Sigma_w, n = k \\ 0, n \neq k \end{cases}, E[v_n v_n^T] = \begin{cases} \Sigma_v, n = k \\ 0, n \equiv k \end{cases} \qquad (3)$$

The distributions of the process and measurement noise are the normal probability distributions

$$p(w) \sim N(0, Q_w), \quad p(v) \sim N(0, R_v) \qquad (4)$$

The covariance matrices Q_w (process noise covariance) and R_v (measurement noise covariance) might change with each time step or measurement, but in our approach we assume they are constant. Due to the process noise injected in the state equation (1), the state vector $x_k \in R^n$ becomes random variable with its distribution approximated by a Gaussian distribution function:

$$p(x) \sim G(\hat{x}, P_x) \qquad (5)$$

By applying a predictor-corrector algorithm, the Kalman filter estimates the process state at current time and then obtains feedback in the form of the noisy measurements. As such, the process of the filter falls into two steps, namely *time update* and *measurement update*. During the time update, the current state and error covariance estimate are projected forward (in time) to obtain *a priori* estimate of the state in the next time step. Next, the new measurement is incorporated into this priori estimate value for calculating the *posteriori* estimate of the corresponding state during the measurement update [1, 2, 3, 4, 5, 6]. However, the process to be estimated and the measurement relationship to the process can be nonlinear in practice. Under these situations the most interesting and successful applications of the Kalman filter is the use of an underlying linearized process model whereby an estimate of the current state of the system is made and then its estimate is corrected by using any available sensor or output measurements. Following the same predictor-corrector mechanism, it is then possible to approximately obtain an optimal state estimate of the state and its covariance. This linearized extension of Kalman filter is referred to as the Extended Kalman filter or EKF [1, 2, 3, 4].

3.2   Unscented Kalman Filter Estimator

The Unscented Kalman Filter (UKF) technique [2, 3, 4, 5, 6] is based on the unscented transformation (UT) and addresses the general problem of state estimation $x_k \in R^n$. The discrete-time process is governed by a nonlinear stochastic difference state-space equation with a measurement $y_k \in R^m$ and is given by

$$x_{k+1} = f(x_k, u_k) + w_k \tag{6}$$

$$y_k = g(x_k, u_k) + v_k \tag{7}$$

The random variables w_k and v_k are representing the process and measurement noise, respectively. They are assumed to be Gaussian, mutually independent, zero mean, white, and with normal probability distributions:

$$p(w) \sim N(0, Q_w), \quad p(v) \sim N(0, R_v) \tag{8}$$

The covariance matrices Q_w (process noise covariance) and R_v (measurement noise covariance) might change with each time step or measurement, but in our approach we assume they are constant. Due to the process noise injected in the state space equation the state vector $x_k \in R^n$ becomes random variable with its distribution approximated by a Gaussian distribution function $p(x) \sim G(\hat{x}, P_x)$. The vital operation performed in the Kalman Filter is the propagation of a Gaussian random state variable $x_k \in R^n$ through the system dynamics. In the Extended Kalman Filter (EKF) the Gaussian random state variable $x_k \in R^n$ is propagated analytically through the first-order linearization of the nonlinear system. This can introduce large errors in the true posterior mean and covariance of the transformed Gaussian random state variable, which may lead to sub-optimal performance and sometimes divergence of the filter. The UKF approach is developed as an alternative to the EKF and addresses this problem by using a deterministic sampling approach. Using the principle that a minimal set of carefully chosen sample points can be used to parameterize mean and covariance, the UKF yields superior performance compared to EKF, especially for nonlinear systems. These sample points completely capture the true mean and covariance of the Gaussian random state variable $x_k \in R^n$, and are propagated through the true nonlinear system dynamics. The cloud of the transformed sample points' distribution captures the posterior mean and covariance accurately to the fourth order while the EKF only predicts with accuracy up to the second order for the posterior mean and fourth order for the covariance [2, 5, 6]. However, the UKF will make more accurate estimates only if the kurtosis and higher order moments in the state error distribution are significant. In some applications the sampling rate could be an important source of degrading in the UKF performance. The main advantage of the UKF is that it does not require the calculation of the Jacobian matrices that could lead to implementation difficulties [2, 5, 6].

3.3   Real-Time Implementation of Extended and Unscented Kalman Filters Estimators

Headings Extensive simulation results for the SOC estimation using the EKF, the UKF, and the RHNF algorithms are explained and compared in this section. Comparative analysis of all the three proposed algorithms in terms of the steady state performance and the robustness capability are also presented in details.

First, the performance evaluation under the nominal battery parameters, namely the battery internal resistance R_i and the battery capacity C_n is presented and then the robustness capability of the algorithms with respect to large initial SOC errors and changes in the battery parameters is illustrated also. The simulations are carried out using the *reference* driving cycle shown in Figure 2.

3.4 Simulation Results in MATLAB Programming Environment

The simulation results in MATLAB programming environment for the both algorithms for the nominal battery parameters and different levels of uncertainty in the initial value of SOC are given in the Figures 2-6.

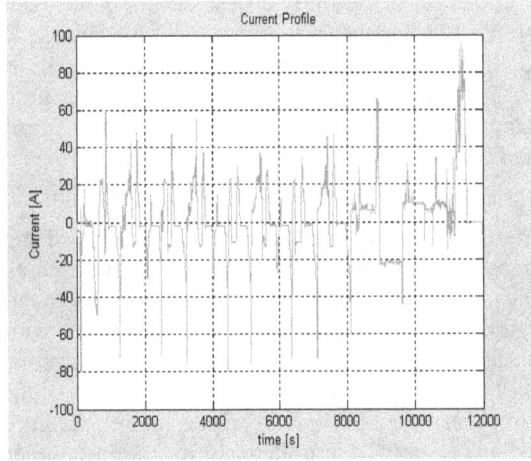

Figure 2. The current profile used in simulations

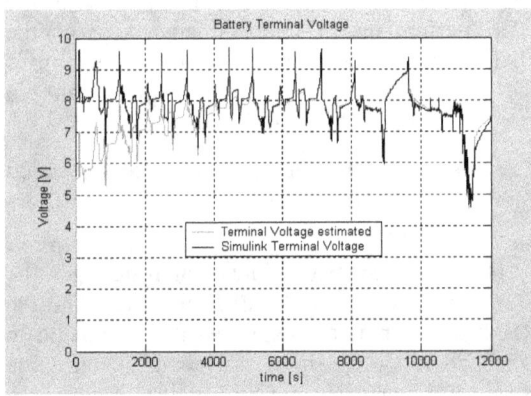

Figure 3. The Battery Terminal Voltage Profile

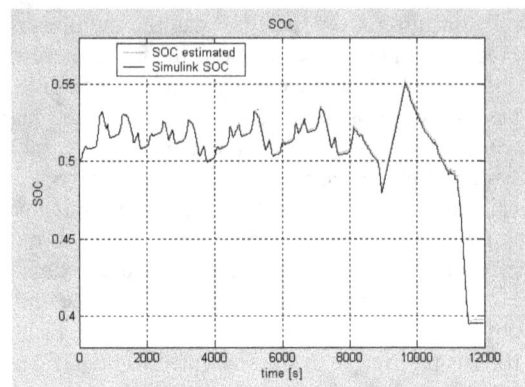

Figure 4. The Battery Terminal Voltage Profile

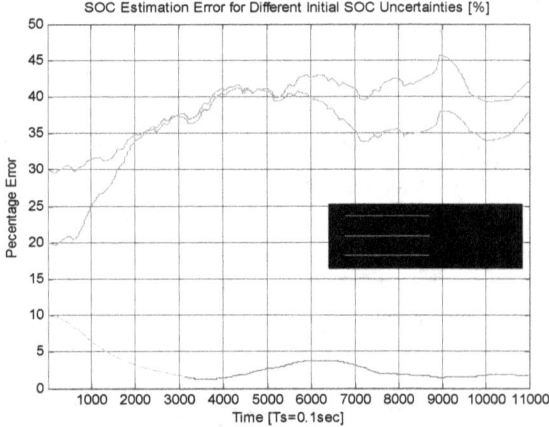

Figure 5. The SOC estimation error using the EKF Estimator

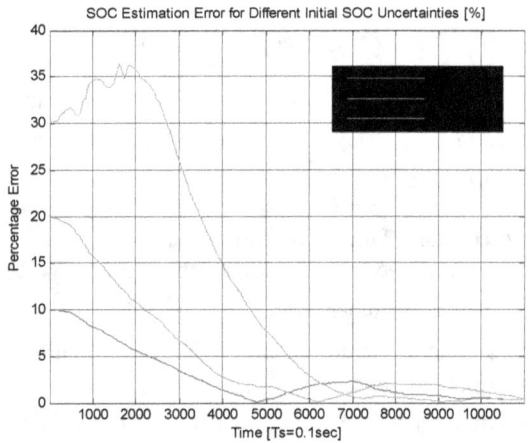

Figure 6. The SOC estimation error using the UKF Estimator

The robustness of the both estimators to the changes in the battery parameters (Resistance and capacitance) is depicted in the Figures 8-9.

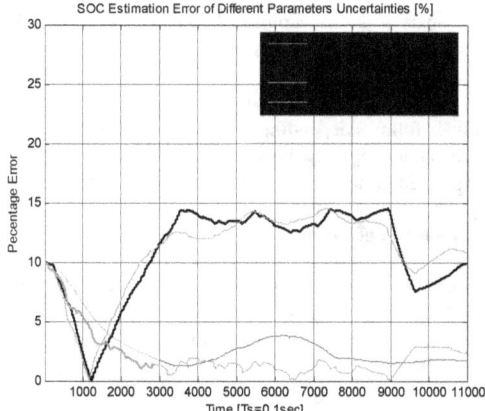

Figure 7. The SOC estimation error using the EKF Estimator

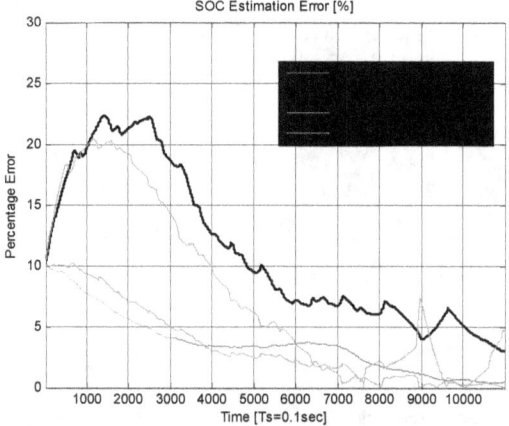

Figure 8. The SOC estimation error using the UKF Estimator

4 Conclusions

For an overall comparative performance of the both algorithms in a more comprehensible framework, the results developed are summarized as follows:

1) The EKF algorithm performs well only for some of the driving cycles
2) The UKF algorithm performs very well for almost all the driving cycles with standard steady state errors sometimes smaller than 2%.

3) The UKF algorithm outperforms the EKF algorithm for almost all the driving cycles in terms of the convergence and the steady state error.
4) The EKF algorithm is not convergent in nearly all cases
5) The UKF algorithm converges very well for majority of the driving cycles
6) The EKF algorithm is not robust in almost all conditions
7) The UKF algorithm is robust in the steady state while it shows a large error in the transient phase

References

1) Plett, G., *Journal of Power Source*, 134 (2) , (2004), 252.

2) LaViola, J.J.Jr, Proceedings of the American Control Conference, 2003, Denver, Colorado, June 4-June 6, 2003.

3) Wan, E. A., and Rudolph van der Merwe, Proceedings of Symposium on Adaptive Systems for Signal processing, Communication and Control (AS-SPCC) IEEE Press, 2000, Lake Louise, Canada, October 2000.

4) Simon Haykin, *Adaptive filter theory fourth edition*, edited by T. Kaillath (Prentice - Hall, NJ, 2003), 936.

5) Tudoroiu, N., Zaheerudin, M.,Cretu, V., and R-E., Tudoroiu, *IEEE Industrial Electronics Magazine* 4(3), (2010), 7.

6) Tudoroiu, N., Zaheerudin, M.,Chiru, C., Grigore, M., and R-E., Tudoroiu, HSI'09 Proceedings of the 2nd conference on Human System Interactions 2009, Catania, Italy, May 2009.

7) Kwo Young, Caisheng Wang, Le Yi Wang, and Kai Strunz, *Power Electronics and Power Systems-Chapter 2*, edited by Rodrigo Garcia-Valle, Joao A Peças Lopes (Springer, Germany, 2013), 325

8) Johnson V.H., *Journal of Power Sources* 110, (2001), 321.

THEORY OF THE ROOT LINES OF THE GENERAL POLYNOMIAL

Nicholas Schmidt, Dipl. Ing. (graduated at the Institutul Polytehnic Timisoara, Romania) E-mail:
nicholas.schmidt007@gmail.com

This paper describes the main characteristics of the root lines of polynomials with complex coefficients. Root lines are defined as the geometric loci of the roots of the real, respective imaginary part of the polynomial. A graphical representation of these lines offers a clear picture of the positions of the polynomial's roots. The coordinates of various points of the root lines can be obtained by use of a computer program. Each root line leads to a given root, this way the polynomial can be solved and graphically represented.
A short description of the rules of variation and properties of these root lines follows depending on the position of the roots of the polynomial. This study contains many numerical examples which show not only how the root lines, but also the polynomials with complex roots and coefficients work, how they can be studied, transformed and solved.

A. Introduction

This study is a continuation and completion of a study about the root lines presented at the 27-th ARA Congress in Oradea, Romania, May 29 – June 2, 2002.
The length of the study was then restricted to 5 pages, therefore some observations in the study could not be mathematically proved or demonstrated.
This paper presents all those observations and new explanations regarding the properties of root lines.

B. Importance of the root lines.

Root lines are characteristic for polynomials with complex roots. A polynomial of degree n has n roots and all roots can be complex. According to some scientific works the concept of complex roots has a particular importance to the physical and engineering sciences.
See Reference [4] Volume II, for the list of such cases.
The same [4], on page 413 and ff. also presents some curves which are root lines, but are called by the author $\mu(x,y) = c1$ and $v(x,y) = c2$. (μ and v are obviously the real and imaginary part of the polynomial).
The name "root lines"of this theory is given by me, and probably is not found in any other book or article.
In the same chapter of that book [4] is also mentioned that the angles between these curves at a multiple root are equal. It explains this with the vector dot product of the gradients. This study gives a more simple explanation for this.
The studies mentioned in [4] are also restricted to polynomials with real coefficients.
My studies have no such restriction, because I found that they can easily be extended to cases with complex coefficients, so they are more general, without restriction.

C. The basic form of a polynomial is

$$P_n(w) = C_n w^n + C_{n-1} w^{n-1} + + C_1 x + C_0 \qquad (1)$$

w is the independent variable of the polynomial which can be a real or a complex number; regarding w see also Par. I.
The coefficients C_n to C_0 are complex or real numbers.

D. Roots of a polynomial.

Roots (or zeros) of a polynomial are those values of the variable w for which the polynomial's value (both the real and the imaginary part) reduces to zero.

E. Real roots of a polynomial.

If the variable w in relation (1) is a real number, and P(w) is represented along a straight reference line, then the roots are points where the curve crosses this line.
The number of real roots sometimes is equal with n, but in other cases less than n (order of polynomial), even 0.

F. Complex roots of a polynomial.

If we compare polynomial x^2-9 with x^2+9 then the condition $P = x^2 - 9 = 0$ can be written also $x^2 = 9$ and hence $x = +3$ or $x = -3$, so this polynomial has two real roots where the curve intersects the Ox line.
The polynomial $P = x^2 + 9$ on the other hand, as we can see from Fig. 1 has the minimum value for x = 0, then P = 9,

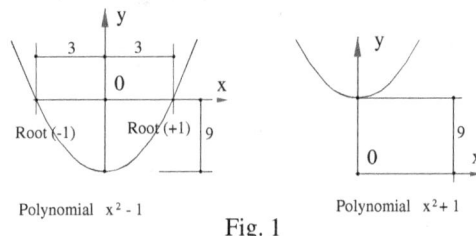

Polynomial $x^2 - 1$

Polynomial $x^2 + 1$

Fig. 1

but it never reduces to zero.

From $P = x^2 + 9 = 0$ results $x^2 = -9$ or $x = \sqrt{-9} = 3\sqrt{-1}$

Because it is not possible to extract radical of -1, this number is called imaginary unit.
So we reach to the notion of complex numbers which have the form a + bi where a and b are real numbers and i = $\sqrt{-1}$ is the imaginary unit.
Then the number of roots of a polynomial is completely solved: Any polynomial has n roots which are either real or complex. Actually the real roots can be considered as complex numbers with b = 0.

G. The second basic form of a polynomial.

If all roots of a polynomial are known, then the polynomial can be written in another basic form

$$P_n(w) = (w - w_1)(w - w_2) \ (w - w_n) \qquad (2)$$

where w_1, w_2, w_n are complex numbers and the polynomial in form (2) obviously reduces to zero if w is equal with one of the roots.

If the mathematical operations are performed, then (2) reduces to (1) because all operations are unique.

H. Independence of the roots and root lines from the reference axes.

This rule is very important for the study of the root lines.

In relation (2), if we choose any point $M(x_p, y_p)$ and calculate the value of the polynomial for this point M, then we choose other two refernce axes (Ox, Oy), the value of the polynomial in *the same point M* will be the same.

Proof.

In relation (2) both w and w_1, w_2, ... w_n are measured from the same reference axes (Ox, Oy). If an axis is moved for ex. by *d* then both w and w_i increase by the same value *d* and *their difference remains the same.*

This rule can be automatically extended to the root lines. A root line constructed in one system (of Ox, Oy) axes remains the same if the axes are moved because each point of the root line is a root. On the other hand coefficients of a polynomial change if the reference axes are moved.

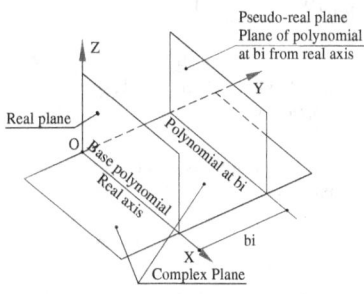

Fig. 2

I. Representation of the roots and root lines of a polynomial.

The real roots of a polynomial are represented along the Ox axis. The imaginary component bi of a complex root has a direction perpendicular to the real axis. So a complex number a + bi is a vector whose real component *a* has direction of the Ox axis and its imaginary component direction of the Oy axis. The complex variable w = a + bi = x + yi then defines a complex plane (See Fig 2)'

Observation.

Fig. 2 has a tri-dimensional representation. The complex variable having two dimensions (x + yi), it is natural to assume that the function P(x+yi) = u+vi has also two components u (real part) and v (imaginary part) which are both represented in a direction z perpendicular to the Oxy plane, so both u and v can be represented as two surfaces, each point of which has a value in direction z. Therefore the complex variable will be noted in this study as w = x + yi.

(Many authors use the notation z = x + yi, but they don't have a figure like Fig. 2 in this paper).

In this study the letter z will be used for a variable whose direction is vertical *and perpendiculat to the complex plane Oxy.* Both u and v have direction of z, but different values.

J. Points of root lines can be obtained by substitution of the complex variable in the polynomial.

If we substitute the complex variable w = a + bi in a polynomial P_n, then in the same point of the complex plane we will obtain two values: u an vi. These values *u* and *v* are assumed to be vertical, i.e. perpendicular to the complex plane Oxy in the direction of the Oz axis (See Fig. 3)

The imaginary unit i only very rarely disappears. Therefore *all terms multiplied by i* added up give an imaginary part noted generally with *v*. The other terms (without i) give the real part noted *u*.

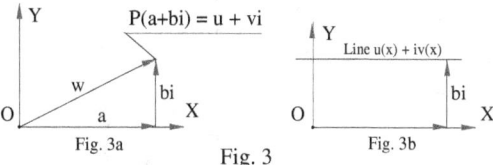

Fig. 3

If we substitute w = a + bi = x + yi in the polynomial and give b a numerical value, but leave x variable, then obviously remain just the terms with x unknown. These have the powers 0 to n. All these terms form than *two polynomials in x, one multiplied by i (imaginary part), and another without i (real part). They* correspond to *the line at distance b* (b can be chosen numerically) one being the real, the other the imaginary part). If these polynomials are solved, they give *n* points for the root lines of the real part and n for the imaginary part. (See Fig. 3b)

Only the real roots have to be calculated, the complex roots are discarded *being outside of the line at distance b.*

K. Different possible positions of the polynomial centre.

The polynomial centre was defined in the first part of this study (in 2002) as the geometrical centre of all roots both in direction x and y. (See [7] Par. (2), Pg. 841)

In case of a single multiple root (i.e. there are no other roots than the multiple root) all root lines are straight lines which intersect where the multiple root is. There then both the real and imaginary part reduce to zero (so this is a complex root) and the polynomial center is also there (all roots are concentated in one point).

So in this case all root lines pass *exactly* through the polynomial center and they are *straight lines* and coincide with the asymptotes.

In case of circular (binary) polynomials in the form $w^n - 1$ the polynomial centre coincides with the center of the circle on which the roots are, this being the *best approximation* for the action of *all roots.*

More about such polynomials see [7], Par 5A2., Pg 843.

For other polynomials the asymptotes correspond to w^n because the other terms for high values of w are negligible against w^n and therefore this case = especially for points far from the center – the polynimial center is similar to (i.e. the best approximation for) the case of a multiple root. It is also an *invariant* of the polynomial and *depends on all roots*. All practical cases which I studied confirmed this hypothesis.

L. Solving polynomials

To solve polynomials even with complex roots is simple if there is an adequate method and computer program.
Such a method is Newton's method in the *complex domain*.
In order to see how this method works, let's see first Newton's method in the *real domain*.
According to the real domain method, for a given x value, the polynomial's value $z = P(x)$ and the value of its derivative is $z' = \dfrac{dP(x)}{dx} = \dfrac{dz}{dx}$ are known (see Fig. 4).

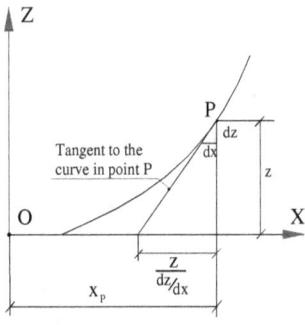

Fig. 4

But z' is the trigonometric tangent of the geometrical tangent to the polynomial's curve. Then the point where the geometrical tangent crosses the Ox axis will be at $x - \dfrac{z}{z'}$.
This is now an approximation closer to a root of the polynomial. In many manuals $\dfrac{z}{z'}$ is noted with $\dfrac{f}{f'}$ (function f and its derivative f').
In the complex domain, the function with a complex variable is

    $f = u + vi$ and its derivative is
    $f' = u' + v'i$ where f'. u' and v' are derivatives of f, both with respect to x.

If we divide f with f' according to the rule of division of complex numbers we obtain

$$\frac{f}{f'} = \frac{u+vi}{u'+v'i} = \frac{uu'+vv'}{u'^2+v'^2} + \frac{vu'-uv'}{u'^2+v'^2}i$$

The obtained two fractions subtracted from an initial value of an arbitrary complex number $w = x + yi$ give a new

approximation of w. This is than repeated until value of $\dfrac{f}{f'}$ becomes negligible.

J.B. Moore in [1] doesn't mention the name of Newton. He calls the above approximations, i.e. the value of $\dfrac{f}{f'}$ 'steepest descent vector' (s d v) but arrives to the same expression as given above.
For the s.d.v. see also [7] Par. 6A., Pg. 845
The above relation inserted in a computer program gives after a few steps value of one root. Then a synthetic division has to be performed, that means the polynomial is divided by $w - w_1$ where w_1 is the value of the calculated root. After this the polynomial's degree reduces by 1.
When n = 1, the root's value is calculated directly.
Observations regarding this method.
1.) The program works well also for polynomials with real coefficients or real roots, but then the imaginary part of the coefficients has to be set to zero. 2.) Whereas Newton's method in the real domain works only if the initial guess is close to a root, there is no such restriction in case of the program in complex domain. 3.) If the program enters in an 'endless loop' (it happens extraordinarily seldom), then increase initial guess, or change sign of the imaginary part (i.e. change the initial guess $1 + i$ to $10 + 10i$, $100 + 100i$, $1 - i$, $10 - 10i$, $100 - 100i$ etc)
Note.
The size of this paper doesn't permit to give a complete listing of this or other programs. If you are interested in more details, please write to my e-mail address.

M. Rotation of the asymptotes of a multiple root multiplied by a complex constant.

A polynomial which has only one *n* times multiple root, i.e.

    $P_n = w^n = (x + bi)^n$ or in the trigonometric form:
    $P_n = \rho^n(\cos n\varphi + i\sin n\varphi)$

If this expression is multiplied by a complex number in the form $a + bi = a(1 + ki)$ where $k = \dfrac{b}{a} = \tan \alpha$ $\alpha = \tan^{-1}(k)$
(using for the argument of the complex multiplier $1 + ki$ the letter α instead of φ in order to make it different of φ of the multiple root) then results

    $(1+ki)\,w^n = \cos n\varphi - k\sin n\varphi + [\sin n\varphi + k\cos n\varphi]\,i$

Equating the real part and the imaginary part separately with zero we obtain

$$\tan n\varphi = \frac{1}{k} = \tan(90° - \alpha) \quad \text{or } n\varphi = 90° - \alpha \text{ or}$$

$$\varphi_r = \frac{90° - \alpha}{n} + k_1\frac{\pi}{n} \tag{3a}$$

and similarly for the imaginary part

$$\tan n\varphi = -k = -\tan \alpha \quad \text{or} \quad n\varphi = -\alpha \quad \text{or}$$

$$\varphi_i = \frac{-\alpha}{n} + k_1 \frac{\pi}{n} \tag{3b}$$

where $k_1 = 1, 2, ...n-1$ (all integer numbers).

Relations (3a) and (3b) show directions of the lines along which the real or imaginary part of w^n is equal with zero, i.e. their root lines.

φ_r and φ_i are measured from the reference axis Ox

The term $k_1 \dfrac{\pi}{n}$ in these relations is called the term of multiplicity, because it adds to φ_r or φ_i $n-1$ more directions.

Proof:

If φ_r or φ_i is substituted in the expression of $Pn = w^n$ then the arguments $\varphi_r + k_1 \dfrac{\pi}{n}$ or $\varphi_i + k_1 \dfrac{\pi}{n}$ become $n\varphi_r + \pi$ resp $n\varphi_i + \pi$, but if for $n\varphi_r$ or $n\varphi_r$ $Pn = 0$ then it will be zero also for $\varphi_r + \pi$ resp $\varphi_i + \pi$.

N. Consequences of relation (3a) or (3b)

Consider a polynomial which has a triple root in the origin and another root at $-3 - 2i$. (See Fig. 5). It can be written as

$$P_4 = x^3(x + 3 + 2i)$$

In the origin $x = 0$ and the parenthesis expression reduces

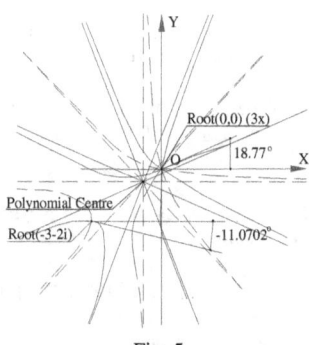

Fig. 5

to the complex constant $3 + 2i = 3\left(1 + \dfrac{2}{3}i\right)$ so in this

case $k = \dfrac{2}{3} = .6666$ $\alpha = \tan^{-1}\left(\dfrac{2}{3}\right) = 33.690067°$

This is exactly the case presented in the previous paragraph (M.), i.e. the complex multiple root $x^3 (= w^3)$ is multiplied by the complex constant $3 + 2i$.

Direction of the real root line will be then given by the angle φ_r equal with (using relations from Par. M.):

$$\varphi_r = \frac{90° - \alpha}{n} + k_1 \frac{\pi}{3} = 18.77° + 60° \quad \text{where}$$

$$\alpha = \tan^{-1}\left(\frac{2}{3}\right) = 33.690077° \quad n = 3 \quad \text{(multiplicity of the}$$

root) $k_1 = 0, 1, 2$

Note that in these relations n is the multiplicity, i.e. the number of roots in the multiple root (in this case $n = 3$) and *not* the order of the polynomial (which in this case is 4!)

Hence results that this polynomial has root lines which in the origin O (where in this case the triple root is) are rotated by φ_r and φ_i and to these directions are added two (in general n-1) more, so the total of directions in which the polynomials real or imaginary part is zero is is equal with n, i.e. the order of multiplicity (in general n and in this case 3) of the root.

The factor $3 + 2i$ is the influence of the other root on the multiple root. It causes a rotation of *all root lines which pass through the multiple (in this case triple) root* by φ_r resp φ_i but the interval between the lines remains always the

same $\left(\dfrac{\pi}{3}\right)$ because of the term $k_1 \dfrac{\pi}{3}$ (in general $k1 \dfrac{\pi}{n}$)

Further it is easy to see that the complex multiplier is $3 + 2i$

$$= 1 + \frac{2}{3} i = a(1 + \frac{b}{a}) \quad \text{and} \quad \alpha = \tan^{-1}\left(\frac{b}{a}\right) \quad \text{where a is the}$$

real part and b the imaginary part of the simple root

In this concrete case (see Fig. 5) it was found that $\varphi_r = 18.77°$

This is the angle of the tangent to the first real root line with

the Ox axis. The other root lines are at equal intervals of $\dfrac{\pi}{3}$

O. Case of several multipliers.

In case that the polynomial has more than one complex roots in different points, these can be multiplied as complex numbers and their product is

$$\rho_1 \rho_2 \rho_3 ... [\cos(\alpha_1 + \alpha_2 + ...) + i\sin(\alpha_1 + \alpha_2 + ...)]$$

The arguments of the roots can be added up to

$$\alpha_{tot} = \tan^{-1}(k_1) + \tan^{-1}(k_2) ...$$

Each argument is measured from the positive Ox axis. In case of a Root Number k, best method is to set this root at the origin, but so that the relative position (i.e. distance) between the roots remains the same. (The simplest solution is then to move the reference axes and not the roots!). All roots have to enter in the sum except the Root k

If there is a multiple root, it is like several simple roots added up, i.e. the *argument* of the multiple root is multiplied by the number of multiplicity of the root.

The total argument α_{tot} is then used in the relations of Par. M. (instead of α).

This theory shows that the root lines at *each root* of a polyno-mial are rotated by a value which depends on *all other roots* of the polynomial. This has some similarity with a system of celestial bodies (like the sun and its planets) where each planet influences the others, but in case of a polynomial the influence depends only on the *angle* of the root, relative to the other root. The angles of all roots are then added up numerically.

P. Rotation of the root lines at a simple root.

With the same relation as in Par. M. to O. we can calculate also the rotation of the root lines at the simple root $3 + 2i$.

The 3 times multiple root is equivalent with three simple roots. (See also observations in the previous paragraph O. regarding multiple multipliers).

For one root we have

$$\alpha = \tan^{-1}\left(\frac{2}{3}\right) = 33.69° \quad \text{For three roots:}$$

$$\alpha_{tot} = 3 \cdot \tan^{-1}\left(\frac{2}{3}\right) = 3 \cdot 33.69 = 101.07°$$

Then $\quad \varphi_r = \dfrac{90^0 - \alpha_{tot}}{n} = \dfrac{90 - 101.07}{1} = -11.07°$

See for this again Fig. 5.

Q. Polynomials multiplied by a complex number.

The first coefficient (i.e. that of x^n) is in most cases of polynomials equal with 1 and $C_{ny} = 0$ (i.e. C_n has no imaginary part). If a polynomial is given in form (2) and the multiplications are performed to obtain form (1), then C_n will result = 1 (real), even if all roots are complex.

So if $C_{ny} \neq 0$ then it can be assumed that the polynomial was multiplied by $C_{nx} + i \, C_{ny}$. This complex factor then produces a rotation of the asymptotes which can be calculated with Rel. (3) where in this case $k = \dfrac{C_{ny}}{C_{nx}}$ $\alpha = $

$\tan^{-1}(k)$

Or, another (better) method is to divide the whole polynomial by the first, complex coefficient $C_{nx} + i \, C_{ny}$ After this division $C_{nx} = 1$ $C_{ny} = 0$ and the polynomial will be a 'regular' polynomial.

The roots remain after division the same, because the polynomial is multiplied just by a constant.

The polynomial centre can be calculated from the divided polynomial. It is the same for both polynomials, because it is a function of *all roots*, and these are the same for both.

As an example consider the following polynomial, given in Ref [5]

$$2x^4 - 30x^3 + 163x^2 - 1773 + (3x^4 + 2x^3 + 472x + 4208)i$$

In this case C_{ny} is not 0, $k = \dfrac{C_{ny}}{C_{nx}} = \dfrac{3}{2} = 1.5$

$\alpha = \tan^{-1}(1.5) = $ therefore all asymptotes are rotated by

$$\varphi_r = \frac{90^0 - \tan^{-1}(1.5)}{4} = 8.422517°$$

φ_r is measured from the Ox axis

If the whole polynomial (i.e. all coefficients) are divided by the first coefficient, i.e. $C_{x4} + C_{y4} \, i = 2 + 3 \, i$ then we obtain the following polynomial:

$13x^4 - 54x^3 + 326x^2 + 1416x + 9078 + (94x^3 - 489x^2 + 944x + 13735)i$ or
$x^4 - 4.15384815x^3 + 25.076923x^2 + 108.923077x$
$698.307692 + (7.230769x^3 - 37.615385x^2 + 72.615385x + 1056.5385)i$

This polynomial has the asymptotess according to the general rule (first real asymptote at $\dfrac{\pi}{2n}$) and the same roots. Hence results the polynomial centre as:

$$\text{PolCtr}_x = -\frac{C_x(n-1)}{nCx(n)} = \frac{54}{13 \cdot 4} = 1.0384615$$

$$\text{PolCtr}_y = -\frac{C_y(n-1)}{nCx(n)} = -\frac{94}{13 \cdot 4} = -1.8076923$$

R. Sum of the angles of the roots of a polynomial related to a point M on a root line of the real part.

If the sum of the angles of all roots of a polynomial related to a point M is equal with 90° then that points is situated on a root line of the real part.

S. Sum of the angles of the roots of a polynomial related to a point M on a root line of the imaginary part.

If the sum of the angles of all roots of a polynomial related to a Point M is equal with 0 then the point is situated on a root line of the imaginary part.

T. Proof of Par. R and S.

If the point M is placed in the origin then the value of the polynomial in point M is equal with

$$P = (-a_1 - b1i)(-a_2 - b_2i) \ (-a_n - b_n i)$$

or written in trigonometric form

$$P = \rho_1 \rho_2 \, \rho_3 \ldots [\cos(\varphi_1 + \varphi_2 + \ldots) + i\sin(\varphi_1 + \varphi_2 + \ldots)]$$

From this expression it is obvious, that if the sum of the angles $\varphi_1, \varphi_2 \dots \varphi_n$ is equal with 90° then the real part of the polynomial is zero and if the sum is 0 then its imaginary part is zero (in point M).

U. Consequence of Paragraph P

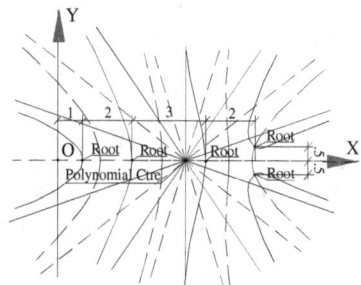

a. The original roots on the Ox axis

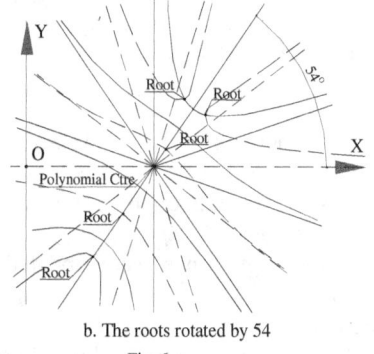

b. The roots rotated by 54

Fig. 6

If all roots of a polynomial are rotated by an angle φ as a rigid body around the polynomial centre, then if φ points in a principal direction then all root lines of the polynomial rotate by the angle φ but their status could be reversed (i.e. a root line which was of the real part becomes of the imaginary part and vice-versa) if the direction defined by φ corresponds to another principal direction.

Proof.

Because of the rigid rotation, the angle at which a root is seen remains the same as before, but the angle φ is added to it n times (for each root). This additional value of $n\varphi$ may cause that the total angle to be a multiple of 90°. In case of Fig. 6 the rotation being 54°, $n\varphi = 5 \cdot 54 = 270 = 180° + 90°$. The 180° doesn't change anything, but the 90° changes. If sum of the angles initially was 0, now it becomes 90°, and if it was 90°, now it will be 180°, that means all root lines will change their status.

As an example consider a polynomial with real roots at x = 1, x = 3 and x = 6 and a pair of complex roots x = 8 + .5i and 8 − .5i. This polynomial has the equation

$x^5 -25x^4 + 225x^3 -903x^2 + 1566\,x - 864$
$+ (-6x4 +106x3 - 622x2 + 1350x - 828)\,i$

If all roots are rotated by 54° around the polynomial centre (at x = 5.2) in positive direction then the roots will be:

2.7313019 − 3.397871i	3.906872445 − 1.779837386i
5.6702282 + .647213595i	6.4412902 +2.55914921i
7.250307204 + 1.971354968 i	

These roots correspond to following polynomial (using Rel. [2])

$x^5 -26x^4 + 276.3176748x^3 -1572.5094075x^2 +$
$4222.1774568x -4701.3827155856$
$+ (-18.2127322x^3 + 288.704436236x^2$
$-1569.03213486x + 2973.05628351)i$

This polynomial has the same root lines as the previous one but all root lines are reversed, i.e. if a root line was before real, now it will be imaginary and vice–versa.

See also the proof at the beginning of this paragraph.

Observe that the rotation of 54° is an integer multiple of

$\dfrac{\pi}{2n}$ = 18 which is the interval between two different root

lines (one of the real part and one of the imaginary part)

See Fig 6. for the initial and rotated polynomial

V. References

[1] Moore, J. B. *A Convergent Algorithm for Solving Polynomial Equations,* Journal of the Association for Computing Machinery, Vol. 14, No. 2, April 1967. pp. 211-315

[2] Garrett Birkhoff, Saunders Mac Lane, *A Survey of Modern Algebra*, MacMillan Publishing Co, Inc, New York, 1977.

[3] A. Ralston, P. Rabinowitz, *A First Course in Numerical Analysis*, 2nd ed, McGraw-Hill, New York, 1978.

[4] F.R. Ruckdeschel, *BASIC Scientific Subroutines, Vol I and II*, Peterborough, NH: BYTE Books, 1981.

[5] W.H. Press, B. P. Flannery, S. A. Teukolsky, W. T. Vetterling, *Numerical Recipes, The Art of Scientific Computing*, Cambridge University Press, Cambridge, New York, New Rochelle, Melbourne, Sydney, 1986-7

[6] Osborne/MacGraw-Hill Science and Engineering Programs Apple II Edition

[7] The 27th Annual Congress of the American Romanian /Academy of Arts and Sciences (ARA), May 29 – June 2 2002, Universitatea of Oradea, Proceedings, Vol. II, Pg. 841, Nicholas Schmidt, Theory of the Root Lines of the General Polynomial (Part 1)

ANESTHETIC CONSIDERATION IN CARDIAC PATIENTS UNDERGOING NONCARDIAC SURGERY

Husham Mohamed Abdallah Zakaria

Universitatea de Medicina și Farmacie "Carol Davilla", Str. Dionisie Lupu nr. 37
Bucharest, Romania; Alshaab Teaching Hospital, Khartoum, Sudan
e-mail: hishca@gmail.com

Abstract: One ofthe biggest challenge today is the safe conduct of anesthesia for patients who might be elderly, have preexisting cardiac diseases and are secheduled to undergo noncardiac suergery. Within financial conestraints of today health services. The aproppriate investigations need to be decided and performed for these patients, in order to inform the anesthetist,surgeon and the patient of the risk of surgery. These should be undertaken only if they will influence management of the patient. The preoperative assessment will help with the formation of a perioperative management plan. Including preoperative optimization and postoperative care, inorder to minimize the risk of an adverse autcome.The most recent guidelines for preoperative evaluation for noncardiac surgery are discussed in details, including assessment of risk factors, andcardiac investigation. Current thinking in preoperative therapy, intraoperative management and postoperative management is discussed, although most patients with cardiac disease have ischaemic heart disease, other specific cardiac conditions and the principles of their management are discussed.

Administration of anaesthesia to patients with preexisting cardiac diseases is one of interresting challenge. Most common cause of perioperative morbidity and mortality in cardiac patients is ischaemic heart diseases (IHD). IHD is number one cause of morbidity and mortality all over the world. Approximately 7 million are considered to be at high risk of IHD.

Goldman et al. reported that 500,000 to 900,000 myocardil infarction occurs annually worldwide with subsequent mortality and mortal other cardiacity of 10-25%. Care of this patients require identification of risk factors, preoperative evaluation and optimization, medical therapy, monitoring and the choice of appropriate anesthetic technique and drugs.

Risk factors -influencing perioperative cardiac mortality are: recent myocardial infarction; congestive heart failure; peripheral vascular diseases; angina pectoris; diabetes mellitus; hypercholesterolemia; dysrhythmias; age; renal dysfunction, obesity; life style and smoking.

Risk factors stratification.

In 1977 Goldman landmark cardiac risk index which was used extesively for preoprative risk assessment for the next tow decades.

Other cardiac risk indices where proposed and adopted. In 1996 a12-member task force of the American College of Cardiology and American Heart Association (ACC/AHA) published guidelines regarding the preoperative cardiovascuar evaluation of patients undergoing noncardiac surgery.

In 2002 these guidlines were updated based on new data:

Evaluation:

Patient having any sort of cardiac ailment need to be evaluated properly preoperatively.

History:

Elicits the severity, progression and functional limitation introduced by cardiac disease. History should include:

1. Exercise tolerance it depicts reserve, it can be exellent. History of participating in sports. Adequate patient able to climb stairs, run a short distance. Poor patient able to do leisure activities only e.g slow ballroom dancing or can walk around in the house only.

2. Angina pectoris: it is symptomatic manifestation of myocardial ischaemia characterized by typical substernal pain which is provoced by physical exertion and relieved by rest or sublingual nitroglycerine.

3. Myocardial infarction: the incidence of MI during the perioperative period is related to time period since the previous MI

4. Co-existing noncardiac diseases:
   a-peripheral vascular diseases
   b-cerebrovascular diseases
   c-chronic obstructive diseases
   d-renal dysfunction
   e-diabetes mellitus
   f-anaemia, polycythemia, thrombocytosis
5. Current medication, awareness about the medication that patient is taking is important during anaesthesia. All medication like Beta blockers calcium channel blockers, nitrates should be continued untill the morning of surgery.
6. Congestive cardiac failure.
7. Dysrhythmias.

Examination:

A carefull general physical examination should be done. It's include assessment of vital signs like blood pressure, pulse rate, rhythm, jugular venous pulse, oedema, pallor, cyanosis, clubbing, jaundice, lymphadenopathy.

In systemic examination, cardiovascular system should be examined for heart sounds, any murmurs. Respiratory system also needs to be assessed in details.

Laboratory investigations:

Cardiac specific tests like ECG, Echocardiograhy, to know ejection fraction, any valvular lesions, wall motion abnormalities, LV function and pressure gradients. Holler monitoring, treadmill test, thallium scintigraphy to detect myocardium of risk, radionuclide ventricuography, dobutamine stress test (DST) for evaluating inducible ischemia in patients who have poor functional capacity. Coronary angiography in patients where DST is positive should be done.

Anesthetic management:

anesthesia goals remain
   1-sable hemodynamics
   2-prevent MI by optimizing myocardial oxygen supply and reducing oxygen demand
   3-monitor for ischemia
   4-treat ischemia or infarction if it develop
   5-normothermia
   6-avoidance of significant anaemia

Management depends upon the type of surgery whether emergency or elective. For emergency surgery proceed for the surgery with medical management of cardiac ailment. For elective surgery perioperative management depends upon, clinical risk factors and surgery specific risk factors.

Clinical risk factors:

Obtained by history, physical examination review echocardiograhy, risk factors are grouped into 3 categories:
   1- major clinical predictors:
      -unstable coronary syndrome
      -decompensated heart failure
      -significant dysrrhythias andsevere valvular diseases
   2-Intermediate clinical predictors:
      -mild angina pectoris, previous MI by history or pathological Q waves, compensated or prior heart failure, insulin dependent diabetes mellitus and renal insufficiency
   3-minor clinical predictors:
      -hypertension, left bundle branch block, nonspecific ST-T waves changes and history of stroke

Surgery specific risk:
   1-High risk surgery: emergent major operation particularly in the elderly, aortic and other major vascular surgery. Anticipated prlong surgical procedures associated with large fluid shift or anticipated blood loss. Cardiac risk is > 5%.
   2-intermediate risk factors:carotid endarterectomy, head and neck surgery, prostate surgery, intraperitoneal and itrathoracic surgery, cardiac risk is <5%.
   3-Low risk procedures: endoscopic procedures, suerficial procedures, catarat surgeries, breast surgeries, cardiac risk is <1%.

Preoperative management:

At risk patients need to be managed with pharmacological and other perioperative interventions that can ameliorate perioperative cardiac events
   1-Optimisation of medical management.
   2-Revascualarization by PCI, revascularizationby surgery by (CABG).

B-blockers have been shown to be useful in reducing periopertive morbidity and mortality in high risk patients and preferably titrated to a heart rate of 50-60 bpm. Alfa 2 agonists can be useful in patients where Beta blockers are contraindicated. Nitroglycerine lowers LVEDP by reducing preload, it improves collateral coronary flow and reduce systemic blood pressure.

Coronary intervention should be guided by patients cardiac condition and by potential consequences of delaying noncardiac surgery for recovery after coronary revascularization.

Preanesthetic considerations:

-Preoperative visit is very important

-Concent obtaining

-Explanation about risk factors

-Continue the medication till the day of surgery. Beta blockers, Calcium channel blockers, Digitalis, Potassium level shoud be normal as hypokalemia can cause digitalis toxicity. Anticoagulants should be stopped.

Premedication:

-To reduce anxiety, to prevent increase in blood pressure and toheart rate, which can disturb the myocardial oxygen supplyand demand and can jnduce ischemia.

Combination of Benzodiazebine and opioids should be given one hour prior to arrival in operating room (Figure 1).

Intraoperative management:

Monitoring:

-ECG should be set on diagnostic mode, monitoring 3 leads improve recognition of ischemia.

-Blood pressure, pulse oximetry, capnography, temperature monitoring, urine output monitoring, control of venous pressure and cardiac output, TEE.

Choice of anesthesia:

Drugs should be selected with the objective of minimizing demand and optimum supply of oxygen and some cardiac drugs shoud be available to maintain hemodynamics to prevent and treat ischemia if it occurs.

General anesthesia

1-intravenous anesthetics

-Thiopental decreases myocardial contractility, preload and blood pressure with slight increase of heart rate, it should be administered slowly.

-Propofol decreases blood pressure and heart rate significantly there is adose dependent reduction in myocardial contractility

-Ketamine is not good with IHD and valvular diseases

-Midazolam decreases mean arterial pressure and increases heart rate, provide exellent amnesia.

-Etomidate causese minimum hehodynamic changes. Excellent for induction in patients with poor cardiac reserve.

Narcotics: morphine is the preferred drug for its relative cardiac stability and very good analgesic effect

-Inhalational agents: isoflurane for patients with good myocardial contractility. Halothane has disadvantage of myocardial depression and potential of dysrhythmias.

-Nitrous oxide provides stabe hemodynamic

-Muscle relaxants vecuronium produces minimum hemodynamic alteration and is a short acting pipecuronium, mivacuriumand dexacurium without any significat cardiovascular side effect.

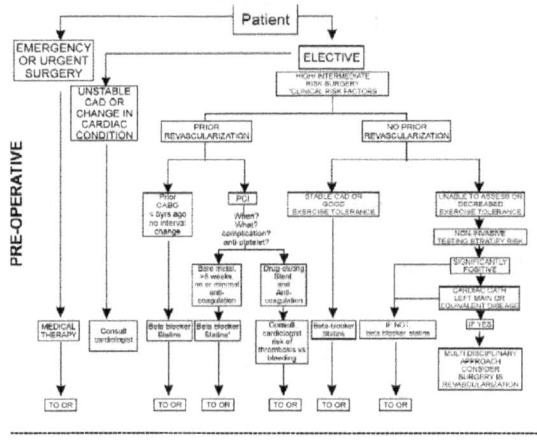

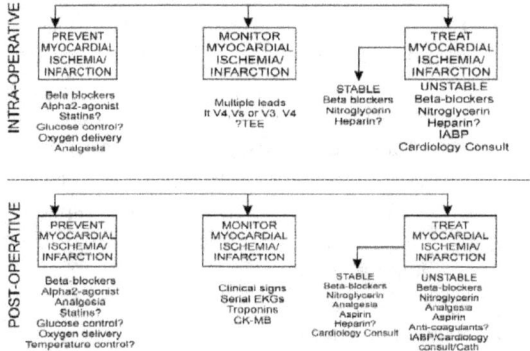

Figure 1 - Scheme of evaluation of the patient

-Glycopyrrolate is preferred over atropine, it producses less tachycardia.

-Regional anesthesia. The potential and well known advantage of RA over GA should be asset in cardiac patients if surgery can performed under RA disavatge of RA include hypotention, care should be taken while giving local anesthetics

Managing intraoperative complications:

1-Ischemia if patient hemodyamically stable Beta blockers and nitroglycerine. Heparin after consultation winth surgeon.

If patient hemodynamicaly unstable support with inotropes.use of intraoperative ballon pump may be necessary urgent consultation with cardiologist to plan for earliest possible catheterization other complication like dysrhythmias, paemaker dysfunction should managed accordingly.

Postoperative management:

Goals:

-Prevent ischemia, monitor for myocardial infarction, treatment for myocardial infarction.

-Although most cardiac events occurs within first 48 hours delayed cardiac event (with 30 days) still happen, and could be the of secondary stress, postoperative stress of extubation, pain, sepsis, hemorrage, anemia,, respiratory problemes can increase demand on the heart and should be minimized and treated.

Valvular heart diseases:

Patients with valavular heart diseases coming for surgery present many challenges .Now it is no longer necessary or even advisable to delay surgery until advanced symptoms are present.The five variables in dealing with the valvular heart diseases are important I-preload II-afterload III-myocardial contractility IV-heart rate V-rhythym.

Hypertension:

Is the commonest cardiac disease all over the world.these patients are decumented to have associted CAD left ventricula dysfunction, renal failure which incease the perioperative risk it is important to control BP preoperatively.But this does not need surgery to be deferred for weeks, to achieve ideal blood pressure control, inpatients with mild to moderate hypertension.it is important to evaluate for target organ damage .it is advisable to continue antihypertensives till day of surgery .any factors of sympathetic stimulus should be avoided.

Dysrhythmias:

May be a marker of severity of underlying CADor left ventricular dysfunction.A symptomatic ventricuka ectopics with stable hemodynamic parameters does not need any treatment preoperatively.prophylactic treatment is not required in supraventricular tachycardia. In a trial fibrillation rate needs to be controlled perioperatively if they occur can be treated by calcium channel blockers beta blockers adenosine.

Patients with conduction delay LBBB do not require pacing unless there is a history of syncope. But in complete heart block patients need to be paced. In patients with pace makers, electro cautary should be used with caution and for minimum period of time. The cautary plate should be as far as possible from the heart. Use of bipolar cautary decreases the risk of pacemaker dysfunction.

5. References

[1] Barash B.G., *Sequential monitoring of myocardial ischemia in the perioperative period.* In: American Society of Anaesthesiology, 2005, p. 411.

[2] Detsky A.S., Abrams H.B., Forbath N. et al., *Cardiac assessment for patients undergoing noncardiac surgery. A multifactorial clinical risk index.* Arch intem Med 1986, 146:2131

[3] Dupius J.Y., Labinaz M., *Noncardiac surgery in patients with coronary artery stent :what should the anesthesiologist know*, Can J Anesth 2005:52:356

[4] Eagle K. A., Brundage B., Chaitman B. et al., *Guidelines for perioperative cardiovascular evaluation for non-cardiac surgery.* AHA/ACC task force report. J Am Coll Cardiol 1996, 27:910

[5] Eagle K. A., Berger P.B., Calkins H. et al., *ACC/AHA guidelines update for perioperative cardiovascular evaluation for noncardiac surgery-executive summary.* A report of American College of Cardiologist/American Heart Association Task force on Practice Guidelines (Committee to update the 1996 guidelines on Preoperative Cardiovascular Evaluation for Noncardiac Surrey).Anesth Analg 2002; 94:1052

[6] Goldman L., Caldera D., Nussbaum S. et al., *Mulyifactorial index of cardiac risk in noncardiac surgical procedures.*N Engl JmED 1977:297:845

[7] Hall M. J., Owings M.F., *Advance Datafrom vital and health statistics* No:329, National Hospital Discharge survey. Hyattsville MD:Department of health and human Services; 2002, No:329

SFÂNTUL IERAH MARTIR ANTIM IVIREANUL

Pr. Prof. Dr. Cezar Vasiliu
Univ. Sherbrooke, Canada
cezarvasiliu@yahoo.com

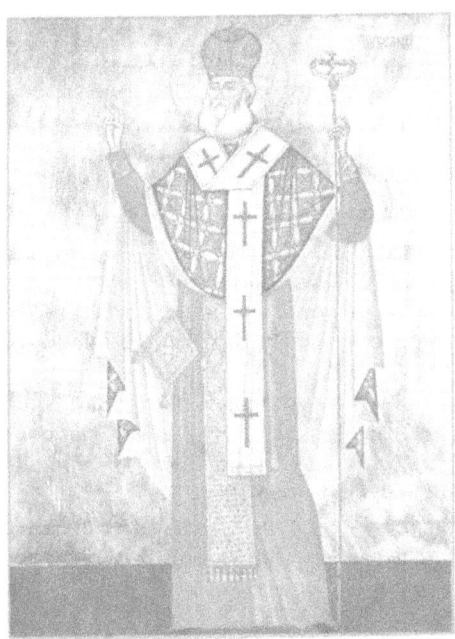

La 27 Septembrie 2016, se împlinesc 300 de ani de la moartea martirică a Mitropolitului **Antim Ivireanul** al Ţării Româneşti. El a fost colaborator pe linie culturală al Domnitorului Constantin Brâncoveanu, martirizat de turci cu copiii, la 15 august 1714.

De-a lungul bimilenarei noastre istorii, conducătorii românilor - fie că s-au numit prinţi, duci, voievozi, domnitori sau regi - au fost personalităţi de mare valoare, prin demnitatea, curajul, omenia şi cinstea cu credinţa în Dumnezeu. Ei au ctitorit salba de mânăstiri şi biserici care înfrumuseţează plaiurile româneşti.

Din ceremonialul înscăunării domnitorilor-noştri nu lipsea obiceiul fastuos al „ungerii" de către mitropolitul ţării, adică primirea binecuvântării divine pentru conducerea poporului. Mitropolitul era una dintre persoanele cele mai respectate ale curţii, fiind, la rându-i, confirmat de domn. Titulatura oficială a domnitorului preciza că domneşte „din mila şi cu darul lui Dumnezeu".

Istoria noastră a păstrat date despre cooperarea dintre domni şi ierarhi. De exemplu, în Muntenia, între Nicolae Alexandru Basarab şi Iachint, între Neagoe Basarab şi Macarie, între Matei Basarab şi Grigore I sau între Constantin Brâncoveanu şi Antim; în Moldova, între Alexandru cel Bun şi Iosif, între Ştefan cel Mare şi Danil Sihastru (duhovnic) sau între Vasile Lupu şi Varlaam. Aceleaşi bune relaţii au continuat şi după instaurarea regalităţii, deşi primii doi Regi - Carol I şi Ferdinand - au rămas catolici.

Sfântul Antim Ivireanul s'a născut la 1650 în Georgia, sau Iviria, numindu-se din botez Andrei. A căzut de tânăr rob la turci şi, după eliberare, a trăit ca monah pe lânga Patriarhia Ecumenică, învătând carte şi arta sculpturii în lemn, a picturii, broderiei şi caligrafiei. Domnitorul Constantin Brâncoveanu l-a adus în Muntenia în 1690, în scopul realizării planurilor sale culturale. Aici a învăţat limba română şi a început opera de tipograf, ca ucenic al lui Mitrofan, fost episcop de Husi.

În 1691 ia conducerea tipografiei bucureştene, prima carte tipărită fiind *"Învăţăturile lui Vasile Macedoneanul către fiul său Leon"*. Au urmat *"Slujba Sfintei Paraschiva şi a Sfântului Grigore Decapolitul"* (1692) şi *"Psaltirea românească"* (1694).

Între 1696 şi 1704 este egumen la Snagov, unde - în primii cinci ani - va tipări 15 lucrări: şapte în grecește, cinci în română, una în slavonă, una în slovo-română şi una în greco-arabă. Dintre cele româneşti amintim : *"Evanghelia"* (1697), *"Acatistul Născătoarei de Dumnezeu"* (1698), *"Învăţături creştineşti"* (1700), *"Floarea darurilor"* (1701), cunoscută pentru conţinutul ei moralizator etc. Şi-a creat ucenici, cei mai importanţi fiind Mihail Ştefan şi Gheorghe Radovici.

Cea mai fecundă perioadă ca tipograf o are la Bucureşti între anii 1701-1705, imprimând tot 15 cărţi: două în romană - *"Noul Testament"* (după ediţia de la Alba Iulia din 1648) şi *"Acatistul Maicii Domnului"* (1703) - una în slavo-română, una în greco-arabă şi 11 în grecește.

EPISCOP AL RÂMNICULUI.

În 1705, egumenul Antim de Snagov este ales Episcop de Râmnic, fiind hirotonit de Mitropolitul Teodosie al Ungrovlahiei. Aici şi-a dovedit aceleaşi bune calităţi organizatorice şi culturale, înfiinţând prima tipografie din oraş şi tipărind - în numai trei ani - nouă cărţi: trei în greceşte, trei în slavo-română şi trei româneşti, prin care începe seria tipăriturilor româneşti menite să ducă la triumful definitiv al limbii române în Biserică. Amintim *"Evhologhion"* sau *"Molitvelnic"* (1706) în două volume, primul un *"Liturghier"* - întâiul tipărit în Muntenia - iar al doilea un *"Molitvelnic"* sau carte de rugăciuni. Tot la Râmnic s'a îngrijit de sporirea bunurilor Episcopiei şi la restaurarea mânăstirilor Cozia şi Govora.

MITROPOLIT AL UNGROVLAHIEI.

În anul 1708 este ales Mitropolit al Ungrovlahiei în locul lui Teodosie, păstrându-se frumoasa cuvântare de înscăunare în care se angaja să-l slujească pe Dumnezeu cu dăruire iar pe credincioşi cu jertfelnicie.

Aici spunea "Şi mă rog bunătăţii Lui şi iubirii Sale de oameni, să-mi lumineze mintea ca să pot propăvădui cuvântul adevărului şi să-mi întărească inima întru frica Lui, ca să pot păstori cuvântătoarea turma Lui cea aleasă, pe care a răscumpărat-o cu prea scump Sângele Său din mâna vrăşmaşului".

A mutat tipografia de la Râmnic la Târgoviste, tipărind 18 cărţi, dintre care 11 româneşti, cea mai importantă fiind *"Capete de poruncă"* (1714), precizând principalele îndatoriri ale preoţilor. Ultima carte tipărită la mânăstirea Antim, în 1716, *"Istoria lui Alexandru Mavrocordat Exaporitul"*, urcă la 63 cărţile tipărite de el într'un sfert de veac - dintre care 22 în româneşte - fiind alături de Coresi, cel mai de seamă tipograf din cultura veche românească – cum scria Pr. Prof. M. Păcuraru.

Pătrunderea limbii române în Biserică a necesitat timp. Primele încercări sunt făcute de Coresi, urmat de Mitropolitul Ştefan, care tipăreşte în româneşte rânduielile tipiconale, de Mitropolitul Teodosie, care introduce lecturile biblice în româneşte şi, în sfârşit, de Antim pentru textul slujbei preotului şi cântarea. În plus, Antim a tipărit în româneşte Liturghierul şi Molitvelnicul, prin care *"datina străină a primit o lovitură de moarte"* - cum scria Nicolae Iorga.

OPERA LITERARĂ

Constă din trei lucrări tipărite pentru nevoile preoţilor : *"!Învăţătura despre taina pocăinţei"* (1705), un mic *"Catehism"* (1710) şi *"Capete de poruncă"* (1714), precum şi unele manuscrise. Dintre acestea amintim: *"Chipurile Vechiului şi Noului Testament"* (azi la Kiev) şi, mai ales, *"Didahiile"* sau *"Predicile"* sale (ed. Gabriel Strempel, 1962), alcătuite din 28 predici şi 7 cuvântări ocazionale, citând din Sfânta Scriptură şi din Sfinţii Părinţi sau amintind filozofi şi poeţi din Antichitate. Ele arată *"hotărîrea mitropolitului de a curăţi o Biserică decăzută în moravuri"* (Nicolae Iorga), aminteşte de asuprirea socială şi naţională a poporului din partea marilor boieri şi a turcilor, afirmând în cuvâtarea de înscăunare *"suntem încongiuraţi de atâtea nevoi şi scârbe ce vin totdeauna neîncetat de la cei ce stăpânesc pământul acesta"*, sau în cuvântarea la Schimbarea la faţă *"suntem supuşi supt jugul păgânului şi avem nevoi multe şi supărări din toate părţile"*. Cât priveşte limba şi stilul, G. Strempel scria : *"Opera lui Antim Ivireanul este un preţios monument de limbă românească, ce se citeşte cu plăcere şi astăzi, după un sfert de mileniu de la data când a fost elaborată"*.

ANTIM-CTITOR.

Între 1713 şi 1715 a zidit mânăstirea cu hramul Tuturor Sfinţilor din Bucureşti, cunoscută ca mânăstirea Antim, după planurile sale. Biserica a fost împodobită în exterior şi în interior, dotată cu odoare, cărţi şi obiecte de cult şi având chilii pentru călugări. A înzestrat-o cu moşii şi alte venituri, din cheltuiala proprie, pentru monahi dar şi pentru săraci, cum reiese din Testamentul său, un model de operă socială a Bisericii.

SPRIJINITOR AL BISERICII ORTODOXE.

Nu şi-a uitat neamul din care s'a născut şi a donat la Tbilisi o tipografie cu caractere georgiene, trimiţând acolo pe cel mai important ucenic al său, Mihail Ştefan. De asemenea, a pus bazele primei tipografii cu caractere arabe instalată la Alep, în 1706. Pentru credincioşii greci din Patriarhiile Constantinopolului, Alexandriei şi Ierusalimului a tipărit numeroase cărţi de slujbă şi teologice în greceşte.

Mitropolitul Antim s'a îngrijit şi de credincioşii din Ardeal, tipărind lucrarea intitulată *"Lumina"*, pentru combaterea prozelitismului catolic al vremii şi a trimis la Alba Iulia pe Mihail Ştefan, care a tipărit aici o *"Bucoavna"* (Abecedar) şi un *"Chiriacodromion"*, ambele în 1699.

ANTIM-PATRIOT.

Pe lângă interesele Bisericii, Mitropolitul Antim s'a dovedit şi un patriot iluminat. El a apărat drepturile sale, ca Mitropolit al Ungrovlahiei, asupra mânăstirilor închinate Sfântului Munte Athos, conduse de călugări greci, care nu mai pomeneau la slujbe pe mitropolitul locului. În plus, a luptat împotriva opresiunii turceşti, militând pentru apropierea de Rusia lui Petru cel Mare. Când Constantin Brâncoveanu a vrut să-l înlăture din scaunul mitropolitan, el s'a apărat în scris - în două rânduri în 1712 -, rămânând ca ierarh până la moartea martirică a lui Brâncoveanu (1714) şi sub noul Domn Ştefan Cantacuzino (1714-1715).

MOARTEA MARTIRICĂ.

Un destin tragic a unit pe domnitorul Constantin Brâncoveanu cu Mitropolitul Antim - martiriul. După ce pe tronul muntean a ajuns, în Decembrie 1715, Nicolae Mavrocordat, primul domn fanariot, au început zile negre pentru Antim. În 1716, cu ocazia războiului ruso-austriac, domnul - impus de turci - a părăsit Capitala, plecând spre Dunăre. La 18 August 1716, după o discuţie aprinsă, la Călugăreni, Antim s'a întors la Bucureşti, la credincioşii săi. Aici a continuat lupta pentru înlăturarea fanariotului, alegând domn pe Pătraşcu Brezoianu.

Reîntoarcerea lui Nicolae Mavrocordat pe tron, în Septembrie 1716, a însemnat şi pedepsirea lui. Mai întâi pe linie bisericească, domnitorul cerând Patriarhiei Ecumenice caterisirea. Aceasta sentinţă injustă a fost corectată, la cererea Patriarhiei Române, de către Patriarhul Athenagoras I, abia la 8 Martie 1966. A urmat exilul, spre Sinai, unde - din nefericire - n'a mai ajuns. Pe drum a fost ucis de turci prin înecare în râul Tingea, lângă Adrianopol ; era 27 Septembrie 1716, dată la care a intrat în calendarul Bisericii noastre. Aşa a sfârşit Mitropolitul Antim Ivireanul, martirizat de turci, ca şi marele sau binefăcător, domnitorul martir Constantin Brâncoveanu.

În **concluzie**, putem spune că Mitropolitul-martir **Antim Ivireanul**, deşi născut pe alte meleaguri, s'a identificat cu poporul pe care l-a slujit, pentru care a tipărit atâtea cărţi, contribuind la desăvârşirea procesului de românizare al slujbelor bisericeşti.

Ca păstor de suflete - Episcop şi Mitropolit - a slujit lui Dumnezeu şi credincioşilor prin pilda vieţii, prin îndemnurile spiritual-morale şi, mai ales, prin Testamentul său, prin care a lăsat întreaga sa avere pentru opere de caritate.

Patriot luminat şi apărător al creştinilor din Patriarhiile răsăritene, Mitropolitul Antim a plătit cu viaţa îndrăzneala de a se fi ridicat împotriva turcilor şi a primului domn fanariot.

Toate aceste considerente au determinat Sfântul Sinod al Bisericii Ortodoxe Romane să-l canonizeze, alături de alte personalităţi religioase româneşti, în vara anului 1992.

Sf. Sinod al Bisericii Ortodoxe Romane a declarat anul 2016 "Anul Comemorativ al Sfântului Ierarh Martir Antim Ivireanul şi al tipografilor bisericeşti"

Veşnică să-i fie pomenirea!

Bibliografie:

N. Dobrescu, **Viaţa şi faptele lui Antim Ivireanu, mitropolitul UngroVlahiei**, Bucureşti, 1910, pp. 119;

D. Teodor, **Despre Antim Ivireanul**, în ST nr. 3-4/1955, p. 236-263;

N. Şerbănescu, **Antim Ivireanul tipograf,** în BOR nr. 8-9/1956, p.690-766;

M. Ruffini, **Il metropolita valacco Antim Ivireanul**, în rev. "Oikoumenikon" III, 1966, p. 357-398;

G. Popescu, **Mitropolitul UngroVlahiei Antim Ivireanul, cârmuitor bisericesc şi propovăduitor al Evangheliei**, în ST 1-2/1969, p. 5-97;

M. Cazacu, **Cum a murit Antim Ivireanul**, în MO nr.7-8/1970, p. 671-691;

I.Ionescu, **Pătimirea mitropolitului Antim Ivireanul**, în "Sfinţi români"...p. 640- 662;

C.Vasiliu, **Mitropolitul martir Antim Ivireanul**, în "Lumea liberă" din 31 Octombrie 1992;

M. Păcurariu, **Sfinţi daco-romani si romani**...p. 109-113.

C. Vasiliu, **Sfinţii Neamului Românesc**, Montreal 2008, p. 185-191

The motivation of the imperative integration of the functional and informatics processing of the economic situational information.

Tudor Stefan Leahu

Free International University of Moldova, Iu. Gagarin bd., 2, 2001, Chisinau, Republic of Moldova, leahu.ts@mail.ru

Abstract: There are revealed in evidence, enumerated and examined the essential factors that motivated the integration of functional and informatics processing of situational information of the unitary economic managerial process. The fundamental ideas of the place, role, interconnections and interactions among the situational (informative, descriptive) information within the framework of the managerial economical system are formulated. It's analyzed the conceptual essence of these factors .

Keywords: motivation, imperative, functional, informatics integration, processing, economic situational information.

Introduction: The gradual evolution of the economic informational domain and the informatics substructures, which ensures it functioning, requires elaboration of the certain fundamental theoretical conceptions of the organization, structuring and transformation of the informative (descriptive, situational) information circulating in the medium of the economical management system in the pressing interconnection and interaction with the milieu informatics subsystem, that damages first.

The necessity in such an approach it's motivated and justified by the existent situation, when the economical informational activities aren't effectuated in the both named areas - "external" and "internal". Because of the permanently pronounced and practically boundless in this sense dynamics of economical material processes, which aren't being evolved without informational matching, there is almost impossible that mentioned situation would someday change in the direction of diminution.

This is the main reason, in the both: functional and informational aspects, the integration of the informational processes and the achievement of their transition from one (the functional) into another milieu (the informatics) and opposite, will probably remain in present.

Therefore,it's significantly valuable and they will dispose the correct estimation and selection in this basis of informatics resources, the following and the guidance of their results indisputable to the contribution and to the efficiency of the activity of economic informatics system (E.Ic.S.) and, not the last, to the increase of general, professional informatics and informational qualification and culture.

Content The problem of integration of the methods of functional and informatics processing is required by a lot of objective and subjective factors and circumstances, which is taking of the present, but particularly in a long term. Virtue of innovation, anticipated by revealing and explaining the essence of the latter (factors, circumstances), it's necessary to be taken in the view by detailed knowledge, deep awareness and taking in permanently consideration of their, convincing confirmations of the need persevering preoccupation of examined problems both the conceptual (theoretical) and practical (applied) plans. In connection with this, first of all, himself demanded to be cleared some new notions derived from the created situation in the domain, to determine the role and the place of such a category of information within the economic management unitary system, in order that afterwards it would motivate the imperative of the named integration, and the contribution to its factors.

Concerning the notions brings the followings. At present and in the foreseeable future it's certified the existence and circulation of economical information in two areas – managerial and informatics. The first it still considered "external", as informational processes are unfurling in the environment of the managerial system, in the outside of the technical means, while the second is considered rightly "internal", the same processes occurring in the interior of these means.

It stressed currently that of all those three informative stages of examined information (1 – primary (initially), 2 – processing, 3 - utilization) only the second is prevalent achieved in informatics modality.

Thus, a lot of procedures and operations of these stages, but also some characteristics for the stage of the data processing, remain fulfilled in manual modality.

Since the economical administration is scoring through certain functions (standardization, regulation, forecasting, accounting, a.s.o.), and lasts thanks to data transformation, these are unfailing through damaging themselves with information products, and with the processing of information within they, in our opinion, can be determined as functional. Such processing is being conditioned and started from the specific of leadership content of the problems of each functions, It's being predetermined and therefore is decisively influencing the informatics processing, while the latter is manifested through various ways of implementing in the physical environment of technical resources. The majority of data processing procedures and operations derived from managerial content of problems are not fulfilled of the subject. In this case, after the structure, composition and sequence of making it, this form the equivalent of those manual, still the informatics – the equivalent of those automatic.

Because of the pronounced dynamics of the terminology of domain investigation and permanently revision, there are enough formulated and used by large notions. Here is the answer to the question why, ,in the time there are enunciation and elucidation of the sense of the new terms, himself requests certain explanations of the existents. In this context, taking it from the beginning, in our opinion, itself the notion of processing requires some determinations and clarifications. Thus frequently in practices and in informatics theory any action exerted on the information is interpreted as information processing, indifferently of the influence or not, the latter on this composition, structure and values. From our point of view, in informational environment, this term can not always be applied to the more generalized sense, replacing any modification of information. Usually, this notion itself implies the calculation processing,the result of this, the changes and the value of the digital informational units.

Drawing a parallel with the material domain, it may come out that the processing of any resource tries to obtain one finished or semi finished product. However any deviously action without fail changes their composition, structure and value of information, after all following with the new product or the prefab. Hence, the more synthesized is the notion, the action exerted on information, is the transformation. Through it are understood any changes in the physical presentation form, composition and sequence distribution of information units within the entities, that includes them, their values as a result of information, structural and calculating processing. Generally, the transformation of information containing in oneself the modification of both the material appearance and the composition, structure, value and their functional content.

Lately, even more affirmative is winning the field the term of processing, which according to DEX means "…. to process the recorded signals on the magnetic tape (support - out)". As any informative action is physically fulfilled, it is thought that the processing is concordant with transformation, but it is achieved in informatics technical environment, with its own respective physical proceedings. So, the transformation itself refers more to the functional framework, but processing – to the informatics, both being of conceptual (imaginary) character.

Since in any domain, inclusively and the informational, of topical interest and better permanently dispose the automatic modality of achievement of works, in present material was to fit the utilization of term of the processing .All the more so that here are elucidated the problems of the constraint of transition and on this basis – the integration of functional (manual) with the informatics (automatic) achievement of the organization, structuring and processing through the selection of respective resources of fulfilling at the firsts

Referring to the categories of information circulating in the economic management system, it can be mentioned that depending on their predestination role, there is standing the situational and decisional information. The firsts, describing (reflecting) the existence and evolution of the managed object (process) in certain moments of time (past, present, future), periods of time (change, day, five days, month of work, a.s.o.), special scales (place of work, crew, sector, section, economic unit, a.s.o.) and the limits of evolution(rates, settlements, tasks). The second (decisional) being the basis of the initiation and guidance of the values during of fulfilled of material activities. Hence the situational information serves as substrate for the formulation of decisional units. At the same time, one of the situational information (of rate – setting, settlement, foreseeing, economic analysis) is disposing a decisional role, since, according to their values, is itself organizing the devolving and the working of the managed object (process), it determined in the measure of its actions.

Holding account of these considerations, it may be concluded whom economical situation it is thought the information what reflecting (describing) the material situation of the careful management (of fabrication, commercialization, consumption), but economic decisional – the information, in basis of which starting and it organized most efficient the functioning of her.

After studying several bibliographic sources we can conclude that after completion of the role in the unitary managerial process it may be one synonymization of the situational information with the informative. Both are recently introduced by

the undersigned. In this connection some memberships of economic informatics domain think that the term of the informative information is tautology so it just means that exists a need to introduce such a term as not informative information, which in their opinion, is a nonsense. By the way, upon this conception is principled, it demand a certain retort to itself. First of all such position is incorrect. In the new dictionary of synonyms (authors: L.Seche, M. Seche, I. Preda, Bucharest, The publishing house VOX, the third revised and enlarged edition, page 185) it comes out that: "INFORMATIVE adj. informational, orientational (Prices)", but in the Explanatory Dictionary of Romanian Language, second edition, page 491 " – INFORMATIVE – A. informativi, e (romanian author), adj. What possessing the role to inform, what serving such informer, of information …"

These definitions demonstrate the fact of existence of two varieties of information – informative and decisional. The first type only informs using description and explanations, but the second type is used for decision making and achievement of certain activities. This differences of information is motivated by the quality of this for the managerial process, which need to include the material and informational compartment. Therefore right source of report may serve not only material object (process), but the information itself. Consequently, in the unilateral descriptive aspect is justified and the notion of the informative information.

Due to the mentioned definitions, the information, which unilateral describe (reflect) the object (process) or action, take to help with molded situation only off on position, it considered informative. It can not serve like a basis of wording decision. For example, the accounting information, taking in isolation, without others categories of economic information express knowing of existence and evolution of the object (process) in the past. Only in combination with the foreseeing or standardization (settlement) information it contribute to the economic analysis, elaboration and wording of the certain decision. These things as standardization, settlement and foreseeing information, take particularly and reflect the limits or amble of the evolution in the future.

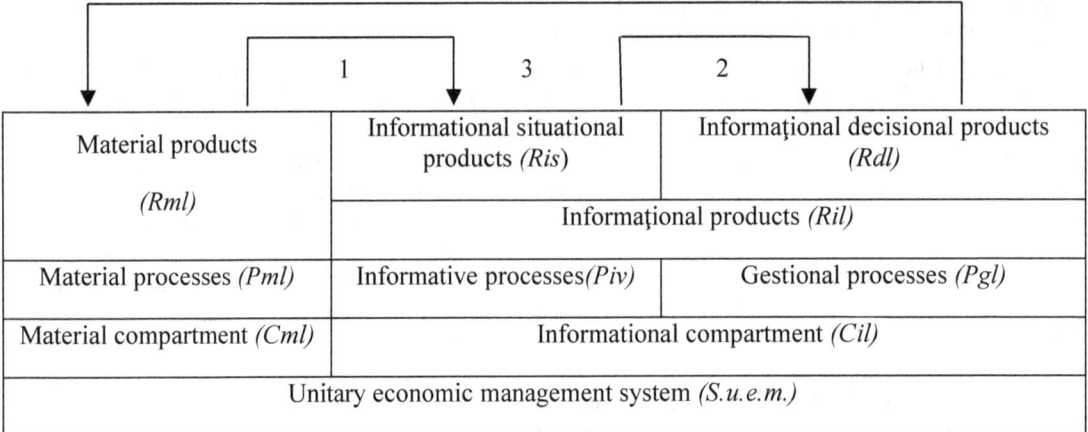

In accordance with the scheme in the fig.1, the tidiness of the descending decomposition on the making of *Su.e.m.* in the constituent component parts can be described through following analytic structural formula in the next succession:

$$Su.e.m. \equiv Cml \ V \ Cil \ (1) \ ; \ Cml \equiv Pml \ V \ Rml; (2);$$

$$Cil \equiv Piv \ V \ Pgl \ (3); \ Ril \equiv Ris \ V \ Rid \ (4) \ .$$

The examined scheme is elaborated starting with the principle of the motivation which predetermine the following consecutivity of functional process of *Su.e.m.*:

$$Su.e.m.: \ Pm \longrightarrow Piv \longrightarrow \widetilde{r} dl \qquad (5)$$

But, it is not excluded on the route of the evolution of each categories of processes so they produced tasks (returns, confirmations) towards the reality of the material, informative or decisional environment. That is why the above order is it's own for the situation when the managed object (process) there is in the dynamics of situation. If it's activities aren't at the initial level, then it is elucidated the presented order:

$$Pdl \longrightarrow Pml \longrightarrow Piv \quad (6)$$

The same scheme and the analytic formula (3) – (5) confirm the fact that in integral gestational process the situational information holds the intermediate position, being formulated in the following evolution of the material activities, after processed in the managerial system with the main goal of obtaining of the values of decisional units, that in continuation are offered to the leaded object for ensure in accordance with the lasts of its evolution of remotely. In this sense, the value of the examined information it's predetermined on their role as "primary matter" of starting of the gestational process. Without their scoring last it can not be initiated, as well as disposed on the "mission of link of the liaison between the managed and gestational under systems of the unitary economic managerial system. From this point of view, in the environment of custom system they are on the decisive predestination, as well as their absence is leading to the interruption and impossibility of the unfolding the indissoluble gestational process.

Conclusions: In the base of affected findings it can be mentioned that any preoccupations of investigative, elaborative and functional order of such information are always current and important.

Once there have been established the functional domains ("external", "internal"), the covering level of the informatics transformative means and methods, the contribution level of the realization of gestational functions, the compliance level of the aspects and the methods of its elucidation (functional – manual, informatics - automatic), as well as the clarification of the transformative, processing and assimilative notions of situational information with the informative one, it starts creating the best conditions for the establishment of the imperative integration of functional and informatics processing of the situational information.

In this context there can be mentioned the following decisive and contributive factors, which requires the objective preoccupation of such problematic integrations as:

1) the permanent trend for automatic processing (not automatization) in interconnection and interaction of both categories of informative resources (situational, decisional) at first stage, and of material and informative in the same integrated modality at last stage. Such mode of approach will contribute in the end to the establishment of the economic system automatically , which due to the Cybernetics father - N. Viener - statement was „ ...the dream of the XX century";

2) such method of conception requires the integration of the functional (manual) and computer (automatic) processing of material and informational works within the framework of unitary management system;

3) such integration of elucidated processes is a preliminary order and contributes to:

a) the highlighting of the entire composition and establishment of the logic of succession of the interactions between operations of the unitary transformational informational process , starting with the organization, structuring and finalizing, and the obtaining informational finished product;

b) the classification of operations of this process in manual and automatic operations ;

c) the highlighting of the modalities of transition from manual to technical operations;

d) the fastening and elaboration of informatics interfaces, which ensure such a transition;

e) supporting of automatic continuity of achievement of integral informational processes.

4) the economic situational informational field is significantly partly covered with informatics means and methods, practically all their workings belonging to the processing stage;

5) this covering will contribute to the supporting of automatic evolution of management functions;

6) since at the moment a lot of operations of the all three transformative stages are manually effectuated, a problem of the present, but also, especially of the future, is the ensurance of the easy, slowly and without suspensions transition from manual to automatic . In this sense primordially it's necessary to establish the composition and possible proceedings of achievement the operations in both environments (manual, automatic) for the integration (inclusion) within the framework of unitary informational process;

7) the automatic thoroughly achievement of management process, contributes its totally approach;

8) in the labour of processing more than 70 – 75% occupy the informational and structural procedures;

9) the rational organization and structuring of information increases the efficiency their processing;

The thorough knowledge, the awareness of the impartiality role, place, essence of the economic situational information, the guidance of the integrated conception of their functional and informatics processing, finally offer the conditions for both material, and spiritual prosperity of the human society.

References

[1] Tudor Şt. Leahu The evolution, the specific and the problems of constitution and functioning of automatized banks of intelligent economic data (A.Bn.Ig.E.D.).. The proceedings of the seventh

international conference on informatics in economy, Bucharest, may 2005. INFOREC Printing House, pp.845-851

[2] Tudor Şt. Leahu The functional – conceptual aspect of the composition, structure and working relations of the composition, structure and working relations of the automatized banks of intelligent economic data (A.Bn.Ig.E.D.). Volume X, 2006. The 2nd supplement of the review "Informatica economică". International Conference "Knowledge management.Projects, systems and technologies". Bucharest, november, 9-10, 2006. Volume II, pp.89 – 96. Review accredited by CNCSIS with B-level.

[3] Tudor Şt. Leahu. The Conception of the General Content of the Composition, Structure and Working of the Informative Components of the Economic Knowledge Base //INFORMATICA ECONOMICA JOURNAL Published with the support of Ministry of Education, Research and Innovation CNCSIS B+,vol.XII, Nr.4(48), pag.105-112.Bucharest, 2008

[4] Tudor Şt. Leahu Imperative integration of functional and informative process of situational economic informatics. Academy of Economic Studies of Moldova. Scientific international conference materials „The Republic of Moldova: 20 years of economical reforms" september 23 – 24 2011, book I, pp. 462 – 465.Chisinau, C.E.P/ al A.S.E.M., 2011

[5] Tudor Şt. Leahu The organisation, structuring and transformation of information of the economic management system. Monograph. Chisinau, E.P.S.M., 2009,
pp.90 – 96, 140 – 144, 169 – 191, 251 – 283.

Negative effects of massive emigration originating from the Republic of Moldova

Doina Guzun[1], Dorin Dusciac[2], Ala Mindicanu[3], Sandra Ayad[4], Frédéric Boisard[5]
[1] University Paris Diderot-Paris 7, France.
[2] President of the Association for the Integration of Migrants (AIM), Paris, France, Deputy Minister - Ministry of Environment of the Republic of Moldova (2014 – 2015).
[3] President of the Moldovan Community of Québec, Member of the Parliament of the Republic of Moldova (1994 – 2001).
[4] Head of the Center for International Research and Documentation of Sexual Exploitation, Scelles Foundation, Paris, France.
[5] Community manager, Center for International Research and Documentation of Sexual Exploitation, Scelles Foundation, Paris, France.
e-mail: doina.guzun@gmail.com

Acknowledgement: This contribution is part of the 4th Global Report Prostitution. Exploitation, Prosecution, Repression, edited by the Scelles Foundation (Paris, France), under the Direction of Yves Charpenel, Deputy General Prosecutor of the Supreme Court of France, President of the Scelles Foundation (ECONOMICA ED., 49, rue Héricart, 75015 Paris, France), 2016. The book will be available for free downloading on: www.fondationscelles.org

1 Introduction

According to the Prosecutor General of the Republic of Moldova, 164 crimes of human trafficking were reported during the first six month of 2015 (compared to 169 in the same period of 2014) along with 38 cases of procuring. The Moldovan Mission to the International Organization for Migration (IOM) estimated that 70% of procurers in Moldova are women (*IOM*, 2013). This can be explained by the fact that most of those who have been accused are already involved in prostitution abroad. For example, a woman between the ages of 20 and 23 who was first a victim of trafficking or a prostituted person, may subsequently become a procurer by the age of 26. The same study provided information on prostituted female witnesses, but there are no statistics on the number of Moldovan nationals practicing prostitution within or outside of the country (*Bulletin of the Supreme Court of Justice of the Republic of Moldova*, 2005). No other study or source to date provides comprehensive data on prostitution in Moldova. The question is whether this lack of data is due to a low rate of prostitution or Moldovan authorities ignoring many cases.

More than 50% of young Moldovans surveyed wish to go abroad, giving rising unemployment and miserable wages as their main reasons. Women constitute 68% of the unemployed and are exposed to longer periods of unemployment than men. Although they are educated to the same level as men, they make up three-quarters of the unskilled labor force and are paid 70-80% of men's salaries, while doing a double work-load since their working day continues at home. That is why they wish to go abroad and find jobs that would pay more than the average wage in Moldova, even though they know a lot about possible risks.

Moldovan women working abroad are mostly active in the caring professions or services (nursing, hospital orderlies, agriculture) and are paid a half or a quarter of the pay rate for locals. One of the most painful features of migration is women's involvement in prostitution and trafficking. According to some national statistics, more than 100,000 Moldovan women were trafficked in 1995-

2002 to countries such as Turkey, Israel, Spain, Portugal, Greece, Germany, Italy, Cyprus, etc. over the last ten years.

Official statistics show a massive number of Moldovan women in Kosovo during the conflict period, accounting for 61% of all the trafficked persons there. Most of them were brought by force and many sold to various owners, without recourse to any legal assistance or any psychological or material support.

However, Moldova remains classified as Tier 2 in the 2014 U.S. Department of State Report on Trafficking in Persons, signifying that the country is still facing serious problems regarding the sex trafficking of women and children. Representatives of the Moldovan authorities and civil society define prostitution as a "sin", therefore prostitutes are presented as "women without morals." This belief takes root in the ideals set by the Church, the government and the society. This way of thinking is actually the main opponent to legalizing prostitution.

2 Prostitution and Easy Money

According to Moldovan psychologists, girls who are involved in the sex trafficking industry have incurred trauma related to sexual abuse. They argue that 40% of these young women were once victims of incest or rape at an early age (*Ziarul de Garda*, February 7th, 2013). These traumas induce serious personality disorders, leading women to agree to have sex for "easy money".

However, over the last few decades, the Moldovan society has evolved enough in the field of prostitution to see its practices diverse considerably. The economic emergence of those practices occurred rapidly in an uncontrolled environment, making easy money the main driving force of economic, cultural and social activities. With this in mind, it's quite apparent to state that the popularity of prostitution increased overtime. Making money quickly and "effortlessly", even if it involves illegal and immoral activities, has become a normal way of life for a significant number of young Moldovans, not only those from vulnerable families and the margins of society. This attitude of abandoning traditional values in exchange for a life of quickly acquired comfort has generated a new phenomenon: the practice of prostitution in public locations. "Library girls" began to appear in the Moldovan press circa 2015

and immediately attracted the public's attention both locally and beyond the Moldovan borders (*Realitatea*, January 7th, 2015). For several years, an alarming number of libraries have been hosting the prostitution of young girls via video conference. This behavior cannot only be explained by economic reasons. In most cases, these girls do not come from very poor families who practice this activity to support themselves. Most of them are actually studying or have another job, but wish to ensure an above-average lifestyle. This shows a clear mutation in the collective mentality of the Moldovan society. Any means are considered acceptable to achieve financial success, resulting from traditional values of work and morality giving way to material values.

In a similar context, there are many cases of Moldovan women and girls looking for rich husbands abroad to get a better life. Sometimes voluntarily and consciously, these girls choose to practice sexual services abroad where there are better financial conditions (*Jail Crunch*, 2014). A new form of prostitution known as "luxury prostitution" or "modeling prostitution" has become a successful industry, run by modeling agencies funded and managed by Moldovan oligarchs. A well-known example of this is the director of Fashion TV Moldova, Corneliu Vidrascu, who was accused of human trafficking and procuring in January 2015. Vidrascu worked for almost five years at the Ministry of the Interior, within the General Directorate for the Fight against Organized Crime and acted as one of the leaders in the Center against Trafficking in Human Beings (*Promotime*, January 3rd, 2015).

Newspapers advertisements play particular a role in attracting clients, with hidden announcements for sexual services or connections with Moldovan or foreign procurers. A study done by the International Center "La Strada" demonstrated that the most words used for attracting customers or future victims of sexual exploitation are either related to job search abroad ("assistance in preparing jobs," "visas for abroad," "employment contracts abroad") or under cover matrimonial advertisements ("mothers who want a good future for their daughters, contact us, we are in contact with wealthy husbands abroad"). To attract clients, women providing sexual services usually post their ads in the "erotic massage services" section. Erotic massage is not illegal in

Moldova, which encourages a number of "providers" to conceal their activity in this way, regardless of where the encounters occur (massage salons, night clubs, personal residences...) (*ProTV Moldova*, July 8th, 2015). The recruitment of girls is also done through coded language, with advertisements for "girl dancers", "erotic masseuses" or through other online job offers (*MoldovaNews*, May 8th, 2015).

3 New Environment – New Challenges

Since the accession of the pro-European government in 2009, the Republic of Moldova has made significant progress in its rapprochement to the European Union. In 2009, Moldova also joined the Eastern Partnership, measure of the EU neighborhood policy to sign an Association Agreements between the EU and each of the six former Soviet republics. By demonstrating its insight and consistency, the Republic of Moldova was able to obtain a free movement regime within the EU so there is no visa required for travel within the EU for up to 90 days every six months. However, Moldovan citizens are not allowed to work in EU member countries without a visa. The visa-free regime came into force on April 28th, 2014, making Moldova the first country in the Eastern Partnership to benefit from this. After four years of intense negotiations, the Association Agreement (including an economic and political component) between Moldova and the EU was ratified by the Moldovan parliament in July 2014. Despite fears expressed by some associations and political parties opposing the pro-European government, the cancellation of visas did not cause a visible effect of any mass exodus of the population. Migration dynamics remain unchanged, and emigration figures have remained stable since 2014. As for forced emigration for sexual exploitation in other EU countries, the situation is still unclear. It is evident that the liberalization of movement to the European Union has facilitated access to the European space to those were not eligible for visas before. This is particularly true for vulnerable unemployed young people living in rural areas – the target population for procurers. Meanwhile, in May 2015, the Moldovan authorities declared that trafficking decreased following the implementation of the visa-free regime (*Moldpres,* May 11th, 2015).

The scale of living in the Republic of Moldova remains one of the lowest and official data regarding the rate of unemployment does not reflect the true state of matters almost at all. Data regarding massive emigration of labor force from the Republic of Moldova to the West varies between 600 thousand and one million citizens. Women constitute more than 60 per cent of this number. This avalanche of illegal migrants (who were never registered· by anyone) was mostly employed in the service's sector: nursing, constructions, seasonal agricultural works, tourism, entertainment, etc. Those employees are paid poorly and aren't protected by the legislation of the host countries.

There are currently no quantitative studies establishing a causal link between the visa liberalization and the flow of trafficking victims from Moldova to the EU. Though one must still consider two aspects: visa liberalization facilitates the departure of victims of trafficking, but it also decreases the number of potential trafficking victims wanting to join family members or work in an EU country. The quantitative value of this effect has yet to be evaluated. In 2013, before the introduction of the visa-free regime, a number of information and communication actions were undertaken by the Moldovan authorities on issues related to the prevention and fight against human trafficking (*Council of Europe*, June 12th, 2014). The new context allows the Moldovan border police to be more effective in identifying and investigating transnational criminal networks.

4 Civil Society: a Mobilization against Prostitution and Human Trafficking

The issue of human trafficking has become increasingly present in the Moldovan society. Recently, topics related to sexual exploitation have appeared more than ever in political discourse. Two legislative initiatives reflect this new recognition: a 2012 bill aimed at criminalizing clients of prostitution, and a 2013 bill on chemical castration of those convicted of pedophilia. However, these two initiatives were not widely discussed in society. Parliamentary debates have been marked by a conflict between MPs from different political parties. The bill on the criminalization of clients has not been approved by the majority of parliamentary representatives (*Parlamentul Republicii Moldova*, June 5th, 2012), and the bill on the castration of pedophiles was initially adopted, but eventually rejected after a few months by the Constitutional Court (*TRM Moldova*, July 8th, 2013).

Even though the subject of prostitution has been present in the speeches and in the programs of Moldovan politicians, this phenomenon remains largely ignored by the upper-political class. Nonetheless, one important manifestation of civil society against human trafficking exists. More and more books and plays dealing with the subject of sexual exploitation have emerged in recent years. This new literature is meant to reflect reality and is largely based on real cases and people. Authors met with victims, listened to their stories and transcribed them in a documentary and poetic fashion to alert the public and authorities on the issues of sexual exploitation. One of the newest and most popular books on prostitution and sexual exploitation is *Bessarabian Nights* (2014, Ed. Aurochs, in English) written by Stela Brinzeanu, a Moldovan writer who immigrated to the United Kingdom. It demonstrates to its readers that sexual exploitation is a reality of migration from east to west that has a significant social, psychological, and emotional impact.

Today, civil society and non-governmental organizations in Moldova conduct various activities for the prevention of human trafficking and the protection of victims. With the help of international and non-governmental organizations (the Moldovan Mission of the International Organization for Migration, the International Center "La Strada," the OSCE Office in Moldova, "Médecins du Monde, " and the Switzerland Fund "Terre des Hommes"), several national and international seminars have been organized for professionals in the field (*Ministerul Muncii, Protectiei Sociale si Familie*, April 22nd, 2015). This training is complemented by the distribution of practical guides and methodical and educational materials.

In general, the cooperative relationship between the state and civil society is satisfactory. Yet, this relationship cannot be called a "total harmonization." The 2014 U.S. Department of State Report on Trafficking in Persons and the report from the Group of Experts on Action against Trafficking in Human Beings (GRETA) of the Council of Europe formulated recommendations on this subject. According to these reports, the

authorities in Moldova must take greater actions to encourage the participation of NGOs and public institutions in the fight against trafficking. NGOs active in the field should be allowed to participate in the decision making. It is also necessary that they support research in this field and adopt practical measures to implement effective cooperation and communication between the police and NGOs at the local level, particularly in the fight against prostitution.

5 A Diligent Government and a Corrupt Legal System

The fight against prostitution and sexual exploitation in Moldova began in the 1990s, when there was a lack of relevant legislation, an institutional mechanism with financial and organizational resources. Since then, the Moldovan government has made efforts in certain aspects of human trafficking, but it has also been negligent of others.

According to the Moldovan Bureau of Statistics, the rate of unemployment in 1998 for women was 17.8%, compared to a mere 10.2% for men. Knowing that in 1994 that number was only 8.9 % for women, it's important to realize that there has been a tremendous increase in women's unemployment. Data from 2004 shows that this number diminished to an acceptable 12%. Nowadays, because massive emigration (more than 600.000 people) and trafficking are not considered to be consequences of unemployment, it is practically impossible to find representative statistics about unemployment in Moldova.

In 2013, the Inspector General of the Police (*Inspectoratul General de Politie*) was established as part of the institutional reform of the Ministry of Internal Affairs and within it, a Center against Trafficking in Human Beings was created. This unit has a multidisciplinary structure, composed of police officers, law enforcement officers, and professionals from the National Anti-Corruption Center, the Service of Intelligence, and the Customs Service and Border Police. In September 2013, the Permanent Secretariat of the National Committee to Prevent and Combat Human Trafficking, a government agency responsible for the coordination of anti-trafficking policy, developed a National Plan including 120 activities for the years to come (2014-2016). One of the most relevant activities is the arrangement of an annual national campaign for a "Week against Human Trafficking", which was created in 2012. This awareness campaign consists of a wide range of anti-trafficking actions, including: public courses, round tables, information and awareness videos, film screenings, photography exhibitions, television programs, and more. At the same time, the Permanent Secretariat monitors these activities organized by the territorial commissions and includes this data in a National Report. This data is usually included in the annual U.S. Department of State Report on Trafficking in Persons where the efforts of these commissions of coordination of actions against human trafficking at the local level are appreciated. A communication strategy was developed by the Permanent Secretariat and put into practice with the specialized website (www.antitrafic.gov.md). According to the GRETA report for 2014 and 2015, Moldova has made considerable progress in information, in education, and in awareness of human trafficking (*GRETA*, March 4th, 2015). During the reference period, more than 1,100 events and 2,100 extracurricular activities focusing on the prevention and fight against human trafficking took place in schools in collaboration with the Ministry of Education. It is still necessary to continue to promote gender equality, the fight against domestic violence, and the de-stigmatization of trafficking victims. According to these same reports, in 2014, 4,229 specialists (judges, prosecutors, psychologists, teachers...) were trained. The most important aspect of this training is the identification of victims, particularly during investigations and judicial proceedings. The Moldovan government has made some improvements in this area, but there is still a lack of resources, not mentioning the deficiency of legal, psychological, and financial services for the victims. So far, there is no compensation system available for all victims. The National Report on the prevention and fight against trafficking human beings for 2014 identified 264 victims of which 116 were trafficked for sexual exploitation (*National Committee for Combating Trafficking in Human Beings, Permanent Secretariat*, 2015). Nevertheless, the government is investing more resources in 38 Health Centers (*Centre de Sanatate Prietenesti Tineretului*) located in all regions of Moldova. These centers were created in 2013 by the Ministry of Health, in collaboration with the

United Nations and are free for those up to the age of 24. There are currently no statistics on the number of young women who have benefitted from these services. Considering the fact that trafficking can be explained by the socio-economic situation of a population, from 2012 to 2014, the Ministry of the Economy developed a series of economic programs through the Organization for Small and Medium Enterprises to prevent trafficking via small and medium investments.

A series of measures to strengthen the fight against prostitution have been included in several strategic documents adopted by the government in the framework of justice reforms (Justice Reform Strategy for 2011-2016, the National Strategy for the Prevention of Organized Crime for 2011-2016, and the Action Plan on Human Rights). Withal, the Moldovan government's efforts have been very minimal from a legal point of view. There are a number of ambiguities in the definitions of trafficking and prostitution in Moldovan legislation. For example, Article 89 of the Contraventions of the Republic of Moldova details the penalties for practicing prostitution (*Parlamentul, Republica Moldova*, January 16th, 2009), but the Moldovan Code doesn't contain a precise definition of prostitution (what it represents and which actions/inactions are considered illegal). Considering that the Moldovan justice system is very sensitive to corruption, judges "play" with terms, allowing different types of fines and penalties that are more or less stringent, making Moldova a true "paradise" for procurers marginalized by laws currently in effect in member States of the EU. Another major problem is the existence of proven complicity of some authorities, which has been reflected in suspended or acquitted criminal cases for no ostensible reason. More and more testimonies of people in prostitution highlight the existence of "collaboration" between themselves and the local police. This has given rise to a new phenomenon, making the fight against prostitution even more difficult. These corrupt prosecutors are the image of a society devastated by the developments of recent decades, constantly in search of an identity and possessing fairly limited means to face social and economic ills (*Unimedia*, July 22nd, 2014).

Most of the recommendations refer to the reform of criminal procedures and the protection of victims and witnesses who are called upon to deal directly with procurers during investigations. Foreign actors have asked the Republic of Moldova to further its efforts in the fight against corruption in the judicial system.

References

1) « Cazul fetelor care făceau videochat în bibliotecă! Trebuie identificate persoanele ce recrutează tinere », *Realitatea*, January 7th, 2015.

2) « Decision of the Plenum of th Supreme Court of Justice of the Republic of Moldova on application of legislative previsions in cases of trafficking in human beings and trafficking in children – no.37 of 22 November 2004 », *Bulletin of the Supreme Court of Justice of the Republic of Moldova*, no.8, 2005.

3) « Investigaţie. Mărturisirile unor prostituate din Republica Moldova », *Unimedia*, vidéo, July 22nd, 2014.

4) « Legea privind castrarea chimică a pedofililor este neconstituţională », *TRM Moldova*, July 8th, 2013.

5) « Masaj erotic cu BONUS: Cat plateau clientii pentru prostitutie. Afacerea, descoperita de oamenii legii », *ProTV Moldova*, July 8th, 2015.

6) « O femeie pentru 60 de minute », *Ziarul de Garda*, February 7th, 2013.

7) « Patru fete din Moldova au facut videochat erotic in bibliotecile din Chisinau. Ce PEDEAPSA incredibila au primit », *Stirile Protv Romania*, January 7th, 2015.

8) « Procuratura: o femeie învinuită de proxenetism riscă 'ani grei de închisoare' », *MoldovaNews*, May 8th, 2015.

9) « Servicii sexuale şi masaj erotic la domiciliu. Deservea câte patru clienţi pe zi VIDEO », *Publika*, October 4th, 2012.

10) « Traficul de fiinţe umane s-a diminuat în urma regimului liberalizat de vize cu UE », *Moldpres,* May 11th, 2015.

11) Council of Europe, *Report submitted by the Moldovan authorities on measures taken to comply with Committee of the Parties Recommendation CP(2012)6 on the implementation of the Council of Europe Convention on Action against Trafficking in Human Beings*, Committee of the Parties to the Council of Europe Convention on Action against Trafficking in Human Beings, CP(2014)8, June 12th, 2014.

12) *Codul Contravenţional al Republicii Moldova*, Parlamentul, Republica Moldova, COD nr.218 din 24.10.2008, CCRMM218/2008, January 16th, 2009, http://lex.justice.md/md/330333/

13) Covrig R., « Director de televiziune, reţinut. Acuzaţii de trafic de carne vie şi proxenetism », *DCNews*, December 19th, 2014.

14) CRIDES/Fondation Scelles, *Revue de l'actualité internationale de la prostitution*, 2013.

15) CRIDES/Fondation Scelles, *Revue de l'actualité internationale de la prostitution*, 2014.

16) GRETA (Group of Experts on Action against Trafficking in Human Beings), Council of Europe, *Reply from the Republic of Moldova to the Questionnaire for the evaluation of the implementation of the Council of Europe Convention on Action against Trafficking in Human Beings by the Parties*, Second evaluation round (Reply submitted on 11 February 2015), GRETA(2015)4, Strasbourg, March 4th, 2015.

17) *Hotline Service – a decade of activity: Aspects of migration and trafficking in human beings in the Republic of Moldova*, International Center for Women Rights Protection and Promotion La Strada Moldova, Chisinau, 2012.

18) Munteanu G., « Video-chat erotic şi PORNOGRAFIC, în capitală. 'Munceau' 19 fete », *Ziarul National*, April 30th, 2015.

19) *National Report on preventing and combating trafficking in human beings 2014*, National Committee for Combating Trafficking in Human Beings, Permanent Secretariat, Chisinau, 2015.

20) Organized Crime and Corruption Reporting Project (OCCRP), « Interviews from the inside - Viorica Ursu, Human Trafficking, 10 years Sentence », *Jail Crunch*, Video, 2014, https://www.reportingproject.net/jailcrunch/video.php?id=0

21) Petrov S., « Modeling sau prostituție de lux? », *Promotime*, January 3rd, 2015.

22) *Raport de monitorizare a procesului de implementare a Strategiei Sistemului național de referire pentru protecția și asistența victimelor și potențialelor victime ale traficului de ființe umane pe perioada anului 2014*, Ministerul Muncii, Protectiei Sociale si Familie, Chisinau, April 22nd, 2015.

23) *Sesiunea a IV-a ORDINARĂ – IUNIE 2012 (session du Parlement)*, Dezbateri Parlamentare, Parlamentul Republicii Moldova de legislatura a XIX-a, June 5th, 2012.

24) *Training on Human-Rights Approach to combating human trafficking for the 2014-2016 National Action Plan (NAP)*, National Committee for Combating Trafficking in Human Beings, Permanent Secretariat, November 2013.

25) U.S. Department of State, *Trafficking in Persons Report*, June 2014.

26) U.S. Department of State, *Trafficking in Persons Report*, July 2015.

27) Vizdoga I., Roman D., Donciu A. et al., *Analytical study on the investigation and trial of cases of trafficking in persons and related offences*, International Organization for Migration (IOM), Chisinau, 2013.

28) Office of the Prosecutor General of the Republic of Moldova: http://procuratura.md/md/newslst/1211/1/6284/

29) http://www.gbv.de/dms/sub-hamburg/385865996.pdf

30) http://www.un.org/womenwatch/daw/egm/eql-men/docs/EP.6_Mindicanu.pdf

31) http://www.wunrn.org/news/2006/06_05_06/061206_moldova_report.pdf

32) https://gramaticamea.wordpress.com/further-reading-analyzing-and-discussing/

Looking through the curtain of history

Ramon Mihai Balogh, Ioana Ionel*, Dan Stepan
Politehnica University of Timişoara, Faculty of Mechanical Engineering, Bv. M.
Viteazu, 1, 300222, Timisoara, Romania,
* ioana.ionel@upt.ro,

Abstract: The Banat region, situated at the geographic crossroads between Eastern and Western Europe, with a special history and a destiny often broken by the vicissitudes of time, is known for many primordialities, all certified without denial. One of them is the first railway track on the present territory of Romania. The present paper, without pretending to be very documented, is an attempt to call to mind certain achievements of the Banat inhabitants related to their creativity, wish of development and entrepreneurial spirit. It briefly presents the railway Oraviţa-Baziaş, inaugurated by 1854 (to carry cargo) and, later, used also to carry passengers, celebrates its 160th anniversary this year. The paper also points out the sad fate of other initial routes of Romanian railways, in retrospect and prospect.
Key words: *Baziaş, Oraviţa, railroad, primordialities, Banat*

1. Introduction

Most Romanians have heard of the town from the Banat, the place called Baziaş, where the Danube enters the country. Moreover, at least the Banat inhabitants know that this is from where the first railway on the present territory of Romania departed, connecting Baziaş to Oraviţa.

The locality is important also for Serbians, because it houses an old convent, founded following the tradition by St. Sava himself. Many elements, thus, gather in order to give great importance to that place. Considering this, it seems hard to believe that Baziaş is about to disappear, maintaining only the memories of its former glory [1].

2. The sad story of the first railways crossing the Romanian territory

In the mid-nineteenth century, the Banat entered into the European rail routes network, which expanded eastward until reaching also our region. According to a very well documented work, written by Şerban Lacriţeanu and Ilie Popescu, *Istoricul tracţiunii feroviare din România* [6] [The history of railway traction in Romania – own translation], vol. I (which deals with the period from 1854 to 1918), Bucharest, 2003, the Banat holds the priority among the Romanian regions having a railway (1854), followed by Transylvania (1858), Dobrogea (1860), Bucovina (1866), the Old Kingdom (1869-1875) and Bessarabia (1871).

The Hungarian State Railways (M.Á.V. – Magyar Államvasutak) were founded only in 1868. Until then, since 1846, a number of private railways had operated on itsterritory. The same happened also in Austria, where St.E.G. (*Österreichisch-Ungarische Staatseisen-bahngesellschaft*) owned the railways until June 1, 1891.

2.1 Railway Oravita – Anina

The most interesting railway track in Romania and in South-East Europe is thus in the Banat region. Built in a mountain region and surrounded by landscapes of rare beauty, the track is entirely an architectural monument, standing as a testimony for the human creativity and technical potential of the time.

The first railway line in Banat was opened at 20.08.1854 between the Danubian port Baziaş and the mountain location Oraviţa. The line length of 62.5 km served to transport coal from the coal region Steierdorf - Anina to the Danube and from there to Austria. The line route was Oraviţa - Răcăjdia - Iam - Iassenova – Biserica Albă - Bazias. This line was named the "line coal" (Kohlenbahn).

For transportation of coal between Oraviţa (Figure 1) and Anina (Figure 2), different technical solutions were adopted that ultimately were not effective technically and economically. Such, at the very beginning, animals were used for the transport in the galleries, after that horse transportation of coal through galleries or on slopes was used.

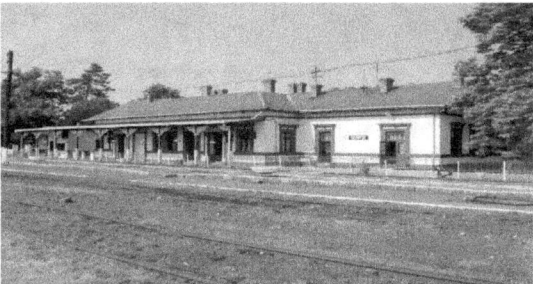

Figure 1. Oravita station

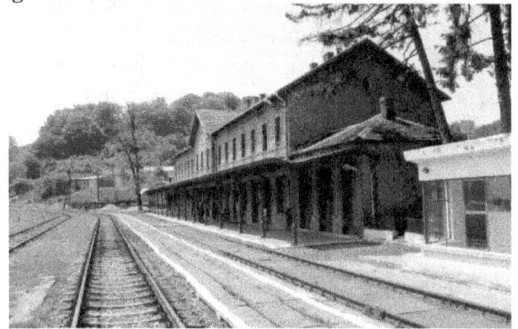

Figure 2. Anina station

Due to increasing quantities of coal to be transported on the Danube, one imposed the construction of a railway line linking Anina directly to the Danube harbor Baziaş.
Track construction began in the spring of 1861 and the opening for freight and passenger traffic was opened on December 15, 1863 (fig 3).

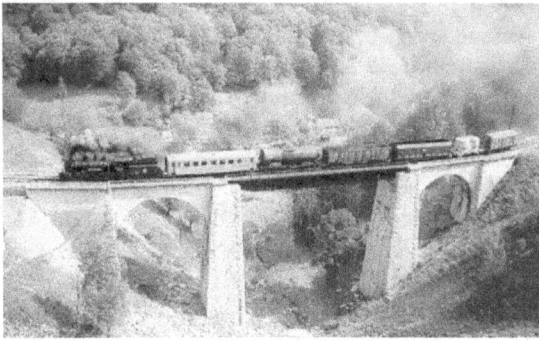

Figure 3. Mixed transport on Oravita-Anina [2]

Presently the line Oraviţa – Anina represents a special and unique engineering achievement. Called also the coal track, the railway Oraviţa-Anina was made by St.E.G. having as model another European monument, the Semmering track in Austria. That had been built only a few years earlier, between 1845-1854 (compared to 1856-1863), between the mountain towns Gloggnitz and Mürzzuschlag in Styria, passing through the Semmering pass. It had 40 km (compared to 33.4) and a level difference of 388 m (compared to our 339 m). The number of tunnels and viaducts of the two tracks is almost identical, fact which made the railway Oraviţa-Anina to be often referred to as the Banat Semmering. Locals and stone specialists brought from the Friuli region of northern Italy participated to its construction. The works were led by engineers Anton Rappos and Karl Dülnig and architects Karl Maniel and Johann Ludwig Dollhoff-Dier. The construction, which according to Georg Hromadka, cost a total of 5 million guilders, was put into operation on December 15, 1863 for freight and on April 4, 1869, for passengers [3, 11, 12].

The line has 33,45 km (Figure 4) with 160 curves within 129 with radius les then 200 m, 14 tunnels from which the greatest is Garliste with total length of 660 m, 10 viaducts the highest has 35 m, Jitin viaduc (Figure 5).

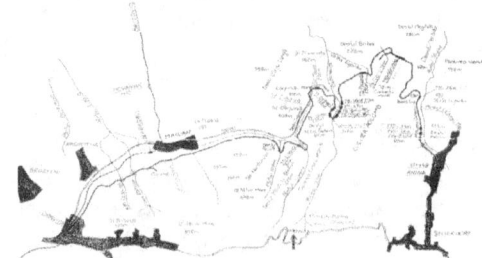

Figure 4. Oravita-Anina 33,45 km railway

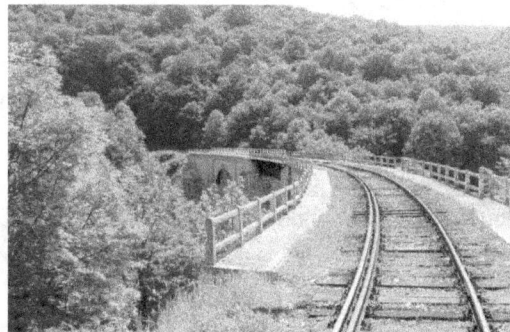

Figure 5. Jitin viaduct

Built under the leadership of engineer Bach and using, from 1851, rail laminated at Reşiţa plants, the track Oraviţa-Bazias was completed in 1854, with a length of 62.5 km. On August 20, 1854, the circulation of freight trains began and on November 1, 1856, that of passenger trains. The coal coming from the surroundings of Anina were loaded in the Bazias port on board of the ships belonging to the large Danube companies navigation, the most famous being D.D.S.G., founded in 1829.

In parallel with the construction of the Oraviţa-Bazias track, in 1847, a project belonging to engineer Anton Rappos was launched, providing for its extension by a normal steam traction track till Lişava. In this case, the coal would have been brought from Anina through a very long underground gallery, called King Ştefan, by horse driven wagons. But after 1852, with the increasing performances of steam traction, the Rappos project was radically altered, permanently abandoning the idea of underground galleries. By 1854, several funiculars were completed for the steeper section between Lişava and Anina. Then, the works to the track itself began. The track covers a distance of 33.4 km, with a level difference of 339 m. It has a total of 143 curves, of a length of 22.027 m, representing 65.9% of the entire route. It must be mentioned that at that time dynamite had not yet been discovered, this being invented only in 1866, three years after the works completion [9, 10].

After the closure of the mining operations on the route, the track was abandoned, as a strategy of C.F.R. (Romanian Railways) on November 1, 2010, being despoiled by scrape iron thieves. This is how another episode from the Banat railway history ended!

Considering its antiquity, the railway Oraviţa-Anina is the fourth track on the present territory of Romania, but, considering its technical and architectural achievements, it is still unrivalled. Due to the track difficulty, the big level difference and the numerous curves crossed, it had always required special locomotives and wagons (Figure 6).

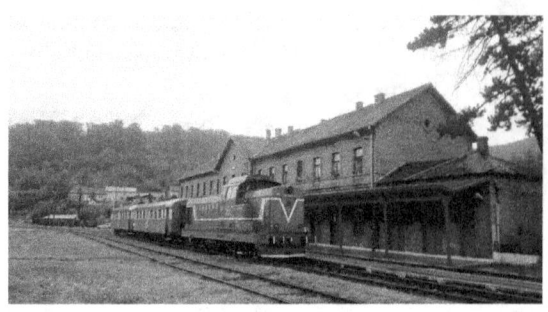

Figure 6. Anina-Oravita train [4]

Even fromits foundation, it had used a locomotive designed for it at St.E.G.'s factory in Vienna by engineer Pius Fink. Called Steyerdorf, it was made in 1861, followed by three others in 1867, namely Karaszova, Gerliste and Lisava. Used until 1891, these first locomotives between Oraviţa and Anina reached a maximum speed of 30 km/h. Their names are worth noting as they belong to the localities situated nearby the route travelled. Finally, the railway ended up having a total of seven halts and stations.

2.2 Railway Oravita-Bazias

In the early years, only one pair of mixed trains circulated per day between Oraviţa and Bazias, as pointed out in The Mixed Railway Timetable on Oraviţa-Bazias route, valid since November 15, 1857, published on April 18, 1858, in the newspaper *Temesvarer Zeitung*. The train used to leave Oraviţa at 7:00 a.m. and arrive in Bazias at 10:02 a.m. From Bazias, it used to leave back at 1:30 p.m., arriving in Oraviţa at 4:38 p.m. Its average speed was of 20.6 km/h. The traction was carried out from the beginning by locomotives built in 1852 by the company Günther from Wiener Neustadt, followed in 1855 by locomotive produced at St.E.G. Factory from Vienna, operating since 1840. The first locomotive distributed to the Oraviţa depot was no. 125 RESICZA and no. 126 ORAVICZA, taken over by St.E.G. in 1858 [15].

Over time, this railway was deeply marked by the political events happening in this side of Europe. After World War I, when the Banat was divided between Romania and Serbia, a 28 km segment of the total of 62.5 remained on Serbian territory. The train circulation was resumed on July 15, 1922, but only on the Romanian sector Oraviţa-

Iam (26.9 km). Bazias could be reached, until 1950, only on the track Timișoara-Jebel-Voiteg-Stamora Moravița-Vârșeț-Iasenova-Biserica Albă-Vračev Gaj. In 1950, due to the conflict between Stalin and Tito, the train circulation on the territory of Yugoslavia was suspended, the section Bazias-Socol-(Vračev Gaj) and the Bridge Nera-(Iasenova)-Iam being closed.

The first train coming from Oravița arrived, thus, in the Bazias station in 1854. After two years, the new rail connection to Vienna by Timisoara turned this place into the *terminus* of European railways. Travelers arriving from Vienna or Paris embarked in Bazias on the vessels of the Danube companies, on which they continued to travel eastward, having as final destinations Constanța, Odessa, Istanbul and Alexandria. Bazias had this important role only until 1879, when the newly built track Timișoara-Caransebeș-Orșova made the connection with the Old Kingdom railways at Vârciorova, the rail route being thus much extended. However, until 1919, Bazias continued to remain connected to Timișoara, facilitating the access of the Banat's capital to the trade on the Danube.

Due to the establishment of the border in its close proximity, Bazias could no longer use neither the rail connection (completely closed in 1950), nor the port, dating from 1795 (Figure 7).

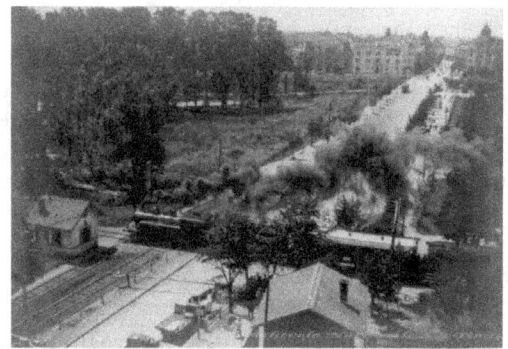

Figure 7: Timișoara-Bazias track [5]

2.3 Railway Seghedin-Timisoara-Bazias

The first railway in present Romania, the one between Oravița and Bazias, was followed shortly by another railway, much more important for the economic development of the Banat, the railwaySeghedin-Timisoara-Bazias, opened in the years 1857-1858. Its builder, as for the majority of the railways in the Banat was St.E.G.

On November 15, 1857, 113.9 km belonging to the section of Seghedin – Jimbolia - Timișoara were inaugurated, out of which only 39 km were in the Romania of today. The first locomotives used on this railway were brought to Timișoara by boats on the Bega channel. They were called KOMÁROM, AUSTRIA, HONT and PESTH, being constructed in the years 1845-1846 at the Belgian factory John Cockerill in Seraing after the American model [6]. In July 1858, the Timișoara depot received the locomotives WARTBERG and LUGOS for passenger trains. It is interesting to highlight the way in which the newspaper *Temesvarer Zeitung* described the arrival of the first train to Timișoara: "When we hear the distant noise of the steam carriage, when we see the high train's chimney full of black smoke, it is as if we were charmed. And when it passes before us, we are almost astonished that we ourselves stand still in one place". This is how the history of the railway in the capital of the Banat began [13].

The historical data were excellently presented by Șerban Lacrițeanu and Ilie Popescu in their work, *Istoricul tracțiunii feroviare din România* [The history of railway traction in Romania – own translation], vol. I, published in Bucharest, in 2003, and will be presented in the following section.

The works started also for the other section, of 94 km, between Timișoara-Jebel-Voiteg-Stamora Moravița-Vârșeț-Iasenova, where it connected to the track Oravița-Bazias. They were led by engineer Ludwig Meyer and began on December 10, 1856, in Vârșeț and on December 27, 1856, in Iasenova and Șag. In order to build the numerous bridges on the route, cement produced from the burning of the limestone extracted from Oravița was used, as well as 7.5 million bricks produced in the seven brickyards placed along the route [13].

The first passenger train on the Timișoara-Iasenova-Bazias route (Figure 8) circulated on July 20, 1858, driven by the locomotive no. 111 WARTBERG, being composed by 11 passenger wagons and 1 luggage wagon.

The railway Seghedin-Timisoara-Bazias connected the Banat region to major European rail routes, as it is here where, for a period of time, the connection between them and the river transport on the Danube was made. The *terminus* point of the railway, Baziaş represented also the beginning of a fluvial journey and then, a maritime journey of great interest to the people of the 19th century, especially those thirsty for new. According to the Train timetable of May 6, 1861, of the St.E.G's southeast network between Vienna and Baziaş, a pair of high-speed trains circulated twice a week, which rode the entire distance of 86 Austrian miles in 28 hours and 18 minutes.

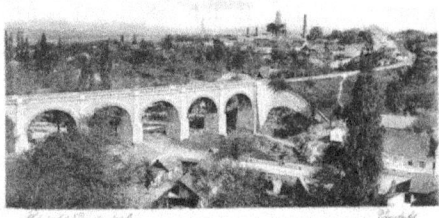

Figure 8: Viaduct, Bazias area [6]

The distance Pesta-Baziaş was covered in 11 hours and 18 minutes, and Timişoara-Baziaş in 3 hours and 6 minutes. The trains from Vienna used to arrive to Baziaş on Tuesdays and Fridays, the journey being continued from there using the boats of the D.D.S.G. [7] Company towards Orşova, Giurgiu, Cernavoda, Odessa and Constantinople. The timetable of these boats was connected to that of the trains belonging to the St.E.G. Company. The fast boats used to leave Baziaş on Tuesdays and Fridays at 8:30 a.m., while those for passengers on Thursdays at 11.30 a.m. The importance of the Timişoara-Iasenova-Baziaş track within the circuit of major international transport relations was maintained for two decades, until the inauguration on June 10, 1879, of the railway Timişoara-Caransebeş-Orşova-Vârciorova. This new railway, linked to those in the Romania of that time, offered the possibility to continue a railway journey to the Orient [8, 15].

During its peak, Baziaş was connected through Vienna, even with Paris and other major European centres. It was situated on the major route Paris-Constantinople, the journey between these cities lasting six days (via Strasbourg, Karlsruhe, Stuttgart, Munich, with the variants Cologne, Hanover, Leipzig, Dresden, Prague, then Vienna, Baziaş, Cernavoda, Constanţa). The boats travelled

the distance Baziaş-Cernavoda in 36 hours and 30 minutes, and back in 54 hours. The route Vienna-Baziaş was justly called, for these reasons, the way to Orient. The years 1858-1879 represented Baziaş most flourishing period, when it was one of the most important points of connection between Eastern and Western Europe [9].

Not used since 1950, the monumental building of the Baziaş station was demolished after 1960, when it was thought that the place would be flooded by the raise of the Danube level once the Iron Gates dam was built. But, the place of the old station is now right on the shore, being occupied by the cottage *Apus de soare (Sunrise, own translation)*. Nevertheless, most of the embankment of the former track between Baziaş and the border with Serbia was flooded by the river. This is how an episode from the Banat history on railway transport ended [15].

3. Conclusions

The sad story of the first railway that crossed also the present Romanian territory ends, although it has existed for almost a century (1854-1950). At present, the circulation is done only on certain sections. Nonetheless, its appearance and existence in this space of the Banat region are milestones in the history of railways and the human potential of the area. No annals on Romanian railways can omit this.

Other lines also had followed the same fate. Now, the only thing we can do is to write about and remember them! The viaducts and tunnels of these historic railways must become models, tailored to modern technology, models of power and perseverance. The constructions that have conquered so many hearts should not be forgotten and the railway Oraviţa-Anina must be kept and introduced in the tourist routes of the Banat region, with care and good local management.

We have no right to forget nor neglect what history has left us, all the more to show indifference towards the current situation, characterized by the lack of maintenance, publicity and visibility or by destruction (moral and material).

References

[1] M. Rusnac – La kilometrul zero al Dunării bănăţene, http://www.timisoaraonline.ro/timisorenii-invitati-intr-o-altfel-de-excursie-cu-trenul-pe-calea-ferata-oravita-anina-una-dintre-cele-mai-frumoase-din-tara/

[3] https://istoriabanatului.wordpress.com/2011/02/16/

[4] M. Rrusnac, Calea-ferata-Oravita-Anina-monument de patrimoniu european/ [http://www.forumtrenuri.com/t114p475-925-oravita-anina]

[5] Timişoara-Baziaş track through ScudierPark on August 8, 1932. Source: ArhiveleNaţionaleTimiş (Mircea Rusnac – Banatul pe marile artere feroviare europene: linia Seghedin-Timişoara-Baziaş, https://istoriabanatului. wordpress.com/2012/02/19/mircea-rusnac-banatul-pe-marile-artere-feroviare-europene-linia-seghedin-timisoara-bazias/accessed April 2016

[6] Ş. Lacriţeanu, I. Popescu, *Istoricul tracţiunii feroviare din România*, vol. I, Bucureşti, 2003

[7] https://istoriabanatului.wordpress.com/2011/09/06/ mircea-rusnac-cea-mai-veche-cale-ferata-din-romania-oravita-bazias/

[8] http://www.banaterra.eu/romana/popoviciu-gheorghe-cateva-date-mai-putin-cunoscute-legate-de-constructia-si-circulatia-trenurilor-pe

http://drumliber.ro/10-motive-ca-sa-mergi-pe-calea-ferata-oravita-anina/

[9] http://www.banaterra.eu/romana/popoviciu-gheorghe-cateva-date-mai-putin-cunoscute-legate-de-constructia-si-circulatia-trenurilor-pe

[10] http://www.banatuldemunte.ro/2009/06/cu-trenu-oravita-anina-33km-in-2h/ http://www.welcometoromania.ro/Oravita/Oravita_Oravita_Anina_r.htm

[11] http://www.intercultural.ro/turismintercultural/Calea-Ferata-Anina-Oravita.html

[12] Banatul în relaţiile internaţionale, Evenimente, Istoria Aninei, Istoria Oraviţei, monumente istorice, https://istoriabanatului.wordpress.com/2011/02/16/ mircea-rusnac-calea-ferata-oravita-anina-monument-de-patrimoniu-european/

[13] https://istoriabanatului.wordpress.com/2012/02/19/ Mircea-rusnac-banatul-pe-marile-artere-feroviare-europene-linia-seghedin-timisoara-bazias/

[14] https://istoriabanatului.wordpress.com/2010/11/17/ mircea-rusnac-pe-urmele-caii-ferate-resita-lindenfeld-delinesti/Mircea Rusnac – Pe urmele căii ferate Reşiţa-Lindenfeld-Delineşti

[15] ***, ANALELE BANATULUI. Serie noua, ARHEOLOGIE-ISTORIEXVIII, 2010, EDITURA MEGA, Cluj-Napoca, 2010, Colegiul de redactie Dan Leopold CIOBOTARU, director al Muzeului Banatului si altii, http://www.muzeulbanatului.ro/publicatii/anale_10_14.pdf, accessed 17.04.2016

IF IT HAD NOT BEEN FOR VUIA, WE WOULD NOT HAVE FLOWN. THE FULFILMENT OF A DREAM

Ioana Ionel*, Sabin Ionel

Politehnica University of Timisoara, Faculty for Mechanical Engineering, Bv. M. Viteazu 1, 300222, Timisoara,

*ioana.ionel@upt.ro

Abstract Now at the anniversary of 110 years from his first flight, the article points out the wish of a Romanian engineer to fly, symbolically, to become a *flyer*. March 18, 1906 – Montesson stands for the date and place registered in human history as the first lift off an aircraft by its own means, situated on board. Traian Vuia's creative genius led him also to other remarkable inventions, such as the forced circulation boiler. In the romantic literature, the term *flyer* is a personification of the longing for the man loved, of intense love for the beloved. The *flyer* is considered the symbol of unrequited love, a man who, in life, was rejected by a woman and who, after death, haunts women on earth, especially the one who refused him. Similarly, Vuia demonstrated that a dream initially declined, can come true and be scientifically proven. His love is directed only towards the Romanian nation, whom he served and to whom he gave all his knowledge and passion!

Keywords: flying under its own power, boiler, priorities, Banat, Traian Vuia

*Motto: "I have never searched for glory, because I know glory often **loses the man**. I do not work for my personal glory, but for the glory of human genius. What does it matter who did these things, it is important that they exist." Traian Vuia*

1. Education

Traian Vuia was born on August 17, 1872, in the commune of Bujoru, Surduc village,at the time, Caraș-Severin county, today, Timiș county. His father, a priest, Simion Popescu, was from Vîrset and his mother, Ana Vuia, was from Lipova, being a relative of the teacher and educator Iuliu Vuia. He had two sisters: Elena (Nina) and Ghizela. The scientist bore his father's name only for two years, 1881-1883, afterwards, the rest of his life, he bored his mother's name [1], [2].

Ghizela Manzur talked about that period of childhood: "Traian Vuia, the child, was working all day long at manufacturing light objects floating in the air, like kites and balloons. He used to do the experiments alone, without calling others - and he loved solitude. The work was bold, tensed, up to self-neglect, he used to enclose himself in the room and work". These words prove what is known about geniuses, that they show their qualities from childhood! According to the child psychology studies conducted, between the age of one and two, the child is a little scientist and needs to explore the world and its environment [3].

At the age of five, he used to go with his sisters to the Romanian confessional school, where, later, he attended two courses. In this robust environment in which he grew up receiving a simple, austere and intellectually sporadic education, it is where his early interests in the technical domain took shape. He learned reading and writing at the school in Bujor, then, he attended the remaining courses at the elementary school in Făget.

From childhood, he began to show passion also for the knowledge of clock mechanisms. During holidays, together with other children, he used to make pinwheels out of wood and tin and fix them on vertical axis in trees in order to show viewers the wind intensity and direction. He also built a large camera, which was original as conception except for the lens, which was bought. The attraction for *everything that flies* and the desire to build flying objects were inducted to him by different events that marked him deeply, namely: (i) his teacher Antal Mahler built for his son Lajos an impressive kite, event accentuating his curiosity for successive building phases and the need to balance this first *flying machine*;(ii) later, the father of another colleague made another device, even more imposing, that two stout individuals barely managed to bridle [5-6].

According to records from that period, from the 2nd to the 7th grade, the child VUIA obtained excellent marks in all subjects. In the 8th grade, Traian Vuia was among the top four students who passed the maturity exam with an exceptional grade. From high school, Traian Vuia remembers a sad episode, when some students of Wallachian nationality were expelled from all the high schools belonging to the monarchy, on the grounds that they infringed the official interdiction to speak Romanian loudly, in public places. The young students received that sanction because they gathered privately to readout works belonging to classical Romanian authors.

He attended secondary school at the State High School in Lugoj, and, in 1892, he had his baccalaureate exam. George Lipovan is the author who devoted a book to Vuia and remained famous for saying that "Traian Vuia owed to the city of Lugoj everything except his work and talent" [12-13].

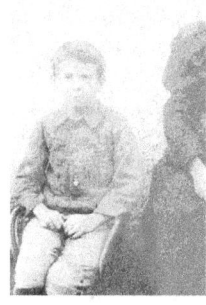 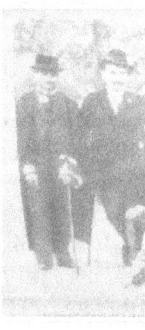 

Figure 1. 1881- Traian Vuia together with his mother and sister
"In the village where I lived with my parents, I built a long series of kites. Later on, as a high school student in Lugoj, I also built several kites, which I used to test on the vast area surrounding the camp." - own translation

Figure 2. Vuia together with his confreres, looking seriously towards the horizon of science

Figure 3. Traian Vuia's PhD thesis in Legal Sciences: *Militarism şi industrialism, regimul de status şi de contractus* [Militarism and Industrialism, State and Contract Regime – our translation], May 6, 1901

Figures 1-2 present photos shoot in the époque.

Throughout the period lived in Lugoj, he spent a lot of time with the family of a well-known lawyer and politician, Coriolan Brediceanu, who later offered him unconditional support in the project that changed human history.

He went to Budapest to continue his studies. In 1892, he attended the coursed offered by the Mechanical Department from the Polytechnic School in Budapest.

But, he also matriculated at the Faculty of Law, where he obtained his PhD. in legal sciences, after defending his paper entitled *Militarism şi industrialism, regimul de statu şi de contractu* [Militarism and Industrialism, State and Contract Regime – our translation] (1901) (see Figure 3). Upon his request, the work was to be written in Latin and not in Hungarian.

Due to material difficulties, after a year, he abandoned his technical studies and transferred to the Faculty of Law, without compulsory attendance, which allowed him to work in order to earn a living.

During those years, the scholarship (stipend)

received from the Foundation Gojdu was a real support. The transfer from polytechnic studies to legal sciences gave him the opportunity to work as an intern in well-known jurists' offices, spending time in several law offices from Banat. Therefore, from 1895 to 1897,heworked as a writer in the office of the lawyer Coriolan Brediceanu, Lugoj; from 1898 to 1899, he worked as a writer for the lawyer Spătaru in Vršac; in 1900, he came back toCoriolanBrediceanu's office in Lugoj as a secretary; from 1900 to 1902 he was an intern and lawyer in George Dobrin's office.

2. Fulfilling the dream of flying

As his great dream was to fly, he began to perform calculations for the design of a flying machine. On July 1, 1902,he arrived in Paris, bringing with him the design of an original "airplane-car", projected as a college student, and its related model constructed during the last 12 months. As Paris was considered at that time the world centre of aeronautics, the young Vuia hoped to manage to fulfill his dream of flying there. He submitted a report entitled "airplane-car project" to the French Academy of Sciences, in which he presented his invention. Although he was given an unfavorable answer, he still received a French patent in 1903, starting the airplane-car's construction right in the autumn of that year (Photos presented in Figures 4-5).

On February 16, 1903, he submitted a report on the "airplane-car" he designed to the Academy of Sciences in Paris, but his project was rejected by the French scientific forum because they considered its realization impossible since aircrafts heavier than air could not fly. Moreover, the machine designed by Vuia had several features that were not understood at first by his contemporaries: it had one propeller instead of two, requirement thought absolutely necessary, and the engine designed seemed unachievable with the technical means of the time. Despite these hindrances, advised by Coriolan Brediceanu, his friend from high school, Vuia patented his invention, obtaining the French patent no. 332.106 of May 15, 1903.

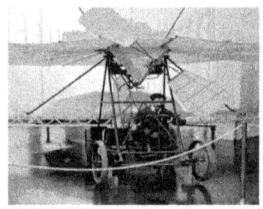 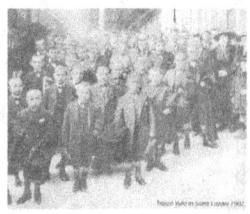

Figure 4. "I go far away, mom, to Paris, but, don't worry, mom, cause I'll come back home from there, either flying or at all." Traian Vuia – own translation

Figure 5. "I left on my own, without any help. The habitants of Lugoj helped me later, contributing to the building of the machine." Traian Vuia – own translation

With the support of his family, who contributed significantly with money, Traian Vuia began building his plane. Vuia was forced to equip his little monoplane (airplane Vuia 1), which he had started to build in France in 1903, with many innovations. Things progressed slowly due to the lack of money. Not affording to purchase an ultra light engine, Vuia resigned himself to invent an engine with compressed carbonic acid, which had the advantage of being simple and inexpensive. However, it could not be accomplished, serving only for demonstrations (its operating time lasted approximately three minutes). In 1905, the airplane was ready. It was a monoplane - almost all attempts at that time were made with biplane aircrafts, Louis Blériot (1872-1937) following Vuia's example, a year later -, it had foldable wings, the same as a fan, carried on a quadri-cycle with pneumatic wheels acting as take-off and landing gear. The entire structure was metallic, made from jointed steel tubes.

At the end of 1905, the plane was ready and was tested for the first time. During this test, the plane rolled along the ground for quite a long distance. This first test was followed by others, to which well-known specialists, such as G. Besançon and Victor Tatin, attended.

On February 5, 1906, Vuia took another test with his plane, but obtained no results due to bad weather. On March 6, a new unsuccessful attempt to lift the plane off the ground was made.

The plane was called the bat due to its shape and had 250 kg, a support area of 14 sqm and an engine of 20 hp. *The bat* flew on March 18, 1906. It accelerated for 50 meters, lifted 1 meter, flew 12 meters, after which the propellers stopped and the machine fell (Figure 6).

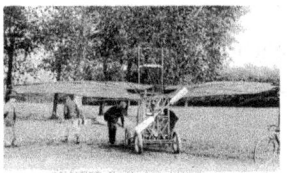 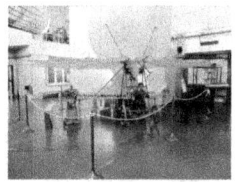

Figure 6. Vuia's machine, called *The bat*, flew for the first time on March 18, 1906, near Paris, and lifted 12 meters. Photography from the museum bearing its name in Montesson

The day of March 18, 1906, will stand for Traian Vuia's first air success, as well as the first flight in human history with a machine heavier than air, which lifted off the ground under its own power. Vuia's plane was made of a steel tubes frame, with bulged cloth wings, which gave it the appearance of a bat. The engine worked with carbon dioxide as fuel and the landing gear consisted of four wheels with tires. This first self-powered flight took place in France, in Montesson. Traian Vuia's plane started moving using its own engine, rolled along about 50 meters, then took off and flew 12 meters at a height of 1 meter. A strong crosswind prevented the machine from continuing to fly, but the premiere happened: the first mechanic flight in history took place.

It is therefore recognized that on March 18, 1906, in Montesson, the Romanian Traian VUIA carried out the first (in the world) take-off from the ground of an aircraft using its own means, situated on the machine's board. At the same time, Vuia also set the monoplane shape and the landing gear (see Photos from Figure 7).

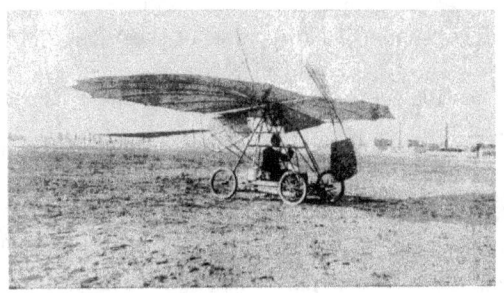

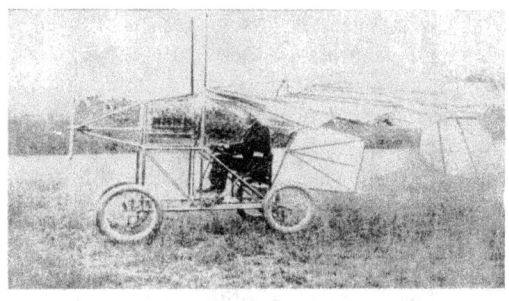

Figure 7. March 18, 1906 - Montesson, the first take-off in the world of an aircraft using its own means, situated on board

The premiere of such a flight is wrongly attributed to the French Santos-Dumont, who achieved the same, but only on September 13 of the same year 1906. Furthermore, after his first successful flight, the one on March 18, 1906, Traian Vuia carried out also other flights. During the one on August 19, 1906, his plane flew 20 meters, at a height of 2 meters and a half.

Vuia further improved his planes; the first used engine was replaced with another, of 24 hp, so that the machine Vuia no.1 became Vuia no. 1-bis, and then Vuia no. 2 [7-8].

On July 17, 1907, Traian Vuia managed to fly over a distance of 60 meters, but, at landing, he damaged the plane. By improving the first model, this became Vuia 1-bis. Later, a new airplane was built, equipped with an engine Antoinette on gasoline, with eight-cylinder V, water-cooled, of 25 hp, created by the engineer Léon Levavasseur. He flew with it in 1907, successfully remaining in the air for about 100 meters (Alberto Santos-Dumont already flew for 200 meters). Thiswas called Vuia 2 and was patented in Belgium. Figure 8 presents stamps celebrating Traian Vuia and his planes.

Traian Vuia continued his research into the technical domain also after 1907. He organized a workshop-laboratory for the research of propellers, and together with Marcel Yvonneau devised some original models of helicopters (considered rotary wing machines), which were presented in public demonstrations (the models realized in 1918 and 1921). The magazines *La Technique Aeronautique* and *L'Atmosphere* published his theoretical studies (see Copies presented by Figure 9).

Among the results of the research conducted by Vuia, some confirmed later, the following ones can be mentioned: he discovered the formula of monoplane aircrafts (subsequently recognized as the most appropriate), made the first airplane wing with variable incidence during flight, applied to his flying machine the principle of a single tractor propeller, made the first airplane with folding wings.

In the following years, Vuia continued to study aeronautics thoroughly, so that in 1918 and 1922, he made two types of helicopters, Vuia I and Vuia II. The machine was equipped with rotary wings, a rudder and a horizontal stabilizer (Vuia II had an Anzani engine of 16 hp). With these machines, Vuia conducted a series of tests and several vertical flights on the aerodromes from Juvissy and Issy-les-Moulineaux [9-10].

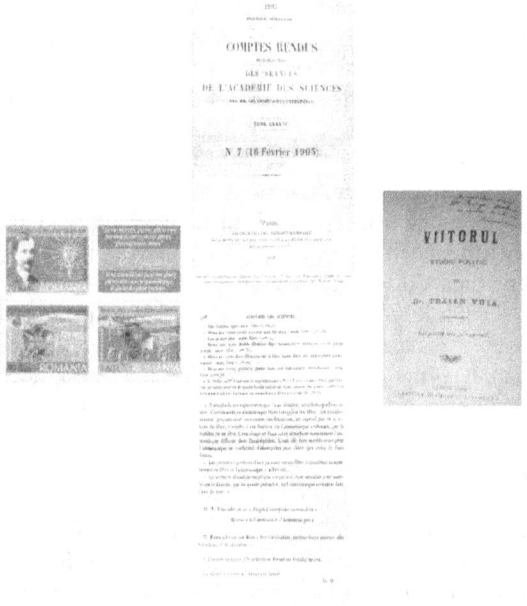

Figure 8. Series of stamps celebrating Traian Vuia and his planes

Figure 9. 1903 - French Academy of Sciences – Vuia's study, accompanied by an invention project for his "airplane-car".

Figure 10. Study published by Vuia, in Lugoj

Traian Vuia, like any visionary, encountered also various problems. By 1923, he intended to start the mass production of his flying machines and helicopter sat Reșița plants. The inventor gave the next message to a friend to deliver back home: "Tell friends back home everything you have seen here. I have proved that a machine 'heavier than

air' can fly. (...) Now, after they are convinced of the possibility of mechanical flight, the number of tests will grow rapidly, specialists will realize special engines, aviation will have a flourishing industry based on my experience, which has become common property. You could see! I did not hide; I ran tests in front of everybody. *Others will continue, more and more.* This is how progress is made ... " – Vuia wrote in a message sent to Professor Constantin Nedelcu, on his return from Paris in 1906 [9-10].

3. Vuia the fighter

The wide range of studied subjects allowed Vuia to practice as a lawyer, but also to publish his studies, among which the study *Viitorul* (see Figure 10) distinguished itself, which was published in the periodical *Drapelul* and printed in a separate booklet. By 1900, he made public his first political opinion saying "The solution to the national problem will be to unite Banat from Transylvania to its mother country." C. Brediceanu gave a reply to this, which said "out of the mouth of these the future speaks."

In the French capital, Traian Vuia continuously tried to influence the destiny of the Romanians in Transylvania, leading a real campaign for the unification with Romania. Thus, together with other Romanian leaders in Paris, on April 30, 1918, he founded the National Council of the Romanians in Transylvania. This new society used as a press channel the magazine *La Transylvania*, in which Vuia published numerous articles supporting the cause of those back home [6], [10].

Vuia was also a counselor in the Romanian delegation at the Paris Peace Conference (1919-1920) where, through the relationships he cultivated, made an important contribution to the unification of the Romanians. On this occasion, he published the paper *Le Banat* to make known the realities to co nationals here. In the same period, he published under a pseudonym a collection of articles appeared in Austro-Hungary, in which the other powers involved in the Peace Conference, i.e. Russia, France and Britain, were defamed. In this context, on his own initiative or following advice from home, he joined freemasonry, together with Alexandru Vaida-Voievod, it seems, to combat the Hungarian element powerfully present in this organization, who acted by all means to prevent the union of the Romanian territories from the former Empire with Romania.

He was the chair of the National Council of the Romanians in Transylvania and Bukovina; he militated for the independence of Transylvania and its union with Romania, in Paris, in Île-de-France, France.

Always active, not only in aviation research, during the Second World War, in 1943, the Romanian engineer created the Romanian National Front in France. This organization acted in connection with the French resistance, which fought against the German occupation.

On Traian Vuia's initiative, on April 30, 1918, the National Council of the Romanians in Transylvania was founded at Paris, an organization militating for the unification of Ardeal/Transylvania/ with Romania. The council published also the magazine *La Transylvanie*, in which Traian Vuia wrote several articles. Vuia published articles on the same subject also in the magazine *La nation tchèque*. He was however very critical on the way in which the union of Ardeal with Transylvania was achieved, considering a big mistake the lack of negotiation regarding the union with the Kingdom of Romania. This would have defended the interests of the locals and would have permitted, in time, a westernization of Romania instead of a balkanization of Transylvania.

In the First World War, Traian Vuia worked for the French Ministry of Defense. Together with the well-known Victor Tatin, he built a torpedo used successfully by the Navy.

In April 8, 1922, in Garches (France), Vuia wrote to his mentor Caius Brediceanu:

"Dear Caius, I received your letter yesterday. I am aware of all that happened in the country. I followed closely the entire electoral campaign. What can I say? I cannot say I was surprised. I had the misfortune to know more closely the Phanariot spirit, not to say, virus. Having no illusion, I could not lose them. I was just surprised when I saw that you believed in a possible cooperation with them. My religion is made long ago.

Do not forget that even if our new masters have no illusion regarding our transformation into upstarts, they have historical evidence of the gentleness and resignation with which we wore so many yokes over so many centuries. They do not ignore our fear of violent resistance either. They

will let us cry and complain until we get tired, like the baby who cries. They admirably know our patience. They know how we have endured without any serious resistance the Hungarian yoke from 1867 onwards [52], not to mention the previous period. They also know that without the European war we would have endured the Hungarian yoke even today. When I read the newspapers from home, I have the impression that the events will take the same course. We can say as Eminescu: "Other masks, the same play" ... "

These are some sections from the thoughts tormenting Vuia.

He was so right! They are still so actual!

An episode that affected Vuia occurred in 1937, when he was invited to give a speech, during the International Exhibition of Aeronautics in the French capital. Because Dimitrie Gusti, the director of Romanian pavilion got sick, and the new director showed no interest for the idea, Vuia could not deliver his prepared speech. Disappointed by the way he was treated, he gave to an acquaintance a note with some written verses from Alfred de Vigny's *La mort du loup*: *Instead you'd fain your weighty task recall: To take, as I, that path that fate decrees, To live, to suffer, and die wordlessly.*

During the Second World War, Vuia was part of the resistance movement in France and was elected chair of the first legal council of the Romanian National Front in France, where he worked hard and published a series of articles in *La Roumanie Libre*. In 1950, Vuia returned in Romania, seriously ill.

4. Memento

Having as a landmark India many centuries before Christ, passing through the Old Testament to the ancient Greeks and the myth of Icarus and the Incas who call themselves the sons of the Sun, an unbowed dream has interlaced with the known history of mankind: the flight. *Memories of the future* about flying have been present in all cultures and civilizations all over the world since ancient times. The dream was to be turned into reality in the year of grace 1906. By a Romanian! Traian Vuia is the man who defied gravity without catapults or other external means and created the first machine heavier than air with which he achieved the self-propelled flight. Among the first Romanian inventors in the field, Traian Vuia (August 17, 1872 - September 2, 1950), has received recognition for his work, even if for a long time the premiere in achieving the mechanical flight was attributed to the Brazilian (naturalized French) Alberto Santos-Dumont (1873-1932).

On May 27, 1946, he became an honorary member of the Romanian Academy. *"Stop praising me. Do not forget that all those who found out something in different areas of the ethnic progress, have no merit; God gave them the mind, they did not do it by themselves; we should be more modest. Boasting about what they did, they do as the Bible says, as the hammer that boasts about having made the stone. When a man discovers something, he should not have a reason for pride, but for humility. The Creator created the universe directly and created also the man, who, as an instrument made by him, executes all that we are aware and know in science, art, techniques. From our incompetence or forgetfulness of this truth, do come all sins. "...* declared Traian Vuia.

Eminescu distinguished between two worlds: the world of dreams and the world of thoughts, taken in antithesis. The world of dreams is *"lumea-nchipuirii cu-a ei visuri fericite"* [the world of illusion with its happy dreams - own translation], *"lumea-nchipuirii cu-a ei mândre flori de foc"* [the world of illusion with its proud flowers of fire - own translation]. The world of reality is *"lumea cea aievea, unde cu sudori muncite / Te încerci a scoate fapte din a stâncei coaste reci"* [the real world, where with worked sweat / You try to take facts from the rock's cold rib - our translation], the world *"unde cerci viaţa s-o-ntocmeşti, precum un faur / Cearc-a da fierului aspru forma cugetării reci"* [where you try to draw up life, like an ironsmith / Tries to give to the harsh iron the shape of a cold cogitation - own translation]. These verses can be interpreted and validated totally through all the work of Mr Vuia, crowning it in the mirror of history!

As in *Luceafărul* [the Evening Star], the genius, Vuia, cannot descend to a middle level, the same as the ordinary man cannot reach a higher level.

That is why, dedicating to Vuia this modest evocation, now, at the celebration of 110 years from his first flight, we fulfill a pious duty towards the one who, set off from Banat to conquer the horizon, he succeed it!

The Banat falcon died on September 3, 1950, in Bucharest, writing his name on the list of personalities who revolutionized modern society. Traian Vuia was a strong creative force, a forerunner spirit of his era, a pioneer of innovative roads in aviation and thermodynamics, as well as a great patriot, who, in difficult moments, served his country.

"Politicians are forgotten, renowned writers are forgotten, but we will not forget him ever. We will say proudly, ... **and we had VUIA**" - paraphrase after Nicolae Iorga, (in original *referring to Aurel Vlaicu*) [14].

References

[1] http://enciclopediaromaniei.ro/wiki/Traian_Vuia

[2] https://ro.wikipedia.org/wiki/Aurel_Vlaicu

[3] "100 cei mai mari savanţi ai lumii", John Simmons, Editura Lider, 1996

[4] http://www.taifasuri.ro/taifasuri/mozaic/6681-traian-vuia-de-la-zmeu-la-avion-nr385-sapt16-22-aug-2012

[5] Mihai-Athanasie Petrescu , "TRAIAN VUIA – VIAŢA ŞI OPERA" Editura Anima, Bucuresti, 2013

[6] Diana Iane, Altermedia, TRAIAN VUIA, UN AVOCAT DEVENIT ZBURĂTOR TEMERAR, „CAIETE DE PROTOCRONISM ROMÂNESC", Revista semestriala de analize si comentarii / NR . 2, https://protocronism.wordpress.com/document ar-altermedia-precursori-romani/precursori-romani-traian-vuia/

[7] I. Cojocaru, I. Iacovachi, „Traian Vuia", Ed. Ştiinţifică şi enciclopedică, Bucureşti, 1988

[8] Valeriu Avram, Paul Sandachi, Ana-Maria Guşă, Bucureşti, "Contribuţii româneşti în aeronautică la începutul sec. XX", , 2001

[9] Dan ANTONIU, Ioan BUIU, Dan HADÎTCĂ, Radu HOMESCU, George CICOŞ, Traian Vuia, Viaţa şi opera (monografie bilingvă, română şi engleză), Editura Anima Bucuresti, 2013

[10] Dictionar de personalitati, www.euroavocatura.ro/dictionar/319652/Traian_Vuia

[11] Seria de filme documentare intitulată „Gândit în România",

[12] G. Lipovan, Traian Vuia realizatorul zborului mecanic, Editura: Tehnica, Bucuresti, 1956.

[13] G. Lipovan , Traian Vuia, un pionier al aviatiei moderne, Editura: Facla, Timisoara, 1972.

[14] N. Iorga, Oameni care au fost, Editura pentru literatura, Bucuresti, 1967.

RISCURILE ŞI ÎNTELEPCIUNEA RAŢIUNII

Stoica Dorin

Master in Philosophy, University of Quebec in Montreal

dstoic2000@yahoo.com

Prin evocarea câtorva mari filosofi si gânditori spirituali, vom demonstra ca ratiunea, în masura în care este fondatoare, poate sa puna bazele unei dezvoltari durabile, consistente si întelepte a planetei. Vom pune în evidenta teorii filozofice, sttintifice si religioase preocupate de statutul ratiunii si mobilizarea ei actuala pentru a atinge dezvoltarea durabila si armonioasa, întemeierea unei lumi mai bune. Concluzia este ca ratiunea este capabila sa înfaptuiasca ceva durabil în masura în care tine cont de integritatea fiintei noastre, inclusiv inconstientul, afectivul, într-o maniera autocritica, vizând întotdeauna perfectiunea, absolutul care îi confera astfel autodepasirea si transcenderea.

Afirmând ca *ratiunea* este facultatea proprie omului prin care se fundamenteaza si justifica cunoasterea teoretica, practica si estetica, îi recunoastem potentialitatea de-a influenta destinul planetar al umanitatii si conformitatea acestuia cu principiile de adevar, bine si frumos. E necesar sa remarcam de la început ca aceasta capacitate specifica omului de a se autodepasi pe sine prin ratiune a fost privita în decursul filozofiei ca o posibilitate *intrinseca* acestuia; Nu toti gânditorii i-au acordat însa capacitati absolute. Unii au admis existenta unei *supraratiuni* intrinseci, altii, extrinseci, de natura transcendenta, care o fundamenteaza si o ilumineaza pe cea strict umana. Dar nici în aceste conditii nu a existat un consens în privinta atingerii absolutului.

Cei care au sustinut posibilitatea ratiunii de a se determina pe ea însasi în chip universal au absolutizat capacitatea acesteia, sustinând ca ea nu este doar iubire de întelepciune, ci *întelepciunea însasi*. Astfel, ratiunea a fost zeificata! Desigur, ei au vazut transcendenta zeitatii ca fiind accesibila cunoasterii transcendentale proprie omului, consubstantiala acestuia. Prin *transcendent* întelegem aici zeitatea, noumenul, suprasensibilul, iar prin *transcendental* domeniul aprioric al cunoasterii, ca laborator al nostru intim, care are ca obiect de cunoastere acest transcendent (obiectul transcendental).

Au existat si filozofi care au negat caracterul de certitudine al ratiunii, încercând sa-i prezinte limitele si contradictiile, mergând în formele lor extreme pâna la agnosticism, atât în ce priveste cunoasterea realitatii înconjuratoare cât si în privinta accesului la transcendenta, la suprasensibil. Acestia au absolutizat limitele ratiunii, riscând caderea în incognoscibil. Ratiunea a fost aproape redusa la ignoranta. În cel mai bun caz, se putea spera doar la *iubire de întelepciune*. Cum am putea gasi un echilibru între toate aceste tendinte ? Se poate spera într-o sinergie a ratiunii astfel încât sa evitam dispersia, uneori extremele?

Fara a adopta în vreun fel o pozitie pesimista asupra situatiei actuale a lucrurilor, deoarece s-ar putea cita totodata o multime de descoperiri si fapte benefice pentru stadiul istoric în care ne aflam, vom spune totusi ca aparitia unor fenomene precum încalzirea planetara, lipsa de autoreflexivitate a stiintei sau tehnicii, problema foametei si a penuriei, lipsa de acces la apa, eugenismul, sclavia sexuala, dictaturile, terorismul, era nucleara sau numerica, clonajul uman, armele biologice, dezradacinarea ontologica, mondializarea-si lista ar putea continua,-sunt stari de lucruri care ne pun în fata unei nevoi de a medita mereu mai profund si mai critic asupra directiei si sensului înspre care ne îndreptam.

Contradictiile sunt inerente gândirii umane, la fel ca si prezenta inconstientului, a irationalitatii, dar tocmai prin re-cunoasterea acestora suntem strabatuti de un fel de *substantialitate universala* care este de natura *rationala* si care patrunde cu lumina sa sufletul si toate trairile noastre. Pentru ca aceste trairi sa poata fi cuprinse de *întelegere* si astfel obiectivate în ceea ce au ele specific, trebuie ca ratiunea sa nu fie rupta de ele, ci sa fie capabila sa le aduca în fata constiintei ca în fata unui focar de lumina prin care sa sesizeze esentialul.

Vorbim în general despre ratiune ca despre o facultate a noastra prin care cunoastem (cu ajutorul unei metode stiintifice sau daca nu cel putin riguroase) temeiul lucrurilor înconjuratoare, ale moralitatii sau ale esteticii. Dar au fost filozofi precum Heidegger de pilda, care au considerat ca ratiunea filozofiei nu trebuie sa urmeze ca fir calauzitor necesitatea celei mai riguroase stiintificitati. Asta nu însemna, desigur, pentru Heidegger, lipsa unei gândiri autentice. Este doar desprinderea de directia pe care o imprimase Husserl ideii de stiintificitate, care era preocupat de gasirea unei metode *strict stiintifice* de apodictica evidenta, care sa fundamenteze toate celelalte stiinte împreuna cu o noua ontologie. Însa pentru a fundamenta o noua ontologie, pentru Heidegger era necesar sa se renunte la pretentia ratiunii de a reprezenta ceva în calitate de *obiecte ale naturii* de plida, ci mai degraba o apropiere a ratiunii de *adevarul fiintei*, fara a carei întelegeri premergatoare nu se pot da ratiuni suficiente ale fiintarilor, nu se pot explica întrebarile ontice: din ce cauza cad corpurile, din ce cauza este asa si nu altfel, din ce cauza foametea, lipsa de autoreflexivitate a stiintei, dezradacinarea ontologica, din ce cauza este de fapt ceva si nu, mai degraba, nimic. Pentru a raspunde la toate acestea, e necesara o întelegere prelabila de natura ontologica, o pre-întelegere a fiintei fiintarilor.

Alaturi de *principiul identitatii* (orice lucru este identic cu sine), *necontradictiei* (este imposibil ca un lucru sa fie si în acelasi timp sa nu fie) si *principiul tertiului exclus* (un lucru exista sau nu exista, a treia posibilitate este exclusa), *principiul ratiunii*, sau mai precis *principiul ratiunii suficiente* constitue capacitatea de baza pentru fundamentarea cunoasterii si moralitatii. În formularea sa ontologica, acest principiu declara: *orice lucru sau fenomen exista în virtutea unui temei*. În logica, principiul spune ca *orice afirmatie sau negatie poate fi acceptata numai daca este dovedita*, numai daca i se indica temeiul. Acest principiu era utilizat înca de catre Aristotel, însa cel care, pentru prima data, i-a dat o formulare explicita, a fost Leibniz. El îl anunta ca pe o mare descoperire: *"Nihil est sine ratione"*. Nimic nu

este fara ratiune. Facilitatea cu care acest principiu este acceptat vine din faptul ca gândirii umane i se pare ca este de la sine înteles ca ceea ce survine sa fie asa cum a survenit si nu altfel. Gândirea noastra este construita în asa fel încât cauta peste tot, în ceea ce este în noi si în afara nostra, ratiunea a ceea ce este. Si totusi gândirea noastra nu este întotdeauna ''în regula''. Acest aspect îl aratase deja Socrate.

Heidegger considera insolit faptul ca au fost necesare doua mii trei sute de ani, de când exista filosofia, pentru ca acest principiu sa-si faca aparitia în forma enuntata mai sus : *Nimic nu este fara ratiune. Tot ceea ce este real are o ratiune a realitatii sale*. Insa principiul ratiunii suficiente formulat astfel nu spune nimic despre ratiunea în sine, în limbaj heideggerian nu spune nimic despre fiinta ratiunii. Nu spune nimic despre ratiunea principiului ratiunii. Principiul ascunde în el ceea ce spune. Dar poate ca acest principiu are în el o ratiune fondatoare care-i asigura caracterul de neclintit.

Filosofia critica a lui Hume ajunsese sa limiteze capacitatile intelectului omenesc printr-un anume tip de scepticism care a fost calificat câteodata drept scepticism moderat. Cunoasterea prin ratiune pur speculativa, asa cum o întelege Hume, este o eroare si o pretentie exagerata. Tot ce putem spera este restrângerea cercetarilor noastre la sfera experientei, desi nici acestea nu arata necesitatea producerii evenimentelor. Observam doar o succesiune care devine pentru noi obisnuinta, care nu indica temeiul acestei obisnuinte. Singurul domeniu de cercetare certa este pentru Hume matematica, unde ratiunea însasi este la ea acasa. Hegel vorbeste mai târziu despre limitele matematicii. Dar nu aparusera înca teoremele inconsistentei si incompletitudinii matematice ale lui Godel, care vor clatina bazele acesteia. Godel reusise sa demonstreze ca exista cazuri matematice în care putem demonstra un lucru si contrarul acestuia: de unde inconsistenta. De asemenea, ca exista adevaruri matematice care nu se pot demonstra în cadru sistemului însusi (de unde incompletudinea). Totusi, Godel nu a negat existenta trancendentei lui Dumnezeu de pilda. Dimpotriva, el încerca sa demonstreze existenta acestuia pe baza transcendentei lui absolute. Dumnezeu exista tocmai fiindca depaseste orice concepere.

Fara a intra în maretul edificiu kantian, dorim doar sa precizam ca ratiunea la Kant este o facultate a ideilor ca principii doar *regulative,* care indica calea spre unitatea sistematica a cunoasterii. Dar ratiunea nu creaza concepte despre obiecte, ci doar le ordoneaza, fiind *neconstitutiva.* Ideile ratiunii pure sunt pozitive fiindca reprezinta regulile pentru îndrumarea metodica si unitara, dar sunt limitative doarece nu fundamenteaza obiectul experientei. Hegel observa foarte bine ca: *"ceea ce Kant a refuzat ratiunii teoretice-libera determinare de sine-el a revendicat-o în mod expres pentru ratiunea practica. Aceasta latura a filozofiei kantiene este îndeosebi cea care i-a câstigat acesteia o mare favoare, si pe buna dreptate"*.[1] Indiferent ca suntem sau nu de acord cu diferentierea ratiunii teoretice de cea practica, meritul lui Kant în domeniul moral poate constitui si astazi un ideal de urmat. Fundamentarea prin ratiunea practica a unei etici stiintifice si ontologice, autonomia vointei practice ca principiu fundamental al libertatii, fundamentarea demnitatii umane pe constiinta morala, îi confera lui Kant merite incontestabile, care pot calauzi si astazi. Caci el ne-a atras atentia asupra raspunderii pe care o purtam în ratiunea noastra pentru actiunile pe care le intreprindem în transformarea lumii în care traim.

Se stie ca Descartes, creatorul filosofiei moderne, a vrut sa aseze toata cunoasterea umana pe *baza neclintita* a *ratiunii fondatoare.* Si pentru a izbuti în demersul sau, el începe prin îndoiala universala, neadmitând ca si cunoastere adevarata decât ceea ce i se prezinta sub forma *clara* si *distincta.* Descartes reia lucrurile de la început dupa ce, asa cum observa Hegel, acestea dormisera peste o mie de ani. Începe acum o noua epoca pentru filozofie, care se afirma cu acest erou al sau. El cauta metoda gândirii *"care ne învata sa conducem bine ratiunea* si sa *descoperim adevarurile pe care le ignoram"*.[2]. Descartes descopera în matematica ceea ce logica sau alte discipline nu puteau oferi, anume o *Mathesis universalis,* care explica toate problemele cu privire la raporturi, proportii, masuri, toate acestea conducând la o stiinta a

ordinii si a masurii. Regulile *clare* si *distincte* pentru conducerea gândirii sunt la Descartes *intuitia* si *deductia.* Prin intuitie, el întelege o forma directa prin care ne dam seama ca gândim si existam, ca un triunghi are trei laturi si asa mai departe. Acestea toate sunt *clare* si *distincte* si nu se mai pot descompune în alte notiuni, fiind în acelasi timp cunoscute prin ele însele, neavând nimic înselator. Acest tip de evidenta dat în intuitie se prezinta constiintei noastre ca o idee clara si distincta. Leibniz sustinuse totusi ca Descartes nu a precizat suficient în ce consta claritatea si distinctia reprezentarii, calitati care erau pentru Descartes însusi atât de importante.

Când vorbim de principiul ratiunii, ne aflam în fata unei proprobleme de fond si atunci, spune Aristotel, este nevoie sa vedem prin discernamânt pentru ce anume trebuie sa cautam probe si pentru ce anume, nu. E nevoie de ceea ce grecii numeau *paideia* pentru a putea vedea evidenta lucrurilor, clare ca lumina zilei, care se prezinta ratiunii. Acestei simple remarci a lui Aristotel i s-a opus de catre unii faptul ca evidenta nu este atât de simpla, deoarece pe masura ce ne fortam sa-i gasim ratiunii sensul cel mai propriu, ne scufundam de fapt într-un abis nesfârsit. Insa pentru Descartes *intuitia* unui lucru apare în evidenta ei nemijlocita ca o idee clara si distincta.

Totodata *deductia* e operatia prin care întelegem toate lururile care sunt consecinta necesara altora despre a care avem cunostinta sigura. Noi spunem astazi ca deductia este un rationament în care se trece de la judecati generale la judecati particulare, sau dela judecati generale la judecati cu acelasi grad de generalitate sau mai mic. Rationalitatea deductiei consta în faptul ca daca premizele sunt adevarate, concluzia este adevarata. Totusi, pentru Descartes, intuitia si deductia nu constitue metoda rationala în sine, ci adevarata metoda consta în *folosirea potrivita* a intuitiei si deductiei. Pentru aceasta folosire chibzuita, e nevoie sa operam cu o alegere initiala: în primul rând, trebuie sa vedem daca dintre toate adevarurile initiale se mai pot deduce altele si aceste adevaruri din care nu se mai pot deduce altele sa fie examinate în lumina usurintei sau dificultatii cu care au fost gasite. În urma acestor operatii, vom sti prin ce examinare va fi potrivit sa începem. Desigur toate acestea Descartes le-a înteles ca suplete spirituala si nu ca aplicare rigida a ratiunii. În discursul sau despre

---

[1] G. W. F. Hegel, *Logica partea I.* Traducere de D.D. Rosca, Virgil Bogdan, Constantin Floru, Radu Stoichita, Editura Academiei, Bucuresti, 1962, p.131

[2] Anton Dumitriu, *Istoria logicii,* Bucuresti 1969, p. 494.

metoda, el enunta doua principii si patru reguli. Primul princiu este *îndoiala metodica*, cea prin care ma pot îndoi de tot: lume, credinta, simturi, Dumnezeu, ratiune. Singurul lucru despre care nu ma pot îndoi este faptul ca ma îndoiesc, deci cuget, caci altfel ar fi absurd. *Dubito ergo cogito, cogito ergo sum* este singura certitudine, este certitudinea dubitativa din care decurge existenta mea, cât si a faptului ca în acest fel am aratat posibilitaea unui adevar indubitabil, deci adevarul îmi este accesibil.

Husserl îl considera pe Descartes ca fiind cel mai mare gânditor al Frantei, cel care, prin inovatia sa, a deschis un drum nou pentru *metoda* si *problematica* transcendental-fenomenologica. Acesta întemeiere prin *eul dubitativ* se va transforma la Husserl în suspendarea privizorie a lumii, stiintelor naturale si chiar spirituale, pentru a gasi *fundamentul absolut* ca punct de sprijin pentru orice judecata corecta. Acest fundament este *constiinta absoluta a lumii*, a tuturor lucrurilor. Se obtine astfel o libertate totala fata de traditia filozofica, stiintifica sau religioasa, cât si o eliminare a oricaror prejudecati. Este o experienta dincolo de care nu se mai poate transcende. Caracterul absolut al unui asemenea obiectiv l-a facut pe Husserl sa afirme posibilitatea acestuia ca înfaptuire concreta, desi numai în limitele devenirii istorice, dar care poate primi concretetea si certitudinea unui realizari efective.

În sensul transcendental în care Husserl concepe accederea rationalitatii mele la fiinta, adevarul sau binele pe care eu le caut nu pot fi conditionate decât de propria mea luare de *constiinta universala*, care prin radicalitatea ei, ma conduce la atingerea adevarului, binelui si frumosului. Chair daca a fost animat de ideea unei stiintificitati a filosofiei, Husserl era constietnt totodata de dimensiunea etica a rationalitatii. De aceea el se exprima într-o conversatie catre Eugen Fink în directia necesitatii unei onestitati a telosului (scopului) oricaror cercetari stiintifice.

Constiinta merge pâna la *fiinta prima si ultima* a oricarei entitati obiective, adica pâna la ceea ce Husserl numeste intersubiectivitatea transcendentala: *o totalitate a monadelor care se unesc* în forme diferite într-o *comunitate*. O asemenea intersubiectivitate nu însemna doar subiecte umane coexistând unele intercalate cu altele, dar mai ales actionând concret unele asupra

altora de o maniera benefica, construind ceea ce înseamna idealul înfaptuit al binelui comun. I s-au adus reprosuri acestui tip de rationalism cum ca ar cadea într-un subiectivism absolut, desi nu transpare clar cum ar avea loc o asemenea cadere din moment ce el nu ramâne doar la cunoasterea de sine universala, ci ajunge la cea *intermonadica*, adica la un fenomen de *constiinta intersubiectiv*.

Renumita reductie fenomenologica cu ajutorul careia, inspirat de catre Descartes, Husserl sustine posibilitatea atingerii adevarurilor apodictice, consta în *punerea intre paranteze a întregii lumi obiective (epohé)*, care conduce la posibila aparitie evidenta a adevarului apodictic. Lumea este ceea ce ramâne din ea dupa ce îndoiala existentei sale a ''desfiintat-o''. Lumea s-a redus astfel la ceea ce este esential. Si aceasta esentialitate este prezenta Fiintei. Este baza de la care va pleca urmasul lui Husserl, Martin Heidegger.

Însa în conceptia lui Heidegger, fiinta se ascunde tocmai prin faptul ca se deschide prin fiintari. Ontologia e camuflata în ontic si acesta este tot pericolul modernitatii si al demoniei tehnicii acaparatoare. Întreaga metafizica a rationalitatii, desi îsi pune intrebari despre ceea ce întemeiaza si face sa fie ceea ce este, nu reuseste sa patrunda adevarul întemeietor, adevarul fiintei, singurul de la care pornind s-ar putea salva starea de lucruri ale fiintarii. Iar fiintarile nu fac altceva decât sa ne transporte dintr-o parte într-alta, fara a avea sentimentul vreunei stabilitati, deoarece totul poate fi transportat si transformat dupa o ordine a eficientei de plida, care pierde orice raport cu umanitatea omului, cu omul ca scop în sine. Tot ceea ce realizam doar în lumina utilitatii si eficientei ramâne în domeniul fintarii, în care se pierde orice rationalitate a adevarului ontologic. Ar trebui sa actionam astfel ca lucrurile create sa exprime adevarul lor ontologic, nu doar sa fie fortate sa apara ca într-un asediu în care le este stoarsa vitalitatea. Vrem ca natura sa nu mai fie doar servitoarea careia i se preia neconditionat energia. Vrem mai mult echilibru si prevedere a ratiunii în raport cu natura.

Daca *tot ce este are o ratiune*, atunci ceea ce este îsi are ratiunea întru *adevar si bine* totodata, într-un fel în ceea ce *trebuie* sa fie. Ratiunea de a fi a tuturor lucrurilor este în adevarul existentei acestora. Acest ''adevar si bine'' nu se prezinta întotdeauna nemijocit în evidenta sa. Pentru a se dezvalui, el trebuie sa fie cucerit printr-o cautare de fond, printr-o gândire fundamentata. Ea este determinata prin inteligibilitatea sa, adica prin adevarul si binele fundamentate ontologic, care sa-i asigure fiintei esenta sa nevatamata.

Ratiunea de a fi a stiintei, de pilda, consta in *"proiectul matematic al naturii"*. Acest proiect este chiar esenta stiintei. Deja Galilei spusese ca esenta naturii este înscrisa în limbaj matematic. Însa pericolul pe care Heidegger îl sesizeaza aici este ca natura este somata sa se dezvaluie astfel pe linia calculabilului si masurabilului, reducând totul la o forma de obiectivitate reprezentabila. Iar tehnica, la rândul ei, scapa dominatiei omului, deoarece esenta ei nu are nimic comun cu umanismul. Tehnica transforma totul în materie prima de dezvoltare, chiar si fiinta umana e fecundata în vitro, modificata genetic. Tehnica produce era informationala a carei esenta nu este înteleasa, creaza ADNuri artificiale, afecteaza biosfera, transforma corpul uman, ameninta de a pune în pericol sistemele simbolice în care vietuim, bulverseaza spatiul reprezentarii lumii si a naturii, raportul cu noi însine si între noi, destinul nostru cosmic. Ratiunea se vede în situatia de a nu mai putea indica sensul, în neputinta de a gândi solutii concrete în fata unei asemenea viteze de transformare. Ratiunea nu poate urma calea dezlantuirii productiviste, energetiste.

Ratiunea cunoaste si poate chiar face proba subiectivitatii sale obiectivate ontologic, dar se vede depasita tocmai de obiectivitatea situatiilor. Ea e somata sa patrunda consecintele fenomenelor actuale în asemenea maniera încât sa poata aplica *principiul responsabilitatii morale* astfel încât sa salveze însasi viata. În fata atâtor probleme, un gânditor de talia lui Heideger afirma *: "Numai un Dumnezeu ne mai poate salva"*. Oricine sesizeaza imediat profunzimea unei asemenea afirmatii.

Dar este Dumnezeu, într-adevar, în afara ratiunii noastre? Este el o entitate suprarationala rupta de capacitatea ratiunii noastre intrinseci. A postula acest hiatus chiar si ca fenomen mundan sau istoric înseamna a nu întelege ca împaratia lui Dumnezeu începe *acum si aici*. Desigur, aceasta este evident pentru credinta. Dar este numai pentru ea? Nu e si pentru ratiune? Nu e si pentru o realitate care se purifica, restabilindu-si temeiul?

De la întruparea Fiului, lumea este locul manifestarii împaratiei lui Dumnezeu si a transfigurarii materiei si omului. Desigur ca starea ultima de transfigurare va fi doar la a doua venire a lui Hristos. Dar pâna atunci, omul coboara cu ratiunea sa in inima sa si se deschide spre a primi ca dar harul Duhului Sfânt, care-i înduhovniceste întelegerea. Omul lucreaza cu ratiunea sa, dar se desavârseste atât în cunoastere cât si moral prin conlucrarea cu transcendenta Duhului Sfant, unica posibiliate de a vedea tot adevarul. Duhul Sfânt înalta umanitatea ratiunii pâna la slava harica a adevarului dumnezeiesc revelat. Despre Iisus Hristos, patriarhul Fotie spunea ca *"ne-a descoperit adevarul în parte, iar Duhul Sfânt ne calauzeste spre tot adevarul"*.[3] Sfântul Vasile cel Mare zicea ca starea de asemanare cu Dumnezeu, starea de desavarsire la care este chemat sa ajunga omul, reprezinta însasi esenta crestinismului. Sfântul Maxim Marturisitorul, cel caruia i se taiase mâna si limba si fusese exilat pe malul rasaritean al Marii Negre, spunea ca *"Duhul Sfânt nu e absent din nici o faptura si mai ales din cele ce s-au învrednicit de ratiune."*[4] Duhul face ca ratiunea naturala sa se transforme dupa har si sa fie ridicata la întelepciune îndumnezeita.

Maxim Marturisitorul spunea ca patrunderea la ultima Ratiune *(Nous)* prin dragoste dumnezeiasca face ca mintea sa se uneasca cu Dumnezeu si, simplu vorbind, sa poarte în sine aproape toate însusirile dumnezeiesti. Iar ratiunea celui care e iubitor de Dumnezeu nu lupta împotriva lucrurilor, împotriva fenomenelor sau a tuturor fiintarilor existente, ci asupra patimilor împletite cu întelesurile. Iar întelesurile sunt dobândite prin bunavointa lui Dumnezeu care a facut tot sufletul dupa chipul Sau si a dat omului *libertatea ratiunii* de a-si alege de buna voie cinstea sa. Dumnezeu ne-a dat cea mai mare valoare cu putinta, punând în noi *libertatea* si capacitatea *naturala de discernamânt*, care primeste lumina absoluta prin Harul dumnezeiesc al Duhului. Discernamântul ne conduce spre *virtute* prin *fapte*, iar luminarea

---

[3] *Dictionar de teologie ortodoxa*, Bucuresti 1994, p.135 (*Fotie, Despre calauzirea duhului. p. 305-309*).

[44] *Sfântul Maxim Marturisitorul, Raspunsuri catre Talasie, Filoc. Rom., vol. 3, p. 48-49*).

ratiunii prin Duh conduce la întelepciunea care înlatura orice culme ridicata împotriva cunoasterii lui Dumnezeu. În acest fel vorbea Maxim Marturisitorul.

Observam ca Maxim Marturisitorul avea o convingere total încurajatoare despre capacitatile *gnoseologice* cât si despre cele *etice* ale omului, chiar în conditia mundana în care ne aflam. Iata ce mai spune el în acest sens: *"...sa nu ne pierdem credinta în virtutea virtutii noastre omorâte. Caci amandoua sunt cu putinta la Dumnezeu: si sa coboare ca sa lumineze mintea noastra prin cunostinta si sa învie virtutea în noi ca sa ne înalte împreuna cu Sine, prin faptele dreptatii"*[5] Multi dintre cei care practica viata crestina cred ca nu ne putem ridica la Dumnezeu decât prin simtire sufleteasca si inima. Dar vorbele sfiintilor parinti ortodocsi nu mai contenesc în a arata virtutile ratiunii. Desigur, ratiunea primeste sens iluminat prin înduhovnicire si astfel preia în ea toata caldura afectiva. Dar chiar si la barbari si nomazi, observa Maxim Marturisitorul, se-ntâmpla sa aflam multi care duc o viata de fapte bune. Astfel, se poate spune în chip general ca în toti este Duh Sfânt. Nimeni nu este privat de atingerea lui, dupa convingerea lui Maxim Marturisitorul.

Sa mai citam câteva virtuti ale ratiunii prin iluminarea tainica cu Dumnezeu spre vedea mai clar care este pozitia crestinismului în acest sens. Iata ce spune în continuare Maxim Marturisitorul: *"Pe aceasta (rautatea) o nimiceste ratiunea prin strategia ei, scrutând cu stiinta duhovniceasca originea si firea lumii si a trupului, împingând sufletul spre tara înrudita a celor inteligibile, împotriva careia legea pacatului nu are nici o putere.*[6] Sfântul Antonie cel Mare, cel dintâi monah care s-a retras în pustie spunea ca rationali sunt doar acei oameni care au *sufletul rational*-si deosebind binele de rau-pun toata grija lor spre cele folositoare sufletului. El credea ca mintea vede chiar si cele din Ceruri si omul dupa partea sa rationala e în legatura cu puterea negraita si dumnezeiasca. Nil Ascetul care si-a petrecut sfârsitul vietii în muntele Sinai, spunea astfel: *"priveste la sufletul care se tine lipit de obisnuinta, cum cade lânga idoli, lipindu-se de materiile fara forma si nu vrea sa se ridice si sa se apropie de*

ratiunea care cauta sa-l calauzeasca spre cele mai înalte". Diadoh al Foticeii, care a fost un mare ascet si adânc cunoscator al tainelor mistice, sustinea ca numai Duhul poate sa curete mintea si nu trebuie sa ne îndoim atunci când mintea începe sa se afle sub lucrarea luminii dumnezeiesti, facându-se stravezie, încât îsi vede cu îmbelsugare propria lumina.

Astfel de cuvinte folositoare din revelatiile vii ale sfiintilor parinti au ca regula de aur curatirea ratiunii umane prin Duhul Sfânt si îndrumarea sufletului de catre o ratiune astfel purificata. Ne întrebam si noi, pastrând o smerita proportie, cum se întreba Husserl în fata Meditatiilor carteziene:*" Merita cu adevarat sa cautam a descoperi în aceste gânduri o importanta eterna? Sunt ele capabile înca sa insufle timpului nostru forte vii?"*[7] Cred ca starea actuala a planetei în întregul ei ar trebui sa ne faca sa meditam mai des asupra întemeierii lucrurilor. În acest sens, alaturi de filozofi, Sfintii parinti sunt de mare folos. Caci într-o lume în buna parte ramasa la suficienta sa naturala si la perplexitatea care o cuprinde, ei vin sa ne trezeasca înca o data încrederea în lumina vie a unei ratiuni sporite de slava înduhovnicirii care o cuprinde, o înalta si îi da sens.

La sfârsitul acestui periplu filozofico-teologic, se impune, fara îndoiala, urmatoarea concluzie: ratiunea autentica este întotdeauna *transcendere*. Aceasta este caracteristica ei de baza. Fie ca este transcendere ca substanta cugetatoare ca la Descartes, ca infinitate pozitiva a ratiunii practice la Kant, ca idee absoluta a unitatii ratiunii teoretice si practice la Hegel, ca si constiinta a intersubiectivitatii transcendentale la Husserl, ca sens al Fiintei fiintarii la Heidegger, peste tot ratiunea transcende, ca într-o depasire de sine universala. A gândi cu adevarat rational, ceea ce implica cuprinderea în sine a simtirii si a iubirii, înseamna a transforma lumea în ceva stabil, fundamental, universal valabil. Insuficientele si amenintarile unei lumi imperfecte, periculoase, cheama catre salvarea vietii, dar viata însasi nu poate fi salvata daca nu primeste în ea adevaratul fundament care o sustine. Acest fundament, fie ca îl numim Fiinta a fiintarii, Spirit Absolut,

---

[5] *Nestemate Filocalice*, Dan Sgârta, *p.73*.

[6] *Nestemete Filocalice*, Dan Sgârta p.36)

---

[7] Husserl, *Meditatii cartesiene*, Traducere Aurelian Craiutu, Humanitas, 1994, p.33

Dumnezeu, este punctul de plecare si de întoarcere pentru ratiunea care întemeiaza cu adevarat ceva concret: sensul si bucuria vietii. Sensul rezida în bucurie, ca iubire a întelepciunii, iar fericirea suprema rezida în gasirea si aplicarea întelepciunii.

BIBLIOGRAFIE :

[1] David Hume, *Cercetare asupra intelectului omenesc.* Editura stiintifica si enciclopedica, Bucuresti, 1987

[2] Leibniz, G.W., *Opere filozofice.* vol. I, Traducere de Constantin Floru, Studiu introdictiv de Dan Badarau, Bucuresti, 1972

[3] René Descartes, *Discours de la methode*, suivi des *Méditations*, preséntation et annotation par François Misrachi, Editions Montaigne, 1951

[4] René Descartes, *Reguli utile si clare pentru îndrumarea mintii în cercetarea adevarului.* Editura stiintifica, Bucuresti, 1964

[5] René Descartes, *Doua tratate filozofice, Viata si filosofia lui René Descartes* de C. Noica

[6] René Descartes, *Méditations métaphysiques*, texte traduction objections et réponses présentés par Florence Khodoss, Presses Univ. De France, 1974

[7] Immanuel Kant, *Critica ratiunii pure.* Traducere de Nicolae Bagdasar si Elena Moisuc. Studiu introductiv, glosar kantian si indice de nume proprii de Nicolae Bagdasar, Editura stiintifica, Bucuresti, 1969.

[8] Immanuel Kant, *Critique de la raison pure.* Traduction de Jules Barni, Flamarion, Paris, 1987

[9] Immanuel Kant, *Întemeierea metafizicii moravurilor. Critica rartiunii paractice*, Edidura stiintifica, Bucuresti, 1972.

[10] Immanuel Kant, *Prolegomene la orice metafizica viitoare care se va putea înfatisa drept stiinta.* Traducere de Mircea Flonta si Thomas Kleininger. Studiu introductiv si note de Mircea Flonta. Editura stiintifica si enciclopedica, Bucuresti, 1987.

[11] G. W. F. Hegel, *Logica partea I.* Traducere de D.D. Rosca, Virgil Bogdan, Constantin Floru, Radu Stoichita, Editura Academiei, Bucuresti, 1962

[12] G. W. F. Hegel, *Stiinta logicii.* Traducere de D.D. Rosca. Editura Academiei, Bucuresti, 1966.

[13] G. W. F. Hegel, *Fenomenologia Spiritului.* Traducere de Virgil Bogdan. Editura Academiei, Bucuresti, 1965.

[14] Edmund Husserl, *Meditatii carteziene, O introducere în fenomenologie*, Traducere, cuvânt înainte si note de Aurelian Craiutu, Humanitas, Bucuresti, 1994

[15] Edmund Husserl, *La crise de l'humaninté européene et la philosophie,* Introduction, commentaire et traduction par Nathalie Depraz, Edition numérique : Piere Hidalogo, La Gaya Scienza, mars 2012

[16] Edmund Husserl *La crise des sciences européenes et la phénoménologie transcendantale*, traduit de l'allemand et préfacé par Gérard Granel, Gallimard, Paris, 1976

[17] Martin Heidegger, *Le principe de raison*, traduit de l'Allemand par André Préau, preface de Jean Beaufret, Gallimard, 1962

[18]Martin Heidegger, *Fiinta si Timp*, Traducere din germana de Gabriel liiceanu si Catalin Cioaba, Humanitas, Bucuresti, 2003

[19] Martin Heidegger, *Repere pe drumul gândirii.* Traducere si note introductive de Thomas Kleininger si Gabriel Liiceanu. Editura politica, Bucuresti, 1988.

[20] Martin Heidegger, *Introducere în metafizica.* Traducere din germana de Gabriel Liiceanu si Thomas Kleininger. Editura Humanitas, Bucuresti, 1999.

[21] Anton Dumitriu, *Istoria Logicii*, Bucuresti, 1969

[22] Ion Bria, *Dictionar de teologie ortodoxa.* Editura Institutului Biblic si de Misiune al Bisericii Ortodoxe Române, Bucuresti, 1964

[23] *Nestemate Filocalice, Vorbe de duh ale sfiintior Parinti*, culegere de texte si maxime alcatuita de Dan Sgârta, Editura Omniscop, Craiova, 2011.

[24] Dumitru Staniloae, *Chipul nemuritor al lui Dumnezeu*, Editura Mitropoliei Olteniei, Craiova, 1987

Nietzsche, Hegel and Historicism: A Polemic With Foucault's Reading of Nietzsche

Paul Catanu, PhD

Humanities Instructor, Champlain College, St-Lambert

1) Introduction

In this paper, we look at how Nietzsche handles the problem of history and historicism. The writing of history and thus indirectly, historicism itself, has a long and complicated history, which can perhaps be dated back to Hesiod, Homer and Herodotus. However, it is useful to distinguish when debating the problem of historicism between the classical historicism of the 19th century German historical school (Ranke, Droysen, von Treitschke and, later, Meinecke and Troeltsch) and the later "new historicism" of thinkers such as Foucault and Derrida. Of course this does not exhaust the problem of historicism which made repeated re-occurrences in philosophical life in the works of Dilthey, Husserl and even Rickert and Windelband, not to speak of Gramsci and Lukács. However, given the limitations of space of this paper, I mainly analyze how historicism occurs in Nietzsche's philosophy and his position vis-à-vis it.

More specifically, I will discuss Nietzsche's concept of history, and his alleged connection to historicism, both by looking back at how he reacted to the German historical school and how his thought influenced the later French thinker Foucault. I will also discuss the connection between the concept of Becoming and Nietzsche's alleged historicism.

There is a tension between Nietzsche's earlier writings on history (*Fatum and History, Free-will and Fatum, The Untimely Meditations*) and the later writings in which a genealogical concept of history is developed. This tension has been noticed by many commentators amongst which the most famous is probably Michel Foucault. However Foucault would have us believe that there is a continuity of the philosophy of history espoused by Nietzsche in the *Untimely Meditations* and in the *Genealogy of Morals*. I believe that this reading of Foucault's, which is especially expounded in

Nietzsche, Genealogy, History, is essentially a misreading and I will try to show why.

I will argue that there is a continuity between the *Untimely Meditations* and *Thus Spoke Zarathustra* and that this continuity sketches an anti-historicistic position within Nietzsche's philosophy. This anti-historicistic position is articulated around Nietzsche's key concepts of the supra-historical, the eternal and the *Augenblick*.

2) Differences Between the Early and the Late Nietzsche's Views On Historicism and History

In the *Second Untimely Meditation,* Nietzsche defines culture as "the stylistic unity of the artistic expressions of a people". This unity alluded to by Nietzsche is present throughout the *Second Untimely Meditation.* Nietzsche identifies the dichotomy between substance and form, between inner and outer, as the problem affecting modern culture. One needs to distinguish Nietzsche's classicism from that of Hegel, as it is most clearly expressed in his *Lectures on Aesthetics.* Hegel still operates with the balance between form and content, Being and appearance that is characteristic of ancient Greek culture. In Nietzsche, there is a tragic rupture between Being and appearance, between form and substance. In the early writings, and this includes the *Second Untimely Meditation,* Nietzsche still experiences the loss of balance between form and substance as a tragic rupture and a problem.

One could define historicism as the belief that all manifestations (ideas, concepts, institutions, texts and, ultimately, Being itself) need to be understood starting from their historical context. This would be the broadest definition of historicism. From there one could proceed to refine the concept of historicism into sub-categories. Hegel would be a historicist in virtue of believing the previous definition to be true, but he also believes that history has a *telos* or purpose. The German historical school is historicistic insofar as it believes the basic definition, but it also believes that one can establish a form of objectivity with respect to historical manifestations. Finally even Foucault and Nietzsche might be historicists according to the general definition. Of course the last two do not subscribe either to the idea of a goal of history or to the idea of an objectivity of history.

As concerns Foucault there is no doubt that he made overtures towards his own special brand of genealogical historicism, but I believe that we need to differentiate within Nietzsche's evolution between the generally historicistic phase of the *Genealogy* and the anti-historicism of the *Untimely Meditations* and of *Thus Spoke Zarathustra*.

The *Second Untimely Meditation* may be said to have a historicist moment. The monumental aspect of history, as Nietzsche describes it, presents the danger of leading to an over-evaluation of the past with respect to the present. There are also Hegelian-dialectical accents to the *Second Untimely Meditation* with its insistence on a German spiritual unity. Nietzsche, although differing from Hegel, is also a classicist insofar as he takes antiquity and more precisely the tragic age preceding Attical classicism as a model that can inform and criticize the present.

Nietzsche is influenced by Schopenhauer when he rejects Hegel's thesis on the identity between Being and appearance. This dichotomy between Being and appearance pushes Nietzsche to partly reject classicism and the idea of antiquity as a model. For Hegel, antiquity is classical insofar as it consists in a perfect coincidence between inner and outer, essence and manifestation, thing in itself and phenomenon. For Nietzsche, following Schopenhauer, the classical consists in the impotence to achieve the balance, unity and identity between inner and outer, essence and manifestation. Antiquity still functions as a model in Nietzsche, albeit as a model characterized by rupture and imbalance.

In the later writings however, history is conceived genealogically and as decadence or nihilism. Nietzsche develops a genealogical concept of history in the *Genealogy of Morals* by referring to the concepts of origin (*Ursprung,*

Herkunft and *Entstehung*).[1] History conceived as genealogy is essentially pluralistic. There is no grand narrative of history as in Hegel. Genealogical history undermines the possibility of a world-historic, universal History. According to the genealogical conception, there is no History, only histories. The unity bemoaned by Nietzsche in the *Second Untimely Meditation* is radically undermined by the genealogical conception of history.[2]

Genealogical history functions by analyzing the hidden, protracted roots of a historical phenomenon. It seeks what lays dormant and what is covered over as the origin of manifestation of the historical phenomenon. However, a historical phenomenon is explained in the German historical school as being the function of a unity and continuity of the Western world that manifests itself in history. For genealogical thought the origin of a historical phenomenon is not manifest and not unitary. A genealogical account of history consists in a genetic account of an event that always seeks a plural and multi-shaped origin.

In the *Genealogy of Morals*, Nietzsche conceptualized history as the struggle between the slave and the master moralities. This can be seen as a covert polemic with Hegel for whom the master-slave dialectic constitutes a fundamental figure in the architectonic of the *Phenomenology of Spirit*. For Hegel, it is ultimately the slave that wins the struggle for recognition because, being more in touch with the earth, he is more *natural* than the master. For Nietzsche, the master is more active as opposed to the slave who is deemed to be reactive.

Nietzsche and Hegel both agree that the history of the West must be understood in terms of the struggle between the master and the slave figures or moralities. But their visions of this history are fundamentally at odds. The struggle between the master and the slave moralities can be re-translated, for Nietzsche at least, and in the *Genealogy*, as the struggle between Rome (master)

---

[1]   *Ursprung*=Origin with the exact meaning "original jump"; *Herkunft*=origin with the exact meaning of "past origin" or "heritage", and Entstehung=origin with the exact meaning of "coming-to-be". See Michel Foucault, Nietzsche, Genealogy, History, in *Language, Counter-Memory, Practice: Selected Essays and Interviews,* Translated from the French by Donald F. Bouchard and Sherry Simon, Ithaca, N.Y., Cornell University Press, 1977.

[2]   *HL*, 4, (*KSA 1*, p.278).

and Judea (slave). For Nietzsche, the relationship between the master and slave moralities does determine the history of the West fundamentally[3], but this determination is not conceptualized through an act of recognition.

The master does not need the recognition of the slave, he does not envy the desire of the slave in order to be essentially that which he is: active. Rather the being-active of the master is not defined in terms of a conscience that opposes him and against which this master must risk his own life. The master is free and active because he does not possess the moral conscience of the slave that determines the slave's essential reactivity.

3) The Young Nietzsche's Anti-Historicism

Gadamer's thesis is that Nietzsche's philosophy did not sprout from the soil of German Idealism as much as in reaction to German historicism.[4] Nietzsche discusses historicism in his famous *Second Untimely Meditation*. There, Nietzsche identifies history and the study of history as an illness that prevents man from acting. Action is opposed to knowledge as in Hamlet's classical dilemma.[5] To know the state and true nature of Being entails being opposed to the action that is associated with Becoming.

In this sense the Dionysian man has similarities to Hamlet: both have had a real glimpse into the essence of things. They have understood the nature of reality, and it disgusts them to act, for their action can change nothing in the eternal nature of reality. They perceive as ridiculous or humiliating the fact that they are expected to set right again a world, which is out of joint. Knowledge kills action...[6] History and knowledge

of history are opposed to action because in order to act, man has to forget. Essential to acting is an ability to forget one's past. For Nietzsche, acting is a projection onto Becoming and Becoming in its turn represents the immediacy of the moment as opposed to the sediments of historical Being.

Nietzsche's generalized perspectivism can be contrasted with Dilthey's positivism. In this way, this perspectivism is understood in relation to the problem of historicism. The *aporia* or contradiction generated by the problem of historicism can be expressed as follows: Human beings are aware that they are historical and that they live in different historical epochs. But the understanding of different historical epochs can only be achieved through the deforming lens of our own present epoch. This means that we can have no objective understanding of past historical epochs since we only perceive them through the lens of our present. Put this way, historicism can be understood as implying relativism. The dilemma can also be put otherwise. We are historical beings, therefore we inhabit a particular historical epoch but in virtue of inhabiting that particular historical epoch we have no access to the past epochs that are constitutive of our historicity.

History is presented as Becoming in the last great attempt to understand it philosophically. This is represented best by Hegel's Idealism. But this Becoming is understood as processuality, development and teleological movement. Nietzsche probably has Hegel in mind when he writes in the *Second Untimely*

A look at Nietzsche's and Hegel's philosophies shows clearly that history comes down to the old metaphysical problem of Being and Becoming. History is not possible without the movement of Becoming, but there are sediments, there are necessities: the flow of Becoming is constantified in Being. The place where the dialectical play of Being and Becoming occurs is precisely this History that we have *been*, that we have *become*, and that we never cease to *become*.

For Nietzsche, history is archaic, critical or monumental but it is also genealogical. Genealogy consists in the study of origins. The origin of history is not hypostasized but studied concretely in terms of its natural origins. The struggle

---

[3]   This is implicit in Nietzsche's theses about the master and slave moralities that are developed in the first essay of *On a Genealogy of Morals*.

[4]   See footnote 229 in *Wahrheit und Methode*, p.130.

[5]   "Thus conscience does make cowards of us all; and thus the native hue of resolution is sicklied o'er with the pale cast of thought; And enterprises of great pith and moment, With this regard, their currents turn awry, And lose the name of action." Shakespeare, W., *Hamlet, Prince of Denmark*, Act III, Scene I.

[6]   *In diesem Sinne hat der dionysische Mensch Ähnlichkeit mit Hamlet: beide haben einmal einen wahren Blick in das Wesen der Dinge gethan, sie haben* erkannt, *und es ekelt sie zu handeln; den ihre Handlung kann nichts am ewigen Wesen der Dinge ändern, sie empfinden es als lächerlich, dass ihnen zugemuthet wird, die Welt, die aus den Fugen ist, wieder einzurichten. Die Erkenntniss tödet das Handeln... GT*, 7, (*KSA*

1, 56-57).

between the master and the slave moralities is to a certain extent the motor of history. The problem is that there is more than one motor of history. To try to understand history solely in terms of a larger, essential process that occurs within it is already to distort history for Nietzsche. Another possible motor of history could be nihilism as has been claimed by Heidegger. But the problem with this description of nihilism is that it is part of the Heideggerian strategy—namely the claim that Nietzsche merely inverts the progress and actualization of freedom in history characteristic of Hegel's Idealism, by proposing a negative development that may be called the history of nihilism. This Heideggerian strategy is ultimately unsatisfactory and untenable since it leads to the radical historicization of Nietzsche's thought.

Some have argued that the paradoxy that characterizes historicism leads to its self-destruction. The problem with historicism is that even though it recognizes the organicity of historicity it still attempts to find a perfectly objective grounding for its understanding of the past. The question of historicism can be reformulated as the question of whether history elevates a claim to truth.

However, for Nietzsche the question is whether history can serve life. We see here prefigured, a theme that is familiar for Nietzsche: the greater importance of life and the value of an evaluation for life over that of truth. Nietzsche does not explicitly apply the concept of force to his understanding of history but the translation of this concept into the realm of history is perhaps appropriate. It is probable that Nietzsche's usage of the concept of play of forces is owed to Fechner (*Elemente der Psychophysik*) or to Büchner (*Kraft und Stoff*).

In this sense, Nietzsche is thinking of forces as physiological and physical entities that have a material application. But the play of forces is also what constitutes the essence of will to power and in this sense these forces can be taken as a metaphysical substrate as well as a material one. Heidegger's reluctance to explain will to power in terms of the play of forces might indicate his awareness of the materialistic implications of such a concept.

But in Ranke's and Hegel's explanations of history, forces are clearly understood as spiritual and metaphysical entities. In Hegel force is related

to its exteriorization. This exteriorization is in turn related to interiorization. Hegel shows that this play of exteriorization and interiorization that is the essence of self-consciousness can be subsumed under the concept of life. Interiorization is given priority in this description of self-consciousness and life but the relationship between exteriorization and interiorization is dialectical.

Life is also related to force for Nietzsche, but his analysis of force does not lead to a dialectic of life. Nietzsche deconstructs the interiorization of force that is the essence of life for Hegel. The interior interiorization and the exterior exteriorization of force cannot be seized dialectically. Rather, the essence of force precedes and destroys the still metaphysical opposition between interior and exterior, between inside and outside. History is understood in historicism as something that has meaning and that makes sense. There is a unity, a continuity, and a necessity behind the events of history. For Nietzsche, these three grounding concepts of unity, continuity and necessity of history have become problematic in the *Genealogy of Morals*.

4) Anti-historicism in *Thus Spoke Zarathustra*

In *Thus Spoke Zarathustra*, Nietzsche inaugurates an original way of thinking time and temporality and thus history. The key element that one has to retain in Nietzsche's *magnum opus* is the concept of the *Augenblick* or the eternity of the instant. What does it mean that eternity can be given in an instant? Eternity has two traditional acceptions: it can mean timelessness or it can mean a temporal endlessness or infinity. Even though both of these meanings are present in Nietzsche's concept of the *Augenblick*, we have to decide which of the meanings is more important.

The Nietzschean *Augenblick* is connected, in its redemptive power, with the thought of eternal recurrence of the same. In the moment when the eternal recurrence of the same is realized and understood, one reaches outside of history and of time. Eternal recurrence is connected to both Being and Becoming. Its aim is to redeem the past and transform and transfigure it into a perfectly willed present and future. Thus eternal recurrence is tied to the Delphic model of historical knowledge of the *Second Untimely Meditation*, which understood history according to inner need and to self-knowledge.

In the interpretation of the tale of eternal recurrence recounted in <u>On The Face and the Riddle</u> of *Thus Spoke Zarathustra*, we must part ways radically with Heidegger's interpretation of Nietzsche. Heidegger assumes that the "schwarze, schwere Schlange" that Zarathustra sees in the throat of the young *Hirt* (pastor) is the snake of nihilism. What the young *Hirt* needs to bite off and that is stuck in his throat is the thought of recurrence that is somehow amalgamated with the thought of nihilism.

If everything repeats itself then nothing is worth anything and all is indifferent (nihilism) and if everything repeats itself then everything matters, everything is worth something (eternal recurrence). The difference is very subtle between nihilism and eternal recurrence: one could call it one of existential nuance.

On the one side the repetition leads to meaninglessness whereas on the other it leads to the creation of meaning. But let us get back, to the "schwarze, schwere Schlange" that is stuck in the throat of the *Hirt* who is none other than Zarathustra himself. According to us this heavy black snake can only be the heavy black snake of *time*. Time is responsible for the will's ill will. The will is still captive because it cannot will backwards in time. What needs to be overcome when thinking the thought of eternal recurrence is time itself and not nihilism. Time is the deeper root of nihilism. It is because of time that the "upmost values lose their value". This is the case because, values are posited historically and later in other historical circumstances they may come to lose their value. But this historical process is inextricably linked to time and its "it was". Were it not for the temporal aspect of the past and for the inability of the will to will backwards and to redeem the historicity of the past, then the problem of nihilism would never occur.

This is why the *Hirt* must bite off the head of the snake, which represents time. If time can be overcome and the past redeemed, then nihilism can be overcome and eternal recurrence can be affirmed. But this interpretation of eternal recurrence also has consequences for Nietzsche's interpretation of history.

So how does the anti-historicism that operates in *Thus Spoke Zarathustra*, sit with the historicism

of the *Genealogy of Morals*? The *Genealogy* is without a doubt one of the writings of Nietzsche that has been the most influenced by Darwinian naturalism. However one can still detect some anti-historicistic moments within it.

In fact the anti-historicistic moment of the genealogical method can be seen in the fact that by looking at the origins of a historical phenomenon, we realize its necessity. Of course, we also realize that the present phenomenon is contingent since it could have been otherwise had the historical, temporal positing been different. But insofar as genealogy opens the present for the future by looking at the past, it transforms and transfigures this present and makes it possible for the future to occur. Insofar as it does this, genealogy redeems both the past and the present by negating them and sublating them. Thus, genealogy has a negative dialectical quality that is also present in the eternal momentariness of the *Augenblick*.

5) Consequences: Nietzsche contra Foucault

It seems clear that the reading of Nietzsche on history that I am developing here is in many ways opposed to Foucault's reading of Nietzsche. Foucault has famously honed in on the concept of genealogy. I cannot hope to do a critique of Foucault's entire philosophical project, but I limit myself to the places where I think he appropriates Nietzsche unfruitfully or unproductively.

There is a danger in reading in Nietzsche a historical teleology of power. I think that despite Foucault's very sophisticated analyses of power he might have fallen pray to this danger to a certain extent. Foucault is certainly opposed to historical teleologies and he has made this clear in many places in his *corpus*. But perhaps Foucault does not distinguish enough between internal historical teleologies and external ones.

Even though Foucault's model of power is non-dialectical and functions from below, there is still a local teleology of power that operates within this concept and I think that this due to a misreading of Nietzsche's notion of power and history. The local, situated notion of power functions so as to perpetuate the regimes of rationality that are described by Foucault, but there is a *telos* that moves power toward its own increase. The emphasis on conditions of enhancement as being more important than conditions of preservation is certainly present in Nietzsche's conception of the will to power.

Foucault is also committed to having enhancement as a condition of power. If not Foucault's concept of power and history would be merely static and that, in itself, would be sufficient to discredit it to a great extent. Thus, it can be shown that there is a local teleology of power in Foucault's concept of power and in his conception of discontinuous history. But locality and local teleology is more fundamentally and clearly understood in terms of the traditional distinction between inner teleology and external teleology. This is the case, at least for Nietzsche if not for Foucault because, the will to power is constituted by *Willlenspunktatione*n.

A *Punktation* is a draft of a contract or of a treaty in Austrian civil law. We say draft contracts or treaties and not treaty points of the will since then we would be coalescing the concept of *Punkt* with that of *Punktation*. The translation of *Willenspunktation* as "treaty points of the will" is also possible however since the draft contract or draft treaty is elaborated by listing points on which agreement is to be reached. In any case, we need to adjudicate as to whether it is the juridical-political nature of the concept of the will to power or its materialist/physicalist aspect that is of greater importance to Nietzsche in his concept of *Willens-punktationen*. The materialist aspects are favored by the fact that we could bring into proximity the concept of *Kraft-punkt* to that of *Willens-punktationen*. Ultimately we may not be able to adjudicate the real nature of *Willens-punktationen* because of the seldomness of its occurrence in Nietzsche's writings.

What is important for the concept of *Willens-punktationen*, insofar as it relates to Nietzsche's alleged teleology of power, is that the *Punktation* builds a global agreement based on a local point upon which agreement is reached. Thus, we proceed from the inside of the treaty through points, which construct externally from the inside out, the agreement/power structure. Thus locality is in fact understood on the model of an interiority of power that then leads to globality and to exteriority. But Nietzsche, takes the further step of deconstructing the *Punktation* and of claiming that the points of agreement have no existence in themselves. This lack of existence is predicated upon the fact that the points are in an eternal and infinite Becoming that can never be arrested in either, interiority or exteriority, locality or globality. *Kraft-punkte*, there is no overall stable increase of the will to power either globally or locally. Instead, the will to power infinitely aggregates and disaggregates itself at the power nodes and configurations of Becoming, perpetually transfiguring and transforming itself.

Thus, if the local-global pair concept is dependent on the inner-outter pair concept and if Nietzsche rejects an internalism of power, it is probable that he would have also rejected a localism of power. Foucault's notion of power is thus built upon and predicated on a misunderstanding of the relationhisp between genealogy and history within Nietzsche's thought.

Thus, it seems rather clear that the Foucaldian concept of an ontology of the present is opposed to Nietzsche's utopian conception of the philosophy of the future. This is the case because Foucault is opposed to a "suprahistorical perspective: a history whose function is to compose the finally reduced diversity of time into a totality fully closed unto itself". But this is precisely what I have been arguing for, in terms of Nietzsche's opposition to historicism. The suprahistorical perspective is to be found precisely there in *Thus Spoke Zarathustra*, when the pastor (*Hirt*) attempts to round off time and close it upon itself in order to digest and overcome it. The serpent of past-time must be bitten off so that the will can will backward and incorporate and redeem the past.

Foucault's presentism (Habermas), the fact that he stubbornly concentrates on the present historical moment in order to think it through lead to his inability to think the future and to his radical historicism. Foucault' suspicion of the utopian tradition is understandable given the pre-Fall of Communism perspective and historical position from within which he was writing. There was no way to think through and conceive an exit out of late capitalism and the dual dynamic of the Cold War.

But this is precisely what is necessary at our present historical juncture. Utopia and utopianism has once more become necessary as a critical function for understanding how to move beyond the present and the past. For these purposes, the Nietzsche of the *Untimely Meditations* needs to be retrieved dialectically and it has to be shown that Nietzsche's dialectic of time and history in the

Second Untimely Meditation is compatible with his later concept of genealogy. Genealogy is not in its essence historicistic. It does possess a dialectical quality insofar as it opens up the present by examining past origins.

By retrieving the past, criticizing and diagnosing the present and opening it up, genealogy makes the future possible. This is Nietzsche's legacy that Foucault has perhaps misunderstood to a certain extent given his suspicion of the dialectic and dialectical thought.

6) Conclusion

The late Nietzsche provides a genealogical and genetical account of history. He pluralizes the essentialized concept of history that is present in Hegel. The conscience of history is re-worked in the *Genealogy* as the tension between slave and master moralities. Through this move, Nietzsche goes beyond Hegel's mere integration of the master-slave dialectic into the movement of consciousness and self-consciousness. By re-conceptualizing the relationship between slave and master moralities in terms of the concepts of reactive (slave) and active (master), Nietzsche advances beyond the merely dialectical conception of history that Hegel possesses. This is because for Hegel history embodies a purposive, unitary *Telos* whereas Nietzsche eliminates teleology from within Becoming (and thus from history).

However my thesis that there is an absolute, infinite and eternal Becoming within Nietzsche's philosophy does cohere with the analyzes presented in this paper. Despite having a historicistic moment, the *Second Untimely Meditation* also speaks of a *supra-historical* moment of World-history. This *supra-historicity* is in fact the seed of the infinite Becoming that we see at play in Nietzsche's philosophy. Becoming is infinite, eternal and absolute and the supra-historicity of the *Untimely Meditations*, their exhortation to stand outside and above our epoch, is definitely something that comes into tension with the genealogical aspect of history of the *Genealogy*. In fact we have to look to *Zarathustra* to effectively decide what is more important for the mature Nietzsche's thought on history.

There we can see that despite the absence of a reference to Becoming, Nietzsche's philosophy wills and desires something beyond all else. What is this something that Nietzsche's philosophy wills and desires? It does not will and desire history and historicity but "all desire wills Eternity, wills deep, deep Eternity" ("*alle Lust will Ewigkeit-, will tiefe, tiefe Ewigkeit*", Zarathustra III, The Other Dance Song).

Notes

Notes

Notes

Notes

www.ingramcontent.com/pod-product-compliance
Lightning Source LLC
Chambersburg PA
CBHW081723220526
45468CB00008B/1949